Stefan Geisen

Soil Protists

Stefan Geisen

Soil Protists

Diversity, Distribution and Ecological Functioning

Südwestdeutscher Verlag für Hochschulschriften

Impressum / Imprint
Bibliografische Information der Deutschen Nationalbibliothek: Die Deutsche Nationalbibliothek verzeichnet diese Publikation in der Deutschen Nationalbibliografie; detaillierte bibliografische Daten sind im Internet über http://dnb.d-nb.de abrufbar.
Alle in diesem Buch genannten Marken und Produktnamen unterliegen warenzeichen-, marken- oder patentrechtlichem Schutz bzw. sind Warenzeichen oder eingetragene Warenzeichen der jeweiligen Inhaber. Die Wiedergabe von Marken, Produktnamen, Gebrauchsnamen, Handelsnamen, Warenbezeichnungen u.s.w. in diesem Werk berechtigt auch ohne besondere Kennzeichnung nicht zu der Annahme, dass solche Namen im Sinne der Warenzeichen- und Markenschutzgesetzgebung als frei zu betrachten wären und daher von jedermann benutzt werden dürften.

Bibliographic information published by the Deutsche Nationalbibliothek: The Deutsche Nationalbibliothek lists this publication in the Deutsche Nationalbibliografie; detailed bibliographic data are available in the Internet at http://dnb.d-nb.de.
Any brand names and product names mentioned in this book are subject to trademark, brand or patent protection and are trademarks or registered trademarks of their respective holders. The use of brand names, product names, common names, trade names, product descriptions etc. even without a particular marking in this work is in no way to be construed to mean that such names may be regarded as unrestricted in respect of trademark and brand protection legislation and could thus be used by anyone.

Coverbild / Cover image: www.ingimage.com

Verlag / Publisher:
Südwestdeutscher Verlag für Hochschulschriften
ist ein Imprint der / is a trademark of
OmniScriptum GmbH & Co. KG
Heinrich-Böcking-Str. 6-8, 66121 Saarbrücken, Deutschland / Germany
Email: info@svh-verlag.de

Herstellung: siehe letzte Seite /
Printed at: see last page
ISBN: 978-3-8381-5157-1

Zugl. / Approved by: University of Cologne, Diss., 2014

Table of Contents

Table of Contents

Abstract

Soil protists occupy key nodes in soil food webs due to their high abundance, fast turnover and functional importance as bacterial grazers. However, methodological drawbacks obscure the knowledge of soil protists, so that many taxa remain unknown. The structure of natural protist communities and taxa-specific ecological functions are also largely unknown. This thesis aims to increase the knowledge on soil protists using a variety of approaches.

In the first part, naked amoebae being presumably the most neglected protist morphogroup were cultivated from several distinct geographical locations across Europe and from high altitude sites in Tibet. During the course of this study, 16 new species and seven new genera were discovered, representing two eukaryotic supergroups (Amoebozoa and Excavata), three classes (Discosea, Variosea and Heterolobosea), and 12 genera (*Cochliopodium, Stenamoeba, Acanthamoeba, Ischnamoeba* n.g., *Darbyshirella* n.g., *Heliamoeba* n.g., *Arboramoeba* n.g., *Angulamoeba* n.g., *Telaepolella, Schizoplasmodiopsis, Allovahlkampfia* and *Parafumarolamoeba* n.g.). This vast number of taxonomic descriptions unveils the tremendous lack of knowledge especially on soil naked amoebae.

The second part aims at deciphering the diversity and community structure of soil protists using four different techniques. First, a modified cultivation technique enabled the quantification of individual morphogroups, allowing determination often up to genus level. It was shown that soil moisture not only impacts the total abundance, but also affects the community composition of soil protists. Second, established cultures of morphologically indistinguishable *Acanthamoeba* spp. were sequenced, enabling differentiation to strain-level. Highly diverse *Acanthamoeba* communities were detected that differed between geographically remote soils. Third, a high-throughput amplicon sequencing approach targeting the protist phylum Cercozoa illustrated that cercozoan communities strongly differed between geographically distant soils and that community composition depended on the type of land management. Finally, a metatranscriptomic approach unveiled the entire active protist community in different soil and litter samples, uncovering a previously unknown diversity. Each of these methods confirmed that soils harbour an enormous diversity of protists and all methods detected differences in soil protist communities depending on geographic origin or treatment. Therefore it is important to understand the respective advantages and disadvantages associated with each of those methods. Further, available skills, equipment and financial resources need to be considered before applying a method to study soil protists.

The third part of this thesis aims to elucidate rarely considered ecological functions performed by soil protists. Generally, soil protists are considered as bacterivores, but diverse protists facultatively consumed fungi. The small amoeba *Cryptodifflugia operculata* revealed another feeding mode – trapping and consuming nematodes. Metatranscriptomic data revealed high relative abundances of both functional groups in all terrestrial soil samples indicating wider ecological functions carried out by soil protists as has been suggested before.

Zusammenfassung

Bodenprotisten nehmen aufgrund ihrer hohen Abundanzen und Reproduktionsraten sowie ihrer Rolle als Hauptkonsumenten von Bakterien eine Schlüsselposition in Nahrungsnetzen ein. Diese funktionelle Bedeutung weist eine deutliche Diskrepanz zur Berücksichtigung in ökologischen Studien und dem bestehenden Detailwissen über Protisten auf. Methoden der Isolierung und quantitativen Analyse der Protistengemeinschaft in Böden detektieren nur einen Bruchteil der Organismen. Die genaue Artenzusammensetzung und individuelle ökologische Bedeutungen bleiben somit im Verborgenen. Diese Arbeit zielt darauf, die Gemeinschaften von Bodenprotisten und deren Funktion durch eine breite Methodennutzung genauer zu charakterisieren.

Im ersten Teil dieser Arbeit liegt der Fokus auf der Kultivierung und Beschreibung bisher unbekannter Amoeben, die aus deutschen, italienischen, tibetanischen und niederländischen Böden isoliert wurden. Dieser Kultivierungsansatz ermöglichte die Beschreibung von 16 neuen Arten sowie sieben neuen Gattungen, die sich in die eukaryotischen Supergruppen Amoebozoa und Excavata, Klassen Discosea, Variosea und Heterolobosea, und 12 Gattungen (*Cochliopodium*, *Stenamoeba*, *Acanthamoeba*, *Ischnamoeba* n.g., *Darbyshirella* n.g., *Heliamoeba* n.g., *Arboramoeba* n.g., *Angulamoeba* n.g., *Telaepolella*, *Schizoplasmodiopsis*, *Allovahlkampfia* und *Parafumarolamoeba* n.g.) aufteilen. Trotz deren hohen Abundanz und Diversität beweist diese hohe Anzahl an Neubeschreibungen die lückenhafte (Arten-) Kenntnis über Bodenamoeben.

Im zweiten Teil der Arbeit liegt der Schwerpunkt auf der Untersuchung der Diversität und dem Vergleich ganzer Gemeinschaften von Bodenprotisten. Es wurden vier grundsätzlich verschiedene Analysewege genutzt. Eine modifizierte Kultivierungsmethode erlaubte die Bestimmung und Quantifizierung von Morphogruppen, oft bis auf Gattungsebene. In Abhängigkeit von der Bodenfeuchte variierte die Gesamtabundanz sowie die anteilige Zusammensetzung der Protistengemeinschaft. Des Weiteren wurden

5

morphologisch nahezu identische Amoeben der Gattung *Acanthamoeba* molekular bis auf (Unter-) Artebene bestimmt, die im Vorfeld kultiviert wurden. Acanthamoeben zeigten eine hohe Diversität innerhalb und zwischen geographisch entlegenen Böden. Zur Bestimmung des Protistenphylums Cercozoa diente eine primerbasierte, high-throughput Amplikonsequenzierung. Die Cercozoengemeinschaft, deren hohe Diversität in allen Böden bestätigt wurde, variierte je nach Landnutzung und Geographie. Eine Analyse der aktiven Protistengemeinschaft erfolgte durch einen Metatranskriptomansatz. Neben der detaillierten Bestimmung aller Protistengruppen der Proben, die sich untereinander stark unterschieden, fanden sich in terrestrischer Umgebung bislang unbekannte Protistengruppen. Jede dieser Methoden zeigt einen unterschiedlichen Ausschnitt der Protistendiversität und geht mit Vor- und Nachteilen einher, die vor Nutzung erörtert werden müssen, da alle eine Option zur Bestimmung der Protistengemeinschaft bleiben.

Im dritten Teil werden funktionelle Nischen der Bodenprotisten in Nahrungsnetzen gezeigt. Eine Vielzahl von Protisten konsumierten neben Bakterien auch Pilze und sind daher eher als omnivor einzustufen. Bei der kleinen Schalenamoebe *Cryptodifflugia operculata* ging der Fraß von deutlich größeren Nematoden mit einer erhöhten Reproduktionsrate einher. Metatranskriptomdatensätze zeigen hohe relative Abundanzen für sowohl omnivore als auch nematophage Protisten. Dies legt nahe, dass Protisten über die bakterivore Rolle hinaus ein weites funktionelles Spektrum in Böden einnehmen.

General Introduction

Soil Protists

Heterotrophic protists are among the most abundant soil organisms with estimated numbers usually ranging between 10^3 and 10^4 per gram soil (Bamforth 1980, Clarholm 1985, Finlay et al. 2000, Domonell et al. 2013). The majority of soil heterotrophic protists (hereafter simplified as "protists") is suggested to be small (< 200 µm), unicellular, reproduce asexually and feed on bacteria, but exceptions to all of these features are the rule rather than the exception (Levine et al. 1980, Darbyshire 1994, Adl and Gupta 2006, Esteban et al. 2006). Soil protists are classically divided into the four most abundant morphological groups, i.e. naked amoebae, testate amoebae, flagellates and ciliates. The taxonomical classification of protists has, however, undergone rapid and fundamental changes in the last decades due to technical developments, such as electron microscopy and most fundamentally by molecular tools (Levine et al. 1980, Cavalier-Smith 1993, Cavalier-Smith 1998, Adl et al. 2005, Adl et al. 2012). Heterotrophic, autotrophic and mixotrophic protists turned out to be highly paraphyletic, being intermingled throughout the eukaryotic tree of life, and represent the vast majority of eukaryotic organisms at least in terms of diversity (Cavalier-Smith 1998, Adl et al. 2005, Adl et al. 2012). Therefore, the term "protozoa" deriving from their nutritional requirements, is diminishing from scientific use (Adl et al. 2014). Flagellates, naked and testate amoebae are composed of paraphyletic species that branch in distant places in eukaryotic phylogenies, while only ciliates form a monophyletic group (Cavalier-Smith 1998, 2003, Adl et al. 2005, Adl et al. 2012). The dominant soil flagellate groups, i.e. small amoeboid cercozoans, free-swimming *"Spumella"*-like and *"Bodo"*-like flagellates (Finlay et al. 2000, Ekelund et al. 2001, Domonell et al. 2013) comprise taxonomically unrelated groups, such as *"Bodo"* in the supergroup Euglenozoa, and both Cercozoa (former supergroup Rhizaria) and *"Spumella"* (former supergroup Stramenopiles) being in the huge supergroup SAR (= S̲tramenopiles, A̲lveolates

7

and Rhizaria (Burki et al. 2007, Adl et al. 2012)). Ciliates are members of Alveolata in the remaining SAR clade. Testate amoebae are found in both supergroups Amoebozoa and SAR (Rhizaria) (Nikolaev et al. 2005). The majority of naked amoebae belong to Amoebozoa, while most of the remaining naked amoebae are members of the supergroup Excavata (class Heterolobosea) and SAR (Rhizaria) (Adl et al. 2005, Adl et al. 2012). Knowledge on most other protists in soils is sparse, but there is evidence that some typically aquatic organisms inhabit soils, such as Foraminifera (Meisterfeld et al. 2001, Lejzerowicz et al. 2010), Heliozoa (Stout 1984), Dinoflagellata (Bates et al. 2013) and Choanoflagellata (Ekelund et al. 2001, Tikhonenkov et al. 2012). Additionally, parasitic Apicomplexa seem to be widespread members of soil protist communities (Bates et al. 2013).

Protist species numbers are difficult to estimate because the species concept differs between protist groups and an enormous cryptic diversity within morphologically similar species has been detected by molecular techniques (Boenigk et al. 2005). The use of sequence data is largely based on the small subunit of the ribosomal DNA (SSU), and this sequence data is useful to supplement morphological information on distinct species or strains (Pawlowski et al. 2012). As a consequence, sequence information for (soil) protists is rapidly accumulating and molecular tools now allow soil surveys without intensive cultivation efforts. Environmental cloning and sequencing and the recent advent of high-throughput sequencing (HTS) techniques therefore provided and keep on providing new insights on the community composition of soil protists, resulting in the discovery of entirely new groups or groups unknown from soils (Fell et al. 2006, Lara et al. 2007a, Lejzerowicz et al. 2010, Bates et al. 2013). Nevertheless, knowledge on soil protists on both, individual and community level is still in its infancy, with new taxa being discovered in basically every soil survey (Urich et al. 2008, Lejzerowicz et al. 2010, Bates et al. 2013, Heger et al. 2013). Classical cultivation-based studies are, however, still essential to increase the reference database in order to assign morphological and eventually functional information to the sequence data.

Naked Amoebae

Testate amoebae, hereafter referred to as "testates" are fairly well investigated due to comparably large body sizes and stable morphological characters (Foissner 1999b, 2006, Wilkinson and Mitchell 2010, Lahr et al. 2013), while knowledge on their non-testate, naked counterparts, hereafter simplified as "amoebae", remains sparse. Amoebae are particularly difficult to identify due to their small size, transparent cell body and the lack of stable morphological characters. Expert knowledge is thus indispensable for their reliable identification (Foissner 1999b, Smirnov et al. 2008). Amoebae cannot be extracted from soils, and their isolation requires establishment of tedious enrichment cultures (Ekelund and Rønn 1994, Foissner 1999b, Smirnov 2003). Consequently, direct enumeration of amoebae is impossible in soils and largely relies on diverse modifications of the most probable number technique (MPN) (Darbyshire et al. 1974, Rønn et al. 1995). MPN is based on a serial dilution of a soil suspension, but protist detection is hampered by high loads of particles remaining especially in low dilutions, which restrict reliable identification of protists. Therefore, MPN studies usually lump soil protists into very rough morphogroups without further taxonomic differentiation, such that paraphyletic amoebae are commonly treated as a single group (Zwart et al. 1994, Scherber et al. 2010, Gabilondo and Bécares 2014). The liquid aliquot method (LAM) uses highly diluted aliquots of the soil suspension and allows a much more detailed deeper determination of amoebae (Butler and Rogerson 1995, Finlay et al. 2000, Domonell et al. 2013), but the drawback are its high work- and time-intensity and reliance on profound taxonomic expertise (Smirnov et al. 2008, De Jonckheere et al. 2012). Recent LAM studies grouped soil amoebae according to morphogroups (Butler and Rogerson 1995, Smirnov and Brown 2004, Domonell et al. 2013), but most morphogroups turned out to contain distinct paraphyletic taxa (Amaral-Zettler et al. 2000, Lahr et al. 2011, Smirnov et al. 2011b). The taxonomic resolution of LAM studies thus differs considerably from the latest taxonomy. This again raises the issue that detailed information on the abundance and community structure of soil amoebae is largely missing.

9

Molecular tools promised to fill that information gap by avoiding many problems involved with cultivation-based methods especially when studying amoebae. However, when applying "general eukaryotic primers" on soil DNA that target a wide range of protists, amoebae, especially the supergroup Amoebozoa, turned out to be largely underrepresented (Berney et al. 2004, Baldwin et al. 2013, Bates et al. 2013, Risse-Buhl et al. 2013). Reasons are inferior SSU amplification due to above-average length of the SSU and common mismatches in primer regions (Berney et al. 2004, Epstein and López-García 2008). Specific amoebozoan-wide primers do not exist, and more specific primers targeting groups within Amoebozoa have not yet been applied in soil surveys. Such focused studies on other protists have, however, revealed a much deeper resolution of the protist diversity (Bass and Cavalier-Smith 2004, Lara et al. 2007b, Lejzerowicz et al. 2010, Vannini et al. 2013). Most of all, a reliable comprehensive reference database to extrapolate more meaningful information from soil molecular surveys is urgently needed.

Diversity, distribution and community structure

The limited knowledge on the diversity of soil protists, and especially amoebae, has resulted in an ongoing debate about the diversity and distribution of these organisms (Finlay 1998, Foissner 1998, Finlay 2002, Foissner 2006). Recent molecular information targeting some protist groups, however, strongly suggests that heterogeneous spatially distinct soil communities exist even within morphologically indistinguishable protist groups (Bass et al. 2007, Boenigk et al. 2007, Heger et al. 2013). For amoebae, one study found different sequences within the morphologically indistinguishable amoeba species *Vannella simplex* that differed between sites of isolation (Smirnov et al. 2002). Sequence data are rapidly accumulating in the genus *Acanthamoeba*. A plethora of sequenced *Acanthamoeba* spp. indicate an enormous hidden diversity within morphologically indistinguishable species in this genus with unique sequences being discovered in basically all studies (Gast et al. 1996, Stothard et al. 1998, Qvarnstrom et al. 2013, Risler et al. 2013). The morphological species concept of *Acanthamoeba* spp. has thus been replaced by sequence types to cope with the enormous sequence diversity in *Acanthamoeba* (Page 1988, Gast et al. 1996, Stothard et al. 1998, Risler et al. 2013). Several of these sequence types include, however, different distinct species that have formerly been described based on morphological features.

Some studies evaluated the community composition of soil protists with a morphotype approach and found that soil protist communities are largely influenced by precipitation, soil moisture and organic matter content of soils (Anderson 2000, Bass and Bischoff 2001, Anderson 2002). Information about the community composition of soil protists acquired with molecular tools are still sparse, but moisture content has also been identified as the key driver affecting the protist community in a large-scale 454 study (Bates et al. 2013). In this study the authors detected an increasing change of the protist community with distance, supporting the notion that deep molecular studies will fundamentally revolutionize the knowledge on soil protist communities. Also plants and even distinct plant species might strongly affect the soil protist community (Turner et al. 2013).

Nevertheless, molecular tools are not without limitations. Amplification efficiency is strongly dependent on primer choice with amoebae are being underrepresented (Berney et al. 2004, Epstein and López-García 2008). DNA of dead or inactive organisms can be transcribed (Pawlowski et al. 2011), while fundamental differences in SSU copy numbers between protists exclude quantitative comparisons (Gong et al. 2013). PCR errors further inflate protist-specific sequence numbers (Medinger et al. 2010, Behnke et al. 2011). Therefore, evaluating the community composition of soil protist with molecular techniques based on DNA is strongly hampered. Using RNA in metatranscriptomic approaches combined with reverse-transcribing RNA with random hexamers is a promising alternative and avoids most problems associated with DNA and primers. Few studies demonstrating the suitability of this approach have successfully recovered diverse protist communities from soils revealing more plausible protist community compositions than comparable DNA based studies (Urich et al. 2008, Tveit et al. 2012, Turner et al. 2013). As expected, amoebae represented a significant proportion of protist

11

sequences in those studies. By capturing the active protist community, metatranscriptomic approaches might therefore even outperform cultivation-based techniques that have been suggested to predominantly enumerate inactive, encysted protists that might be less important in soil functioning (Foissner 1987, Berthold and Palzenberger 1995). Depending on the question being addressed, all methods seem to find their eligibility in soil surveys, but (dis)advantages of the respective methods have to be considered.

Ecological importance

Protists and especially amoebae occupy key positions in soil systems. They are considered to represent the major bacterial grazers, thereby linking nutrient flow to higher trophic soil organisms (Hunt et al. 1987, de Ruiter et al. 1995, Crotty et al. 2011). They also liberate nutrients from the consumed bacterial and stimulate plant growth (Clarholm 1985, Bonkowski 2004). However, most information on the impact of protist grazing on bacterial communities has been obtained in studies with just one or few model protists. Since the diversity of protists in morphology and phylogeny is enormous (Cavalier-Smith 1993, Cavalier-Smith 1998, Adl et al. 2012), it seems inevitable that at least certain protists differ in their ecological functions. Glücksman et al. (2010) investigated the impact of cercozoan flagellates and found that the bacterial community was changed in a species-specific manner. Most strikingly, they found substantial feeding differences even between closely related species. Similarly, different feeding impacts on the bacterial community were shown for related ciliates (Weisse et al. 2001), while amoebae-specific feeding differences have not yet been studied. More natural conditions investigating interactive effects of different protists and potential feeding differences are even more rarely approached. Recent studies showed that an increased diversity of the protist community enhanced bacteria community diversity by reducing bacterial competition in favour of subdominant species, while total bacterial abundance was reduced (Saleem et al. 2012, Saleem et al. 2013).

Protists are generally treated as bacterivores, but other functional feeding groups and respective ecological functions have rarely been addressed. Cryptic information, however, strongly suggests that trophic interactions among protists are common, as successful cultivation of various protists is only possible in presence of other protists. For instance, *Paradermamoeba levis* has been cultivated on small vannellid amoebae (Smirnov et al. 2007), *Deuteramoeba algonquinensis* needed small amoebae or flagellates for growth (Mrva 2010) and *Thecamoeba* spp. has been co-cultivated on other amoebae or ciliates (Page 1977). Not surprisingly, also algae serve as prey for a variety of amoebae (Smirnov and Brown 2004, Mrva 2010, Smirnov et al. 2011a, Hess et al. 2012, Berney et al. 2013). Several protist taxa are further known to feed on fungi, for example flagellates (Hekman et al. 1992, Ekelund 1998, Flavin et al. 2000), ciliates (Petz et al. 1985, Petz et al. 1986, Foissner 1999a) and a variety of amoebae (Old and Oros 1980, Chakraborty and Old 1982, Chakraborty et al. 1983, Mrva 2010). Among soil fauna, nematodes have been shown to be affected by protists. Small flagellates have been shown to kill, but not feed on nematodes (Bjørnlund and Rønn 2008), while larger testate amoebae (Yeates and Foissner 1995), ciliates (Doncaster and Hooper 1961) and amoebae (Sayre 1973) directly feed and grow on nematodes. In addition to direct feeding, non-trophic interactions, both positive and negative have been reported between protists and fungi (Rønn et al. 2002a, Vohník et al. 2011, Koller et al. 2013), nematodes (Neidig et al. 2010, Bjørnlund et al. 2012) and earthworms (Bamforth 1988, Bonkowski and Schaefer 1997, Winding et al. 1997, Tiunov et al. 2001). Detailed knowledge on all of these interactions remains, however, on a very crude taxonomic level. It seems that we only grasped the tip of the iceberg on deciphering the importance and the multitude of ecological functions carried out by soil protists.

Further, some soil protists pose direct risks to human health. Among the facultative pathogens are amoebae of the genus *Acanthamoeba*, which are ubiquitous and very abundant in soils (Page 1988). Several *Acanthamoeba* spp. cause the eye infection Amoebic Keratitis (Schuster 2002, Siddiqui and Ahmed Khan 2012), and also fatal Granulomatous Amoebic Encephalitis (GAE)

(Schuster 2002, Khan 2006, Qvarnstrom et al. 2013). Similarly, GAE can be caused by *Naegleria fowleri* and *Balamuthia mandrillaris* (Visvesvara et al. 1993, Schuster 2002, De Jonckheere 2004). While *N. fowleri* has so far only been reported from hot water environments (De Jonckheere 2004), soils are likely to represent environmental niches for *B. mandrillaris* (Dunnebacke et al. 2004, Ahmad et al. 2011). In addition to these directly hazardous protists, endosymbiotic pathogenic bacteria are common in amoebae, such as *Mycobacterium*, *Legionella*, *Salmonella* and *Listeria* (Anand et al. 1983, Abu Kwaik 1996, Greub and Raoult 2004, Lamoth and Greub 2010). Protists are therefore considered to be "Trojan horses" for pathogenic bacteria to evade the human immune defence (Greub and Raoult 2004, Horn and Wagner 2004). Taken together, soil protists are highly diverse members of the soil food web and their ecological importance is commonly being accepted, but detailed knowledge on a wide range of species, the community compositions and taxon-specific ecological functions of soil protists is largely missing.

Aims

The main objective of this thesis was to increase the knowledge on abundant taxa, community structures and functions of soil protists. Therefore, a variety of distinct approaches were applied that are summarised in three major parts.

Part 1

This part, aimed at increasing the knowledge on soil protists focusing on amoebae, through extraction and cultivation from geographically distant soils across Europe and Tibet. These cultures form the basis for the description of new species and genera from four distinct clades of amoebae, summarized in four chapters;

Chapter 1 – Two *Stenamoeba* species, *S. sardiniensis* n. sp. and *S. berchidia* n. sp. isolated from the same soil on Sardinia, Italy are described, revealing morphological characters that help in taxonomic re-classification of the class Discosea.

14

Chapter 2 – *Cochliopodium plurinucleolum* n. sp. also extracted from Sardinian soil is described and the genus *Cochliopodium* spp. is revised.

Chapter 3 – Six new genera of the class Variosea cultivated from diverse soils (and freshwater) are erected, enabling a deep re-structuring of Variosea.

Chapter 4 – The new heterolobosean genus *Parafumarolamoeba* and six new *Allovahlkampfia* strains are described cultivated from a range of soils from Tibet, Sardinia and the Netherlands.

Part 2

This part aimed at amending existing information on the diversity, abundance and community composition of soil protists. Four different methods are applied to target distinct questions on soil protist communities, each depicted in a single chapter;

Chapter 5 – The abundance and diversity of soil protists is investigated microscopically using a modified enrichment cultivation technique, to analyse the impact of changes in soil moisture conditions on the protist community.

Chapter 6 – Cultures of *Acanthamoeba* established in the first part of this thesis (Chapters 1 - 4) are subject to more focused molecular analyses followed by phylogenetic analyses to investigate potential cryptic diversity of *Acanthamoeba* spp.

Chapter 7 – The protist phylum Cercozoa is targeted in a HTS approach applying specific primers to compare cercozoan community compositions in a range of geographically distant soils and between different levels of soil treatment in order to evaluate main drivers determining cercozoan soil communities.

Chapter 8 – The entire soil protist community is analysed using a metatranscriptomic approach to investigate and compare active protist communities between soils and to explore protist clades uncommon for soils.

Part 3

The last part aimed at deciphering ecological functions performed by soil protists focusing on feeding interactions other than with bacteria. Using predominantly cultures obtained in Part 1, interactions between protists and other soil eukaryotes are being investigated, to evaluate whether the current classification of protists as mainly bacterivorous holds true.

Chapter 9 – A wide range of cultivated soil protists were tested for facultative fungal-feeding potential. Further, the presence and relative abundance of known obligate fungal feeding protists is determined using data-mining approaches, e.g. from sequences obtained in Chapter 8.

Chapter 10 – The interaction of the small testate amoebae *Cryptodifflugia operculata* with soil nematodes are investigated and the presence and potential importance of this interaction in soil investigated in terrestrial ecosystems examined using the metatranscriptomic data obtained in Chapter 8.

Part 1

Cultivation and descriptions of new protist species and genera

Part 1 – Chapter 1

Two new species of the genus *Stenamoeba* (Discosea, Longamoebia): cytoplasmic MTOC is present in one more amoebae lineage

Geisen Stefan[2], Weinert Jan[2], Kudryavtsev Alexander[1], Glotova Anna[1], Bonkowski Michael[2] and Smirnov Alexey[1]

[1] Department of Invertebrate zoology, Faculty of Biology and Soil sciences, Saint Petersburg State University
[2] Department of Terrestrial Ecology, Faculty of Zoology, University of Cologne

Abstract

Two new species of the recently described genus *Stenamoeba*, named *S. berchidia* and *S. sardiniensis* were isolated from a single soil sample on Sardinia, Italy. Both share morphological features characteristic to *Stenamoeba* and form in phylogenetic analyses together with other *Stenamoeba* spp. a highly supported clade within the family Thecamoebidae. The ultrastructural investigation of *Stenamoeba sardiniensis* revealed the presence of cytoplasmic microtubule-organizing centres (MTOCs), located close to one of several dictyosomes found inside the cell. This is the first report of cytoplasmic MTOCs among Thecamoebidae. The presence of MTOCs is now shown in five of nine orders comprising the class Discosea and potentially could be a phylogenetic marker in this group. We re-isolated *Stenamoeba limacina* from German soils. This strain shows a similar morphology and an almost complete SSU rDNA sequence identity with the type strain of *S. limacina* originating from gills of fishes, collected in Czech Republic.

Introduction

Naked lobose amoebae (gymnamoebae) belong to the phylum Amoebozoa Luhe 1913, which comprises a number of phylogenetic lineages, covering all groups of lobose amoeboid protists (Cavalier-Smith 1998, Smirnov et al. 2011b). Among them, gymnamoebae are distributed among three groups currently recognized in the rank of classes - Tubulinea, Discosea and Variosea (Smirnov et al. 2011b). While Tubulinea are well outlined and proven to be monophyletic, and the same is probably true for Variosea, the monophyly of Discosea is more ambiguous, since they appear paraphyletic in most analyses (Cavalier-Smith et al. 2004, Smirnov et al. 2005, Tekle et al. 2008, Kudryavtsev et al. 2009a, Pawlowski and Burki 2009, Lahr et al. 2011). Discosea usually segregate into a number of relatively independent lineages with unstable position; a remarkable exception is the stable coupling of the orders Vannellida and Dactylopodida. However, Dermamoebida, Thecamoebida and Centramoebida showed weak statistical support in recent analyses (e.g. Kudryavtsev and Pawlowski 2013), but a surprisingly stable tendency to form a clade, recognized as the subclass Longamoebia in Smirnov et al. (2011b). Morphologically Thecamoebida, Centramoebida and Dermamoebida are rather different and until now there is no evident character that could be considered as a synapomorphy of this clade.

The genus *Stenamoeba* (order Thecamoebida, family Thecamoebidae) was established in 2007 to accommodate the former *Platyamoeba stenopodia* Page 1969 after the genus *Platyamoeba* was abandoned based on molecular data by Smirnov et al. (2007). It comprises oblong, linguiform amoebae with a large anterior area of hyaloplasm, usually occupying half or more of the cell. During locomotion several longitudinal ridges similar to those in *Thecamoeba* occasionally appear, but they never remain stable such as in *Thecamoeba*. A single nucleus is usually positioned at the border separating hyaloplasm and granuloplasm. Amoebae with *Stenamoeba*-like morphology are common in various environments, but *Stenamoeba* was monotypic until *S. limacina* and *S. amazonica* isolated from organs of freshwater fish hosts added (Dyková et al. 2010b). Both of these species display characteristic morphological features unique to *Stenamoeba*, supported by molecular data revealing a solid monophyly of the genus *Stenamoeba* (Dyková et al. 2010b). This was the first indication that the genus *Stenamoeba* may be rather species-rich, but many new species may have not been recognized earlier due to morphological similarities with *S. stenopodia* and the shortage of discriminating characters. Recent add-ons of sequences, probably belonging to unnamed members of the genus *Stenamoeba* to the GenBank database further supported this idea.

In the present paper we describe two more species of *Stenamoeba* and show the presence of cytoplasmic microtubule-organizing centres (MTOCs) in one of the newly described isolates. We also report a new strain of *Stenamoeba limacina* isolated from soil of the Hainich-Dün region (Germany) that shares profound morphological and almost complete sequence identity with the type strain described by Dyková et al. (2010b).

Materials and Methods

Sampling sites

Samples were taken from the upper 20 cm of mineral soil of an intensively managed grassland plot at Berchidda-Monti long term observatory, managed by the University of Sassari, on the island of Sardinia (Italy), Berchidda district, 40°46′N, 9°10′E (Lagomarsino et al. 2012), the upper 20 cm of an ex-arable field in the central part of the Netherlands (52°06′N, 6°00′E), and the upper 10 cm of a beech forest in the Hainich-Dün region (Germany), 50° 56′ 14.5″-51° 22′ 43.4″ N, 10° 10′ 24.0″-10° 46′ 45.0″ E. Sampling was performed in the course of 2 different projects, EcoFINDERS (http://ecofinders.dmu.dk/) in 2011 (Sardinian site), and Biodiversity Exploratories (Fischer et al. 2010), http://www.biodiversity-exploratories.de/1/home/, in the year 2008 (German site).

Isolation and cultivation

Soil samples were incubated in 90 mm Petri dishes with 0.15 % wheat grass (WG) medium, made by adding vacuum-dried wheat grass powder (Weizengras, Sanatur GmbH, D-78224 Singen) to PJ medium (Prescott and James 1955a) to the weight concentration of 0.15 %. The medium was autoclaved and sterile filtered through

0.45 µm Whatman filter paper. Parallel samples were inoculated with the same medium and agarised by adding non-nutrient agar (AppliChem, Darmstadt, Germany) at a final concentration of 1.5 % (WG agar). The agar was autoclaved (122°C, 20') prior to use. Enrichment cultures were maintained for 10-14 days at room temperature and ambient light. To establish clonal cultures, amoebae were picked manually with a tapered glass pipette using an inverted phase-contrast microscope, and transferred to 60 mm Petri dishes with WG medium. Amoebae were subcloned once or twice until free from other eukaryotes, and fed on the accompanying non-identified bacteria. Observations and measurements of live cultures in Petri dishes were made using Leica DMI3000 inverted microscope equipped with PhaCo (Phase contrast) and IMC (Integrated modulation contrast) optics. Observations and photographs of amoebae moving across the glass surfaces were done using either a Leica DM2500 microscope equipped with PhaCo and DIC (Differential Interference Contrast) optics or a Nikon Eclipse 90i equipped with PhaCo and DIC optics.

Electron microscopy

For transmission electron microscopy (TEM) amoebae were fixed at +4°C with a 2.5 % solution of glutaraldehyde in a 0.05M sodium cacodylate buffer (pH 7.4) for 40 min followed by postfixation with 1 % osmium tetroxide prepared with the same buffer for 1 hour. Cells were washed with buffer (3x5 min) between fixation steps. Fixation was initiated in Petri dishes by a quick replacement of culture medium with glutaraldehyde. During buffer washes cells were scraped away from the Petri dish bottom and collected by gentle centrifugation. After osmium tetroxide treatment, cells were washed with buffer (2x5 min) followed by a gradual decrease in buffer concentration down to glass-distilled water. Amoebae were finally embedded in 2 % agar prepared with glass-distilled water; blocks of agar (ca. 1 mm^3) containing cells were subsequently cut out, dehydrated in a graded ethanol series followed by 100 % acetone, infiltrated and embedded in Epon 812 epoxy resin (Fluka). Silver to light gold sections were cut with a diamond knife on a Leica Ultracut 6 ultramicrotome and double-stained with 2 % uranylacetate prepared with 70 % ethanol and Reynolds' lead citrate. Sections were observed using a JEOL JEM1400 transmission electron microscope operated at 80kV.

DNA Extraction

DNA was extracted with guanidine isothiocyanate buffer (Maniatis et al. 1982b). To collect cells, 60 mm Petri dishes containing amoebae were washed 2-3 times with fresh medium. The medium was always completely discarded form the dish. Subsequently, 100 µl guanidine isothiocyanate buffer was added to the dish and distributed across its entire bottom. Cells were scraped off using a disposable cell scraper, and the buffer with floating cells was collected with a Pasteur pipette and transferred into an Eppendorf tube before being precipitated with isopropanol and ethanol according to above cited protocol.

PCR

SSU rDNA gene fragments were amplified using universal eukaryotic primers (Table 1). The cycling conditions included a 5 min initial denaturation at 95°C followed by 30 cycles (each comprising 95°C for 30 sec, 50°C for 60 sec, and 72°C for 120 sec), and a final elongation step at 72°C for 5 min. Subsequently, 8 µl of the PCR product were purified by adding 0.15 µl Endonuclease I (20 U/µl, Fermentas GmbH, D-68789 St. Leon-Rot), 0.9 µl Shrimp Alkaline Phosphatase (1 U/µl, Fermentas GmbH) and 1.95 µl water to a final volume of 11 µl. This mixture was incubated for 30 min at 37°C, and for another 20 min at 85°C. Afterwards (partial) PCR products were sequenced using the Big Dye Terminator Cycle sequencing kit and an ABI PRISM automatic sequencer. Primers used for sequencing are listed in Table 1.

Table 1. Primers used for amplification, lengths of fragments and accession numbers at GenBank as well as the Culture Collection of Algae and Protozoa (CCAP). Primer sequences are as follows: RibA (5' > acc tgg ttg atc ctg cca gt <3'), RibB (5'> tga tcc atc tgc agg ttc acc tac <3') S12.2 (5'> gat cag ata ccg tcg tag tc <3') S20R (5'> gac ggg cgg tgt gta caa <3') (Cavalier-Smith and Chao 1995, Pawlowski 2000).

Species	Primer pair	Sequence length	GenBank accession number
S. berchidia	RibA-RibB	2124 bp	KF547921
S. sardinensis	S12.2-S20R	772 bp	KF547922
S. limacina	Rib A-S20R	698 and 963 bp	KF547923 and KF547924

Phylogenetic analysis

For phylogenetic analysis newly obtained sequences were aligned with an extensive alignment covering all major groups of eukaryotes using Seaview 4 (Gouy et al.

2010); the alignment was manually polished. To increase sampling of *Stenamoeba*-related sequences, sequences obtained in the present study and all named *Stenamoeba* sequences downloaded from GenBank were used as a query for BLASTn search using default search parameters in GenBank; top 10 hits were downloaded; mounted in the same alignment and analysed using PhyML (1148 sites, GTR + Γ model with 4 rate categories, optimized proportion of invariable sites). Sequences that robustly grouped with *Stenamoeba* species were added to the analysis; the rest were removed.

The phylogenetic analysis to obtain maximum-likelihood (ML) tree was performed using a subsample of the alignment containing all Stenamoeba-related sequences and a representative amoebozoan sample. RaxML (Stamatakis 2006) was used with the following parameters: 1605 sites; GTR+Γ model, 25 rate categories (Lanave et al. 1984), optimized proportion of invariable sites. The tree showing the best likelihood value was bootstrapped with 100 replicates. Bayesian analysis was performed using MrBayes 3.2 (Huelsenbeck and Ronquist 2001) with the GTR+Γ model, 8 rate categories and the covarion model. The analyses were performed as two separate runs of four chains each with default parameters, until they ceased to converge (final average standard deviation of the split frequencies less than 0.01); this required about 1 million generations; the first 30 % of generations were discarded as burnin. Calculations were performed using the facilities of Bioportal of the University of Oslo (Kumar et al. 2009).

GenBank numbers of newly obtained sequences are represented in the Table 1; type cultures of *Stenamoeba sardiniensis* and *S. berchidia* as well as a strain of *S. limacina* isolate 61 are deposited with CCAP under the accession numbers CCAP 2571/1 - CCAP 2571/3

Results

Morphological observation

Stenamoeba berchidia n. sp.

23

Light microscopy (Figs. 1-4): Active, moving trophozoites were oblong, or linguiform (Fig. 1). The hyaloplasm was very pronounced, covering up to 2/3 of the cell. *S. berchidia* was broader than *S. sardiniensis*, occasionally exhibiting a nearly fan-shaped appearance. The granular part of the cell was narrower than the hyaline part, often pointed at the posterior end. No distinct uroidal structure was observed. Pronounced longitudinal surface ridges were formed during active locomotion (Fig. 2, black arrow) and transverse waves running over the hyaloplasm were observed occasionally. A single contractile vacuole was always located in the posterior part of the cell.

Average length of the locomotive form was 18.9 μm (range 12-24 μm), and the average breadth 8.6 μm (range 6-12 μm). The average length/breadth ratio was 2.3 (range 1.4-3.7).

Non-directionally moving cells adopted variable shapes, often forming dactylopodia-like projections or assuming an irregular crescent-like shape (Fig. 3). As *S. berchidia* was very active, amoebae virtually never formed a stationary phase during observations. Floating amoebae produced several small pseudopods, not reaching far out of the central cell mass; cells floated not readily and only for a short time under our conditions of observation, so these forms might not have represented developed floating forms.

A single, vesicular nucleus was always located close to the border between the hyaloplasm and the granuloplasm (Figs. 2 and 3, white arrows). The average diameter of the nucleus was 1.7 μm (range 1.3-2.2 μm); the average diameter of the nucleolus was 0.75 μm (range 0.5-0.9 μm). No inclusions such as crystals were observed. Amoebae produced spherical, double-walled cysts of ca. 10 μm (Fig. 4) in diameter in old cultures.

Stenamoeba sardiniensis n. sp.

Light microscopy (Figs. 5-9): Trophozoites in active locomotion were oblong or linguiform, with a distinct frontal area of the hyaloplasm, occupying about 1/2 of the cell (Fig. 5). The granular part of the cell was in general narrower and thicker than the hyaline part, almost pointed at the posterior end in some individuals. No distinct

24

uroidal structures could be observed. At times small longitudinal surface ridges occurred on the dorsal surface of moving cells (Fig. 5, black arrow), but they never stayed stable for a long time. The single contractile vacuole was always located in the posterior part of the cell.

Average length of the locomotive form was 17.0 μm (range 13-24 μm), and the average breadth 6.1 μm (range 5-7 μm). The average length/breadth ratio was 2.8 (range 2.2-3.9).

During non-directional movement, cells sometimes produced very short dactylopodia-like cytoplasmic projections, or became narrow and stretched, or even temporarily branched; the latter was especially characteristic when amoebae tried to adhere both to the object slide and the coverslip (Fig. 6). Stationary amoebae usually adopted an irregular shape with no distinctive characteristics or most often floated just above the cover slip surface, without adopting a pronounced floating form. Notable was also the extremely low ability of trophozoites to attach to the glass slides, making it difficult to observe the typical locomotive morphology. Fully developed floating forms produced up to three thin, blunt and radiating pseudopods of different length (Fig. 9).

The cells observed contained a single vesicular nucleus, always located at the border of hyaloplasm and granuloplasm (Fig. 7, white arrow). The average diameter of the nucleus was 1.7 μm (range 1.5-2.1 μm), the average diameter of the nucleolus 0.65 μm (range 0.5 -0.9 μm). Cells contained no crystals or any other remarkable inclusions. Amoebae in old cultures formed smooth, spherical, double-walled cysts ca. 10 μm in diameter (Fig. 8).

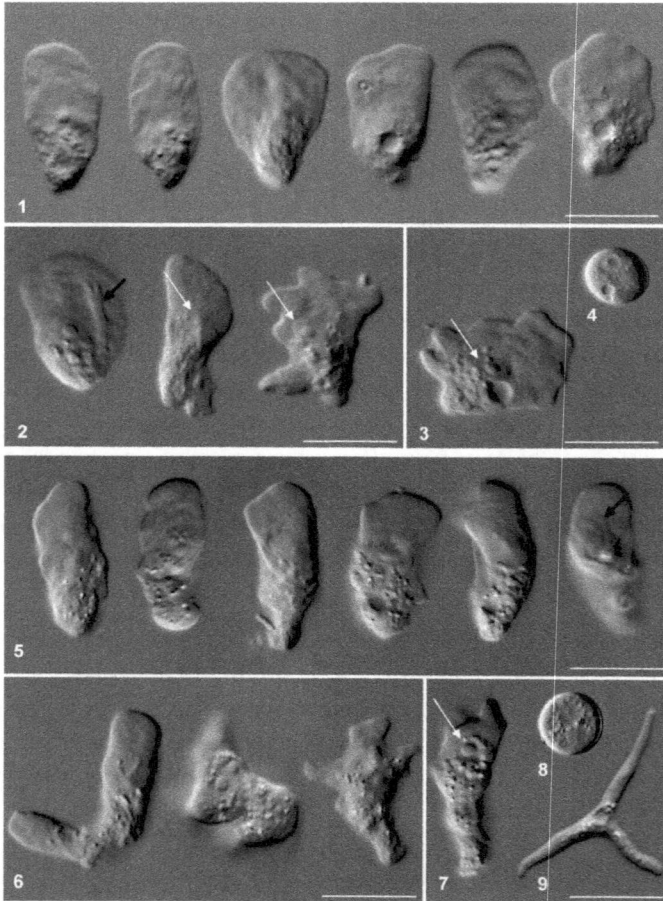

Figs. 1-9. Light microscopic images of newly described *Stenamoeba* species; Scale bars: 10 μm

1-4. *Stenamoeba berchidia*. 1. Locomotive form, ranging from oblong and linguiform to almost fan-shaped. 2. Pronounced longitudinal surface ridges, occasionally formed in few cells (black arrow); single nucleus located at the border between granuloplasm and hyaloplasm (white arrows). 3. Stationary amoebae of irregular shape. White arrow shows nucleus. 4. Mature cyst.

5-9. *Stenamoeba sardiniensis*. 5. Locomotive form, demonstrating characteristic oblong, linguiform shape. some cells show short-lived small longitudinal surface ridges (black arrow); 6. Cells in non-directed movement. 7. The single nucleus located at the border between granuloplasm and hyaloplasm (white arrow). 8. Mature cyst. 9. Floating form.

26

Electron microscopy: The majority of observed cells were covered with the plasma membrane without any visible glycocalyx on its surface (Fig. 10). However, in a few cases a layer of fuzzy material 20-30 nm thick was observed in a limited part of the cell surface area that looked like a food cup under formation (Figs. 11-12). The peripheral part of the cytoplasm contained filamentous material that looked like a microfilament network, and some of these filaments were seen to form bundles extending into the cytoplasmic projections (Fig. 13). The nucleus was irregularly rounded in sections, and contained a rounded, central electron-dense nucleolus that sometimes contained more transparent areas visible in sections, and numerous patches of electron-dense heterochromatin scattered in the nucleoplasm (Fig. 14). Mitochondria in sections were rounded or oval, containing an electron-dense matrix and tubular cristae. Associations of mitochondria with cisternae of rough endoplasmic reticulum were regularly observed (Fig. 15).

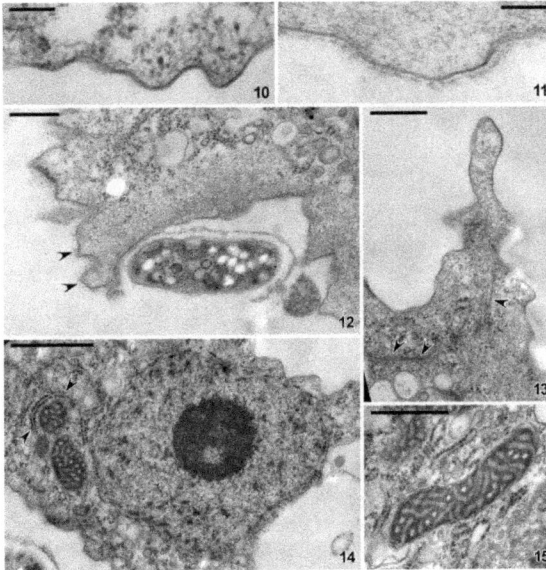

Figs. 10-15. *Stenamoeba sardiniensis*.
General ultrastructure. Scale bar: 0.2 µm in 10-11, 1 µm in 14, 0.5 µm in others.
10-11. Plasma membrane and cell coat (note a layer of fuzzy material over the plasma membrane in 11). 12. Formation of a food cup for ingestion of a bacterium. Arrowheads indicate glycocalyx-like fuzzy layer over the plasma membrane. 13. Cytoplasmic projection with bundles of microfilaments (arrowheads). 14. Nucleus, mitochondria and cisternae of rough endoreticulum (arrowheads). 15. Mitochondrion and associated cisternae of rough endoreticulum.

Large dictyosomes consisting of ca. 10 flattened cisternae were scattered in the cytoplasm (Fig. 16). In all cells numerous microtubules were observed around the dictyosomes expanding into the cytoplasm. In few sections an electron-dense body was seen close to the dictyosomes with numerous microtubules radiating from it (Figs. 17-18). Therefore, it was interpreted as a microtubule-organizing centre (MTOC). This MTOC was most probably cylindrical as it looked in sections either as a rounded body ca. 100nm in diameter, or had an elongated shape ca. 600nm in length. No internal structure of a MTOC could be seen in sections. The very low frequency of occurrence of MTOCs in sections suggests that each cell had a single MTOC that was always located close to the dictyosomes. Not all dictyosomes in the cytoplasm were, however, associated with the MTOC.

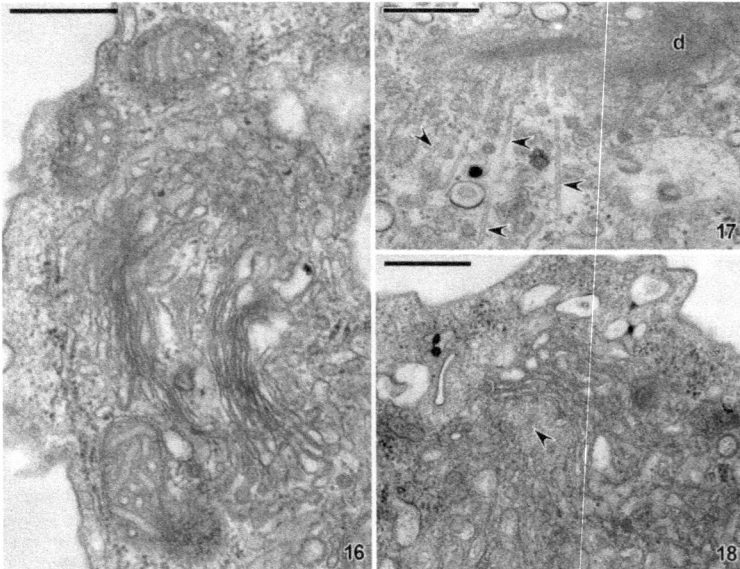

Figs. 16-18. *Stenamoeba sardiniensis*. Dictyosomes and MTOC. Scale bar = 0.5 μm.
16. Part of the cytoplasm showing dictyosomes, mitochondria and cisternae of rough endoreticulum. 17. Microtubule-organizing centre adjacent to a dictyosome (d) in a longitudinal section; arrowheads indicate microtubules. 18. Dictyosome and a putative microtubule-organizing centre in a cross-section (arrowhead).

Stenamoeba limacina strain 61

Light microscopy (Fig. 19): Locomotive morphology and other features of trophozoites were very similar to those presented in the original description of *S. limacina* (Dyková et al. 2010b). Following this description, we also noted the poor adhesion of amoebae to the glass surface, which made light microscopic observations very difficult. Amoebae detached from substratum started to move only after several hours. Average length of trophozoites was 19.9 μm (range 15-29 μm), breadth 6.9 μm (range 4 -11 μm) and length/breadth ration (L/B) 3.0 (range 1.6-4.9). The average diameter of the nucleus was 2.7 μm (range 2.0-3.5 μm), the average diameter of the nucleolus 1.3 μm (range 0.9 -1.7 μm). Cells contained no crystals or any other remarkable inclusions. We have not seen cysts in our cultures.

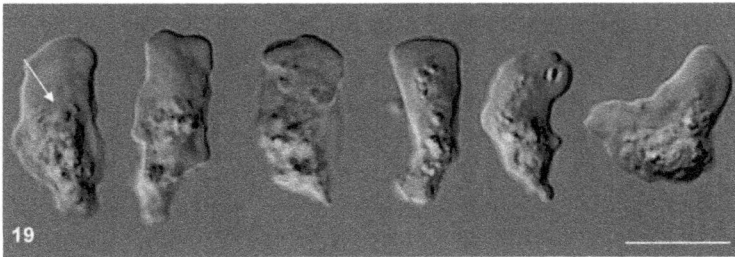

Fig. 19. Light microscopic images of *Stenamoeba limacina* strain 61 in locomotion. The single nucleus is located at the border between granuloplasm and hyaloplasm (white arrow). Scale bar = 10 μm.

Molecular phylogeny

The resulting phylogenetic tree (Fig. 20) revealed three large clades, corresponding to the classes Discosea, Variosea and Tubulinea; the class Discosea divided into two subclasses - Flabellinia and Longamoebia with high PP but negligible bootstrap support. The same pattern occurred across the entire tree with high posterior probabilities for nearly all branches, and with often much lower bootstrap values. The present analysis revealed a well-supported clade unifying the genera *Thecamoeba*, *Sappinia* and *Stenamoeba*, which corresponded to the order Thecamoebida *sensu* Smirnov et al. (2011b) and a clade unifying the orders Dermamoebida, Centramoebida as well as environmental sequences of a so-called

29

LKM74 clade. All three *Stenamoeba* sequences investigated in the present study robustly grouped within the clade corresponding to the genus *Stenamoeba*. Among these, *S. limacina* strain 61 in two sequenced fragments, 698bp in the 5' and 963bp in the 3' region of the SSU rDNA gene (including a fragment of the V4 region) showed nearly complete sequence identity (1 bp difference at position 601 of *S. limacina*) with the sequence of *S. limacina* studied by (Dyková et al. 2010b), which is congruent with their morphological similarity. Two new species, named here *S. sardiniensis* and *S. berchidia* showed distinct sequence differences from all yet known *Stenamoeba* strains. Although these species form short branches within the *Stenamoeba* clade (Fig. 20), this should not be interpreted as evidence of close sequence similarity, because sequences belonging to distinct species and isolates mostly differ in variable regions not included in the phylogenetic analysis, while the sequence differences within the set of sites used for the Amoebozoa-wide analysis were minimal.

Manual analysis of the alignment revealed that the V4 region of the SSU rDNA gene sequence in *S. berchidia* possessed a unique 56bp long signature sequence ranging from position 686 to 742 (AGGGAGAGGGGCGGGGAGGGGCGACCCTCCTCGT TTTCCTTCTCCCTTCGTGCGGC). The V7 region in *S. berchidia* also included a unique signature motif, i.e. GTTCCTTTC (position 1849-1857). The SSU rDNA gene sequence of *S. sardiniensis* has a very distinct V7 region with characteristic motifs at position 595-622 (TTTATCGAACACCGTCTCTTCTTCCTTC) followed by a G-rich area in position 623-641 (GCGGGAGGGGGGGGCGGCG), differentiating it strongly from all other *Stenamoeba* strains.

30

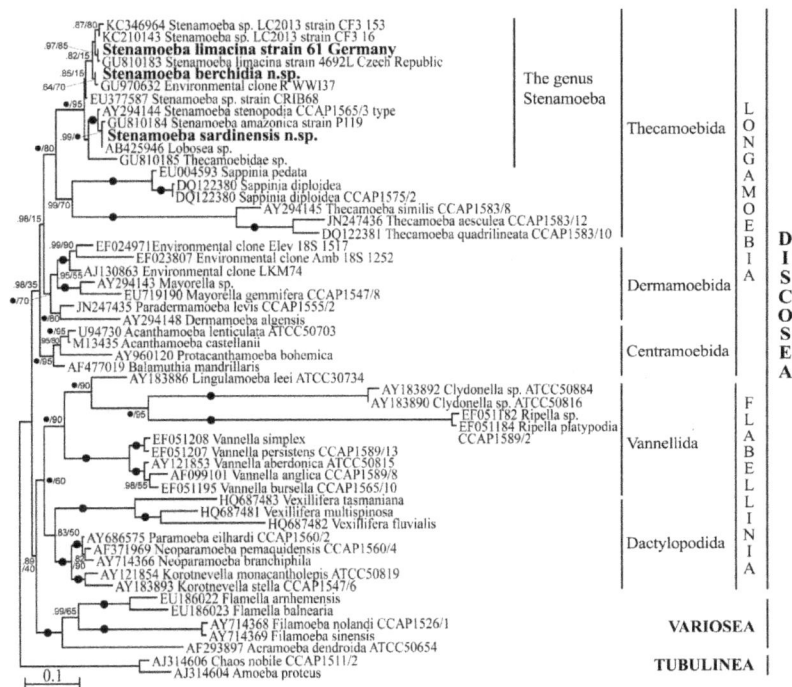

Fig. 20. Phylogenetic tree based on the SSU-rDNA gene. Strains described in this study are in bold. Support values at each node indicated as RAxML / PP value. Black dots on nodes indicate 1.00 / 100 support level; smaller black dots show 1.00 PP values. Indications in nodes without strong support (< 60 % (RAxML) and < 0.60 (BI) omitted). All branches are drawn to scale.

Discussion

Identification of *Stenamoeba limacina* strain 61 and justification to establish two new species - *Stenamoeba berchidia* n. sp. and *Stenamoeba sardiniensis* n. sp.

Stenamoeba limacina strain 61. The *S. limacina* strain 61 in morphology and size is very similar to *S. limacina* Dyková et al., 2010. The limax-like locomotive form reported for *S. limacina* was less pronounced in our strain as the posterior end seemed to narrow more strongly as originally described for *S. limacina*. Our strain was slightly larger than the type one (length 15-28 μm against 15.4–19.2 μm

31

reported for the type strain) but maintained similar length to breadth ratio. Trophozoites overloaded with phagocytosed bacteria noted by Dyková et al. (2010b) were not observed in our strain, which probably depends on the culture conditions. We have not seen cysts in our cultures, but this also may be a physiological property of a particular strain and cannot serve as a distinctive character. Based on the sequence similarity we have to conclude that these differences represent intraspecific polymorphisms. We provide a slightly revised diagnosis of *S. limacina* to incorporate our data on this species and to remove some characters repeated in the diagnosis of the higher taxon (the genus *Stenamoeba*) (Smirnov et al. 2007).

The type strain of *S. limacina* was isolated from kidney tissue of the freshwater fish *Gobio gobio* Linnaeus (Cypriniformes), which was caught in the Lužnice River (Czech Republic, South Bohemia), whereas our strain was isolated from a mineral soil of a beech forest in the Hainich-Dün region (Germany, Thuringia). This is a remarkable (and quite rare) case of reliable amoebae species re-isolation from two very different and relatively distant locations (ca. 550 km linear distance), confirmed both at the morphological and molecular level. The finding of *S. limacina* free-living in the environment supports the idea on its amphizoic nature (Dyková et al. (2010b) and show that this species is not necessarily a freshwater fish-associated one, but can live in soil as well.

Stenamoeba berchidia: Cells of S. berchidia were on average slightly longer (18.9 μm) and explicitly broader (5.9-11.6 μm) than those of *S. sardiniensis* and *S. limacina*. Sometimes *S. berchidia* demonstrated an almost fan-shaped appearance, similar to the shape of *S. amazonica*, but only for short periods. A remarkable character distinguishing *S. berchidia* from all other *Stenamoeba* species is the strong tendency to attach and move on coverslips; trophozoites of *S. berchidia* were always very active, and nearly never stationary. Wave-like surface ripples passing from the anterior to the posterior edge of the cell during locomotion are characteristic for *S. berchidia*. Similar ripples were only documented for species of the genus *Clydonella*, but waves in this genus move in the opposite direction from the posterior to the anterior edge of moving cell (Sawyer 1975).

Stenamoeba sardiniensis: Cells of *S. sardiniensis* shared several characteristics with *S. limacina* such as shape, ratio of hyaloplasm to granuloplasm, low attachment to coverslips, size and length/breadth ratio. However, cyst morphology allows a differentiation as *S. sardiniensis* forms double walled cysts, while cysts in *S. limacina*, if present, appear single-walled. Further, the average cyst diameter of ca 10 μm is larger in *S. sardiniensis*. However, the most reliable way to differentiate these two species remains the SSU sequence, detailed below.

Cytoplasmic MTOC as a phylogenetic marker among lobose amoebae

In contrast with many other groups of amoeboid protists, naked lobose amoebae move predominantly with their acto-myosin cytoskeleton; cytoplasmic microtubules are rare and never form cortical networks or organized bundles (Grębecki 1994). However, a number of amoebae species are known to possess cytoplasmic microtubule-organizing centres (MTOCs) showing a characteristic pattern of microtubules, radiating from an amorphous or lamellar central mass (e.g. Bowers and Korn 1968). Normally MTOCs in naked amoebae are associated with dictyosomes, but in lobose amoebae dictyosomes were never shown to be specifically associated with the nucleus unlike in most other organisms; moreover, in amoebae MTOC may be associated with only one of numerous dictyosomes, as it as shown in the present study. Among lobose amoebae, cytoplasmic MTOCs are known in Centramoebida (Bowers and Korn 1968)(Bowers and Korn 1968); members of the genera *Stygamoeba* (Smirnov 1996), *Gocevia* (Pussard et al. 1977), *Endostelium* (Bennett 1986), *Cochliopodium* (Kudryavtsev 2004); *Thecamoeba* and *Pellita* (Kudryavtsev et al., unpublished). The present study adds the genus *Stenamoeba* to this list. All these genera belong to different phylogenetic lineages comprising the class Discosea and represent five of nine orders, currently included in this class. Therefore, the presence of MTOC theoretically could be a synapomorphic character, unifying several lineages of Discosea.

Among the known discosean lineages, cytoplasmic MTOCs were never shown in the clade comprising Vannellida and Dactylopodida despite these groups were the subject of close attention and various TEM studies (e.g. Page and Blakey 1979; Page 1980; Dyková et al. 2000; Dyková et al. 2003). Therefore, it seems that MTOCs are absent in this clade. Among other discosean lineages, MTOCs are not yet reported in the orders Trichosida and Dermamoebida. However, EM studies of both orders never specifically aimed at recovering cytoplasmic structures. Among various TEM data available for *Trichosphaerium* (Angell 1975, 1976, Rogerson et al. 1998), only Schuster (1976) provided detailed TEM data. His study did not reveal MTOCs, but a single study of a single species might miss to identify these delicate structures and thus provide no conclusive answer to such a minute detail in the ultrastructure. For example, a rather detailed study of the ultrastructure of the genus *Stenamoeba* (Dyková et al. 2010b) revealed no MTOCs (also we cannot reject the hypothesis that it might be absent in some species, such as *S. limacina* or *S. amazonica*). Another group not showing MTOCs is Dermamoebida, but, similarly, there are few studies dedicated to the general ultrastructure of their cells - *Mayorella pussardi* was studied by Hollande et al. (1981), and although no MTOCs were detected, the authors noted a microtubular layer "surrounding the nucleus", which was never recovered in any other amoebae species. Other TEM studies of Dermamoebida were either aimed to the cell coat structure (Page 1983b, 1988, Smirnov and Goodkov 2004, Smirnov et al. 2007) or to the general overview of the ultrastructure (Cann 1981; Smirnov et al. 2011b). Cytoplasmic MTOCs are tiny structures, often visible only in a few TEM sections of many examined ones and requiring good fixation quality to be preserved. They can easily be missed if a study not specifically aims at recovering cytoplasmic structures, and from this point of view both mentioned lineages require more detailed study to prove or disprove the presence of cytoplasmic MTOCs.

The above analysis shows that the presence on MTOCs potentially may be a shared character of Discosea with the exception of the clade containing Vannellida and Dactylopodida. The later clade is also unified by the presence of unique surface structures - pentagonal glycostyles that led Cavalier-Smith et al. (2004) to establish a taxon Glycostilida. Later we have abandoned this group as a taxon (Smirnov et al. 2005, Smirnov et al. 2011b), but the present data indicate that this might have been premature. Distribution of MTOCs along with the accumulation of molecular data and increasing quality of phylogenetic analysis support the hypothesis that Discosea might split in two groups, differing in composition from the present Flabellinia and

Longamoebia (Smirnov et al. 2011b). Those would be the previously mentioned "glycostilida" (Vannellida and Dactylopodida) and a group combining the remaining lineages within Discosea. This suggestion is weakly supported by some molecular data, where Stygamoebida and Himatismenida tend to group with other members of the subclass Longamoebia, thus reducing numbers of independent discosean lineages (e.g. Kudryavtsev and Pawlowski 2013) and shaping a sort of larger clade within Discosea. Support for this remains, however, weak and more data and analyses are required to prove or reject this hypothesis.

MTOCs are also known in amoebae of the genera *Corallomyxa* and *Stereomyxa* (Benwitz and Grell 1971a, b, Grell and Benwitz 1978), classified by Page (1987) in a separate suborder Leptoramosina within the order Leptomyxida. Leptomyxida are members of Tubulinea, however in modern systematics both Stereomyxa and Corallomyxa are placed as Lobosa incertae sedis (Smirnov et al. 2011b). The reason for this is that none of those strains were studied using molecular methods, and no reliable re-isolations of these amoebae are recorded. A strain, resembling *Corallomyxa* and named *C. tenera* (Tekle et al. 2007) was shown to belong to the rhizarian genus *Filoreta* (Bass et al. 2009a). These two genera share a very specific morphology and there is no agreement on the phylogenetic position of both. Page (1983a) lists the genus *Stygamoeba* in the family Stereomyxidae together with *Stereomyxa* and *Corallomyxa* based on certain morphological characters, especially for the species *Stygamoeba polymorpha* (Sawyer 1975). Smirnov (1996) showed that *Stygamoeba* differs from *Stereomyxa* and *Corallomyxa* by the shape of mitochondrial cristae, but the value of this character remains uncertain since all phylogenetic relatives of *Stygamoeba* share tubular cristae. However, in the system by Page (1987) *Stygamoeba* was missing, while *Stereomyxa* placed in Leptomyxida (together with *Gephyramoeba*), and *Corallomyxa* formed a separate order Loboreticulatida (Page 1987 p. 206). All this indicates that the position of *Corallomyxa* and *Stereomyxa* is uncertain, and unless the opposite is proven we can assume that cytoplasmic MTOCs are not present in Tubulinea.

Finally, MTOCs are present in all lineages of Conosa (Cavalier-Smith 1998). This suggests that the presence of cytoplasmic MTOCs is a plesiomorphic feature of

Amoebozoa, while its absence in certain lineages such as Tubulinea and "glycostilida" probably is a secondary loss.

Stenamoeba - a widely distributed and probably species-rich genus

Stenamoeba stenopodia was long believed to represent a single species of this genus, because even before this genus was erected, no other platyamoebian with similar morphology was described. To a certain extent, the marine species *Lingulamoeba leei* Sawyer 1975 somehow resembled *S. stenopodia*, but molecular studies clearly place *L. leei* in a separate lineage (Peglar et al. 2003, Kudryavtsev et al. 2005, Smirnov et al. 2005). More recently, *Stenamoeba stenopodia*-like species were recorded at Valamo Island (North-West Russia) by Smirnov and Goodkov (1999) and meanwhile detected in soil samples from many different locations, including various sites in Russia, Switzerland, Denmark and Canada (Smirnov, unpublished observations), as well as The Netherlands, Germany and high altitude soils from Tibet (Geisen, unpublished observations). *Stenamoeba limacina* isolated from gills of fishes (Dyková et al. 2010b) and the present isolation of this species from a soil habitat, as well as a number of unnamed environmental sequences and sequences from strains identified at the morphological level as *Stenamoeba* spp. (see Fig. 20), suggest that this amoeba genus may be another widely distributed taxon consisting of no less than 10 different species hard to identify by light microscopic characters. Molecular markers will be helpful to discriminate species and estimate the true diversity within this amoeba genus.

Diagnoses

Stenamoeba berchidia n. sp.

Diagnosis: Length in locomotion 12-24 µm (average 18.9 µm), breadth 6-12 µm (average 8.6 µm), L/B ratio 1.4-3.7 (average 2.3). Pronounced frontal hyaloplasm covering up to 2/3 of the cell. Sometimes nearly fan-shaped appearance in addition to the typical oblong or linguiform locomotive form. Posterior part of the cell narrower than anterior hyaloplasm, often pointed. Pronounced longitudinal surface ridges formed during active locomotion, sometimes transverse waves running over the hyaloplasm. Single nucleus usually located at the border of hyaloplasm and granuloplasm with average diameter of 1.7 µm (range 1.3-2.2 µm), nucelolus 0.75 µm (range 0.5-0.9 µm). Smooth, spherical double-walled cysts ca 10 µm in diameter.

Food: bacterivorous; habitat: soil; etymology: the species name refers to the geographic area of the origin (Berchidda); type location: soil, Sardinia region (Italy), Berchidda district, 40°46′N, 9°10′E; type material: type culture *Stenamoeba berchidia* strain 22 deposited with CCAP, accession number CCAP 2571/2.

Differential diagnosis: Morphologically typical for the genus *Stenamoeba*, with size dimensions similar to *S. limacina*. High activity even on cover slips, the absence of a pronounced floating form and the large hyaline area (up to 2/3 of the cell) are unique to *S. berchidia*. A new feature in the genus *Stenamoeba* present in *S. berchidia* are transverse waves running over the hyaloplasm as has been observed in the genus *Clydonella* (Sawyer, 1975a). Distinct SSU rDNA gene sequence and nucleotide pattern in region V4 of the SSU rDNA gene sequence starting in positions 686-742 and a short V7 region differentiate this species from any other known *Stenamoeba*.

Stenamoeba sardiniensis n. sp.

Diagnosis: Length in locomotion 13-24 µm (average 17.0 µm), breadth 5-7 µm (average 6.1 µm), L/B ratio 2.2-3.9 (average 2.8). Posterior granular part of the cell is thicker than the frontal hyaline part, narrowing to the pointed end. Longitudinal surface ridges on the dorsal surface are occasional and briefly present on the moving cell. Floating form with up to three thin, blunt and radiating pseudopods of different length. Single nucleus located at the border of hyaloplasm and granuloplasm with an average diameter of 1.7 µm (range 1.5-2.1 µm), nucleolus 0.65 µm (range 0.5 -0.9 µm). Smooth, spherical double-walled cysts ca. 10 µm in diameter.

Food: bacterivorous; habitat: soil; etymology: the species name refers to the geographic area of the origin (Sardinia); type location: Sardinia region (Italy), Berchidda district, 40°46′N, 9°10′E; type material: type culture *Stenamoeba sardiniensis* strain 17 deposited with CCAP, accession number CCAP 2571/1.

Differential diagnosis: Locomotive form typical for members of the genus *Stenamoeba*. Size and length/breath ratio are only within the values reported for *S. limacina*. Also the tendency to detach from the substratum is a shared feature of *S. sardiniensis* and *S. limacina* but cysts differentiate both; *S. sardiniensis* has larger (10 μm), double walled cysts, while those of *S. limacina* are smaller (5.6 μm) and single walled. A unique SSU rDNA gene sequence with a distinct V7 region is characteristic for *S. sardiniensis*.

Stenamoeba limacina Dyková, Kostka et Pecková 2010, emend.

Flattened amoeba of elongated limax-like or linguiform shape with anterior part round or truncate; anterior hyaline area covering up to 1/2 of the cell, posterior part granular; single contractile vacuole; length in locomotion 15-28 μm (average 19.9 μm), breadth 4-11 μm (average 6.9 μm), L/B ratio 1.6-4.9 (average 3.0); transitory folds or wrinkles on the surface of trophozoites, floating form with blunt radiating pseudopodia; cysts spherical, single walled. *Stenamoeba limacina* deposited with CCAP (accession number CCAP 2571/3).

Acknowledgements

Supported with research grants involved in the EU-project 'EcoFINDERS' No. 264465, the DFG-Priority Program 'Biodiversity Exploratories' (Bo 1907/2-1, Bo 1907/2-2), research grants from Saint Petersburg State University and RFBR 12-04-01825 grant. The present study utilized equipment of the core facility centres "Development of molecular and cell technologies" and "Culturing of microorganisms" of Saint Petersburg State University. We further thank the managers of the three Biodiversity Exploratories, Swen Renner, Sonja Gockel, Kerstin Wiesner, and Martin Gorke for their work in maintaining the plot and project infrastructure; Simone Pfeiffer and Christiane Fischer giving support through the central office, Michael Owonibi for managing the central data base, and Markus Fischer, Eduard Linsenmair, Dominik Hessenmöller, Jens Nieschulze, Daniel Prati, Ingo Schöning, François Buscot, Ernst-Detlef Schulze, Wolfgang W. Weisser and the late Elisabeth Kalko for their role in setting up the Biodiversity Exploratories project.

Part 1 – Chapter 2

Discrepancy between species borders at morphological and molecular levels in the genus *Cochliopodium* (Amoebozoa, Himatismenida), with the description of *C. plurinucleolum* n. sp.

Geisen Stefan*[2], Kudryavtsev Alexander*[1], Bonkowski Michael[2], Smirnov Alexey[1]

[1] Department of Invertebrate Zoology, Faculty of Biology and Soil Sciences, Saint Petersburg State University, Russia
[2] Department of Terrestrial Ecology, Zoological Institute, University of Cologne, Germany

Abstract

Amoebae of the genus *Cochliopodium* are characterized by a tectum that is a layer of scales covering the dorsal surface of the cell. A combination of scale structure, morphological features and, nowadays, molecular information allows species discrimination. Here we describe a soil species *Cochliopodium plurinucleolum* n. sp. that besides strong genetic divergence from all currently described species of *Cochliopodium* differs morphologically by the presence of several peripheral nucleoli in the nucleus. Further, we unambiguously show that the Golgi attachment associated with a dictyosome in *Cochliopodium* is a cytoplasmic microtubule organizing centre (MTOC). Last, we provide detailed morphological and molecular information on the sister clade of *C. plurinucleolum*, containing *C. minus*, *C. minutoidum*, *C. pentatrifurcatum* and *C. megatetrastylus*. These species share nearly identical sequences of both, small subunit ribosomal RNA and partial Cox1 genes, and nearly identical structure of the scales. Scales of *C. pentatrifurcatum* differ, however, strongly from scales of the others while sequences of *C. pentatrifurcatum* and *C. minus* are nearly identical. These discrepancies urge for future sampling efforts to disentangle species characteristics within *Cochliopodium* and to investigate morphological and molecular patterns that allow reliable species differentiation.

Introduction

The genus *Cochliopodium* Hertwig et Lesser, 1874, comprises lens-shaped amoebae covered by a flexible layer of carbohydrate scales (tectum) located only on the dorsal surface of the cell (Bark 1973, Kudryavtsev 2004, Kudryavtsev et al. 2005, Kudryavtsev 2006). When the cell is viewed from the top during locomotion, it shows a thick central granuloplasmic mass surrounded by a broad peripheral sheet of hyaloplasm. It is widely accepted that the size and shape of scales comprising the tectum of *Cochliopodium* are species-specific, so that details of scale structure allow the unambiguous identification of morphospecies (Bark 1973, Kudryavtsev 1999, 2004, 2005, 2006). However, it was demonstrated previously that there is a group of morphospecies in *Cochliopodium* that can be distinguished based on size, formation of cysts and several characteristics of the locomotive form such as uroidal filaments and shape of the hyaloplasm (Kudryavtsev et al. 2004, Kudryavtsev 2006), while sharing nearly identical scales. Namely, these are *C. barki*, *C. minutoidum* and several strains morphologically resembling *C. minus*. *C. megatetrastylus* and *C. pentatrifurcatum* were also recently added to this group of morphospecies (Anderson and Tekle 2013, Tekle et al. 2013), although scales of the latter species are surprisingly different from the rest of the group. The distinction of these morphospecies may be facilitated by using gene sequence data (Kudryavtsev et al. 2005, Kudryavtsev et al. 2011,

Anderson and Tekle 2013, Tekle et al. 2013). Therefore it seems unavoidable to decipher how morphological and ultrastructural differences are related to the genetic divergence between strains. Yet, the number of sequenced species is relatively small compared to the total number of around 20 morphologically defined *Cochliopodium* spp., and this is the main obstacle for understanding the evolutionary relationships within the genus and borders between morphospecies.

In the present paper we revise the phylogenetic relationships in a clade of *Cochliopodium* comprising closely related species *C. megatetrastylus*, *C. minus*, *C. minutoidum* and *C. pentatrifurcatum* (hereinafter referred to as *C. minus*-clade) and describe a new soil species *C. plurinucleolum* whose scale structure is very similar to the members of this clade. However, molecular trees show only distant relationships between this species and members of the *C. minus*-clade.

Results

Microscopic observations of *Cochliopodium* strains

Cochliopodium plurinucleolum n. sp.

Trophic amoebae were highly variable, showing most frequently oval, fan-shaped, or crescent-shaped locomotive forms (Fig. 21A-B). Many amoebae temporarily adopted a triangular, drop-shaped locomotive form with length greater than breadth caused by adhesion of the posterior end to the substratum (Fig. 21C). The granuloplasm in above view was entirely surrounded by a thin hyaloplasmic sheet (Fig. 21A-C), which never exceeded $1/5^{th}$ of the total body length and was often equally broad at the anterior and lateral parts of the body. The dorsal surface of the hyaloplasmic sheet was often completely covered by the most peripheral scales of the tectum (Fig. 21A-B). Sometimes the anterior margin of the hyaloplasm extended beyond the border of the scale layer; it was then

41

smooth or slightly irregular (Fig. 21C). Subpseudopodia were never seen. Few short trailing adhesive filaments were visible in the posterior end of the cell during fast, directed locomotion (Fig. 21A-C). Amoebae in non-directed movement (Fig. 21D) were oval, triangular, or had an irregular shape, often angulate and stretched over the substratum, with the hyaloplasm split into several flattened projections on distinct sites of the cell margin. The stationary form was nearly rounded, sometimes slightly wrinkled or oval without any extensions of the peripheral hyaline area, which was equally broad but retracted in comparison to the locomotive form.

A differentiated floating form was rarely developed, and only when the cell was artificially disturbed. Amoebae contracted shortly after detachment from the substratum; the central granuloplasmic area became spherical, and the peripheral hyaloplasmic sheet contracted and produced several hyaline pseudopodia. Those pseudopodia were mostly short, but extended and became longer than the central cell body in amoebae getting closer to the substratum and attaching to it. Amoebae started locomotion immediately after settling down on the substratum.

A single spherical nucleus (Fig. 21E) was located in the central part of the granuloplasm close to the dorsal surface of the cell. It had finely granular contents and several spherical pieces of nucleolar material at the periphery (1-2 of these pieces were usually visible simultaneously). The nucleus was inconspicuous and often concealed by the organelles and inclusions in the granuloplasm such as abundant food vacuoles or crystals of varying sizes and shapes. Several asynchronously working contractile vacuoles, capable of fusing together, were present in different areas of the granuloplasm. Cysts (Fig. 21F) were always formed in older cultures. The cyst wall consisted of two layers; the outer one incompletely enclosing the cell was made up of the scales comprising the tectum in a trophic amoeba. No cyst pores were observed. Cysts could survive at least several weeks of complete drought in culture. Amoebae were feeding on bacteria and multiplied both in liquid and on agar media.

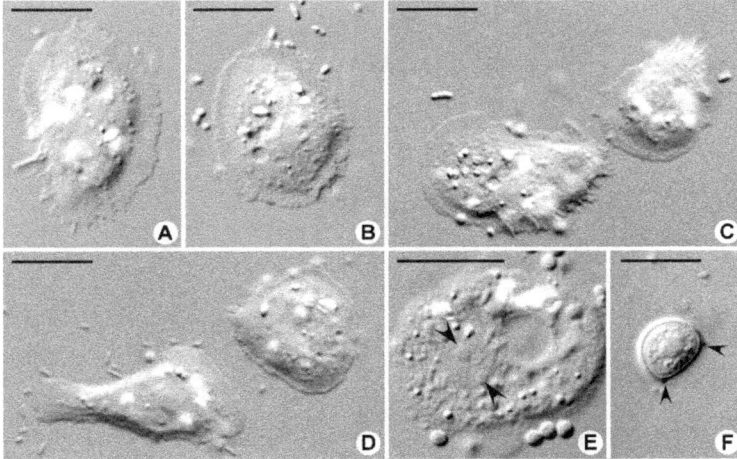

Fig. 21. *Cochliopodium plurinucleolum* n. sp., light micrographs, DIC. A-C. Locomotive forms on the glass surface. D. Amoebae during non-directional movement. E. Nucleus in an amoeba partly squeezed with a coverslip (arrowheads indicate nucleoli). F. Cyst (arrowheads indicate margins of an outer cyst wall formed by scales). Scale bar = 10 µm in all Figs.

Transmission electron microscopy showed that the tectum consisted of scales of uniform structure (Fig. 22) located on the dorsal surface of the cell (Fig. 23A). Scales had flat, circular base plates with a grid-like structure formed by a square mesh (mesh size 0.015 µm; Fig. 22, 3B-D). Four vertical stalks were rising from the centre of the base plate and terminated with a funnel-shaped apical part consisting of ca. 15 fine radial spokes and a dense outer rim. The material between radial spokes was organized in poorly discernible concentric rings thus giving an apical part a somewhat spiderweb-like appearance (Fig. 23C-D).

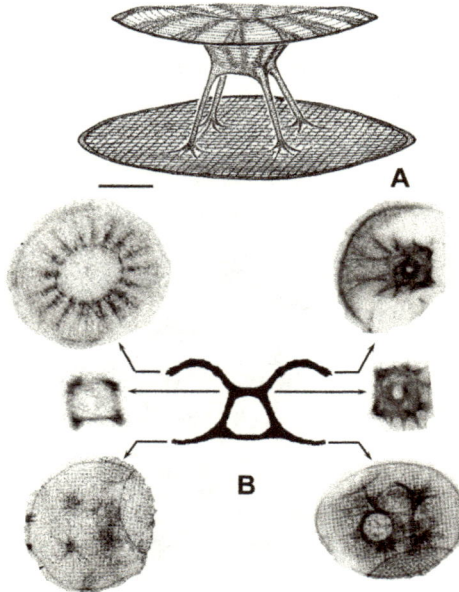

Fig. 22. Diagram representing scale reconstruction in *C. plurinucleolum* n. sp. and *C. minus* CPE based on the ultrathin sections. A. Drawing of a scale, scale bar = 0.1 µm. B. Scheme of the scale appearance at different levels of sectioning in *C. minus* CPE (left) and *C. plurinucleolum* n. sp. (right), not to scale.

The nucleus in sections was oval or irregular in shape (Fig. 23A, E). It contained numerous dense patches of heterochromatin. In most of the sections one or two rounded nucleolar pieces were visible at the periphery of the nucleus (Fig. 23E). A single large dictyosome was located between nucleus and dorsal surface of the cell (Fig. 23E). An elongated granular structure, the "Golgi attachment", was located close to the side of a dictyosome opposite to the nucleus, and numerous microtubules often emerged from the Golgi attachment into the cytoplasm (Fig. 23F). Mitochondria in sections were rounded or ovoid with tubular cristae (Fig. 23G). Cysts in sections were ovoid, with the denser and more poorly preserved cytoplasm than in the trophic amoebae. The cyst wall consisted of an inner layer of medium electron density that was 0.1-0.2 µm thick and an outer layer consisting of scales (Fig. 23H). The scale layer occupied the part of the cyst surface where the tectum on the surface of a trophic amoeba was located, in accordance to what was seen with light microscopy (Fig. 21F, 3H).

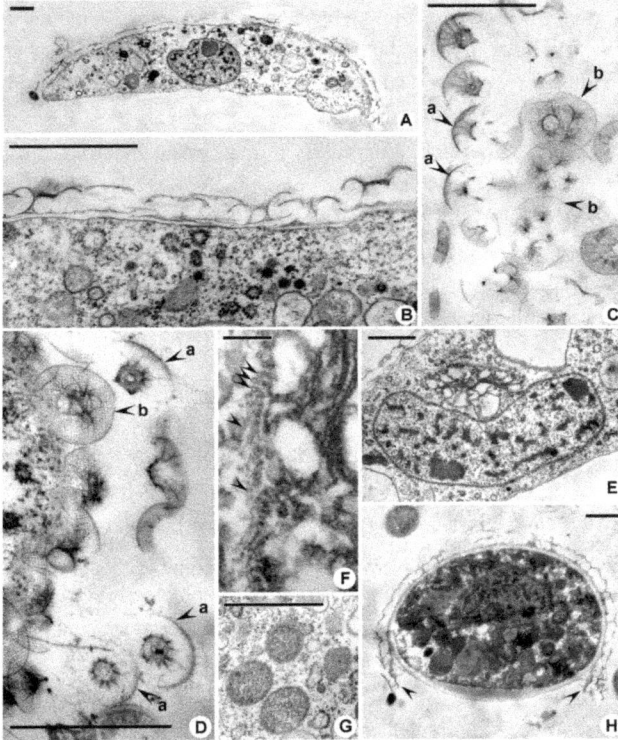

Fig. 23. Transmission electron micrographs of *Cochliopodium plurinucleolum* n. sp., strain 8. A. Cross-section of the cell adhering to the substratum showing dorsal surface covered with tectum (above) and ventral, naked, surface (below). Nucleus is seen in the centre. B. Vertical section of scales on the dorsal cell surface. C-D. Details of the scales in tangential sections: grid-like base plates (b) and apical parts (a). E. Nucleus (note 2 peripheral nucleoli) and a dictyosome. F. Detail of the dictyosome: note microtubules (arrowheads) adjacent to the dictyosome's side opposite to the nucleus. G. Mitochondria. H. Cyst (note inner cyst wall underlying tectum; the latter covers only part of the surface; arrowheads indicate margins of an outer cyst wall formed by scales). Scale bar = 1 μm in all Figs. except F, where it is 0.2 μm.

Cochliopodium minus strain CPE

Length of the locomotive form in this strain was 12-30 μm (average 20 μm), breadth 18-46 μm (average 28 μm), length: breadth ratio was 0.42-1.08 (average 0.73) (n=76). Amoebae, in above view, were generally broadly triangular, oval or fan-shaped during locomotion, and changed their outline very quickly (Fig. 24A). The central mass of granuloplasm was located posteriorly and surrounded by a peripheral hyaline sheet (Fig. 24C-D). Anterior and lateral parts of the hyaline sheet were usually much wider than the posterior one, their width was equal to approximately one-third of the cell length. The margin of the hyaline sheet was very uneven, producing numerous waves, irregularities and short subpseudopodia (Fig. 24C-D), sometimes deeply cleaved into several lobes. Occasionally, the anterior area of the hyaloplasm retracted, and lateral parts became wider. In this case amoebae temporarily expanded in two opposite directions before resuming locomotion in one direction. Rear parts of the lateral hyaline margins, as well as the posterior end of the cell often adhered to the substratum producing a number of trailing filaments (Fig. 24C-D), occasionally becoming as long as the remaining cell body. Non-directionally moving and stationary amoebae (Fig. 24E) showed either irregular shapes with the cytoplasm split into several lobes, or were rounded, with the central granuloplasmic mass completely surrounded by the hyaline sheet of approximately equal breadth from each side. A floating form was only occasionally seen in cultures and occurred more frequently when amoebae were placed on a glass slide. When floating, amoebae usually contracted and their central granuloplasm became a spherical compact mass. The peripheral hyaloplasm folded and often produced several hyaline pseudopodia (Fig. 24F). The floating form was usually maintained for a maximum of 20 minutes until amoebae settled back to the substratum and started locomotion.

The majority of cells possessed a single vesicular nucleus (Fig. 24B), but cells with 2 or 3 nuclei were occasionally seen, that were usually larger than the uninucleate ones. The nucleus was spherical, with a conspicuous envelope and a large central nucleolus that sometimes contained one or several small cavities. The nuclear envelope was sometimes outlined with a coarse layer consisting of fine granules. The diameter of the nucleus was 4.3-9.2 µm (average 6.3 µm), that of the nucleolus 1.9-5.1 µm (average 3.6 µm) (n=32). Amoebae possessed several asynchronously working contractile vacuoles. The granuloplasm contained 5 to 15 refractile crystals of a bipyramidal shape 1-2 µm in length (Fig. 24C, E). Several food vacuoles and a number of spherical or elongated granules below 1 µm in size were also present. The tectum was clearly visible in phase contrast or DIC optics covering the dorsal surface of the granuloplasm (Fig. 24E) and the hyaloplasmic surface in non-directionally moving and stationary amoebae, as a number of granules over the surface. In cultures amoebae encysted regularly producing spherical or ovoid double-walled cysts ca. 15 µm in diameter.

Scales reconstructed from TEM sections (Fig. 24G-J) appeared to be identical in structure and dimensions to those of *Cochliopodium plurinucleolum* (Fig. 22). A very slight difference in their appearance was a less dense periphery of the top part, with concentric rings being better visible, but this may vary between individual fixations. Dimensions of scales are given in Table 2. The nucleus was rounded in sections containing a central, electron-dense nucleolus (Fig. 25A). Rounded or oval mitochondria had electron-dense matrix and tubular cristae (Fig. 25A-B). Each cell possessed a single dictyosome located close to the nuclear envelope between nucleus and the dorsal surface of the cell. A Golgi attachment (Fig. 25C) visible as a dense granular structure was located close to the dictyosome, opposite to the nucleus. Cysts of this strain had the same structure as those of *C. plurinucleolum* (Fig. 25D-E). The dense inner wall was fibrous and contained a number of cavities filled with granular material in a young cyst that also occupied the space between the folded plasma membrane of the encysting cell and the cyst wall (Fig. 25E).

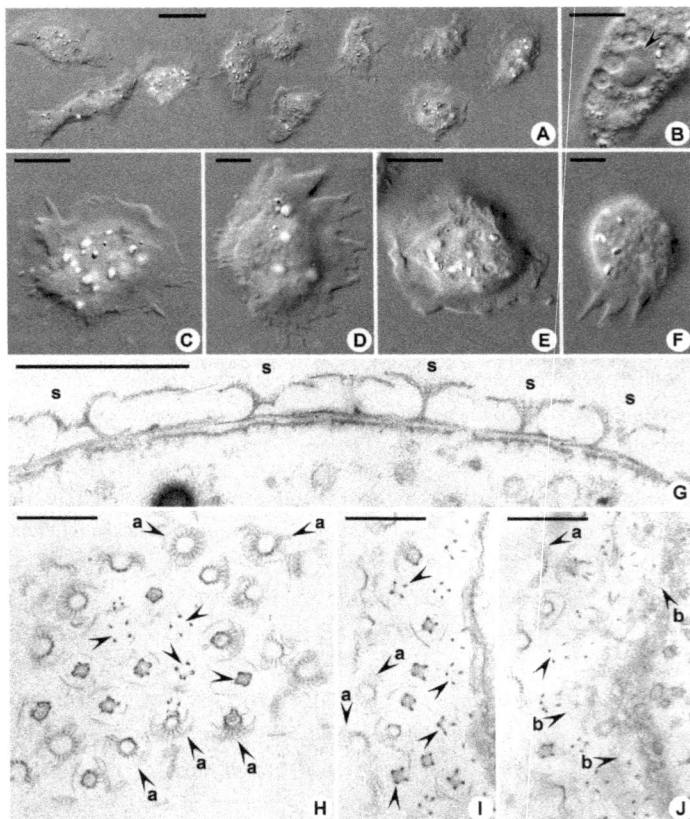

Fig. 24. *Cochliopodium* sp. CPE. Light (A-F) and electron (G-J) micrographs. A, C-D. Locomotive forms on glass surface. B. Nucleus (indicated by arrowhead). E. Amoeba in non-directed movement. F. Floating form. G. Scales (s) on the dorsal surface of a cell in a vertical section. H-J. Serial tangential sections showing scales cut at different levels. Note tangentially sectioned base plates (b), apical parts (a) and central columns (arrowheads). Scale bar = 20 μm in A, 10 μm in B-F, 1 μm in other Figs.

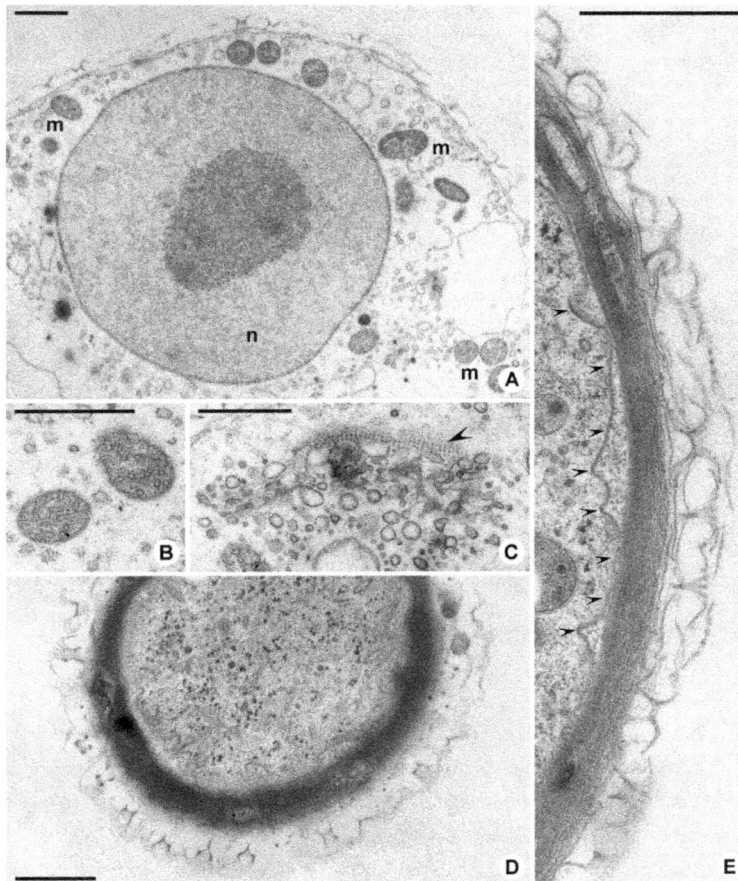

Fig. 25. *Cochliopodium* sp. CPE, electron micrographs. A. Section of a trophic amoeba showing nucleus (n) and mitochondria (m). B. Mitochondria at a higher magnification. C. Dictyosome and a Golgi attachment (arrowhead). D. Cyst wall in oblique section. E. Transverse section of a (young) cyst wall; arrowheads indicate underlying plasma membrane. Scale bar = 0.5 µm in A, 1 µm in other Figs.

Table 2. Comparison of the scale characteristics among *Cochliopodium* species with scales of "Category 1" (Anderson and Tekle 2013). Source references are in brackets after species names, numbers in brackets show average values where available

Species	Base plate diameter (μm)	Top part diameter (μm)	Scale height (μm)	Number of spokes in top part
C. plurinucleolum n. sp.	0.465-0.76 (0.627)	0.445-0.577 (0.5)	0.185-0.3 (0.25)	ca 15
Cochliopodium minus CPE	0.538-0.788 (0.671)	0.475-0.654 (0.56)	0.185-0.338 (0.25)	15-17
Cochliopodium sp. NYS strain (probably *actinophorum*; Nagatani et al. 1981, Yamaoka et al. 1984)	0.64	0.8	0.7	24
C. barki (Kudryavtsev et al. 2004)	0.7–1 (0.81)	0.6–0.85 (0.67)	0.35-0.45	20
C. minutoidum CCAP 1537/7 (Kudryavtsev 2006)	0.49-0.63 (0.56)	0.47-0.6 (0.54)	0.26-0.3	14-17
C. minus CCAP 1537/1A (Kudryavtsev 2006)	0.64-0.77	0.58-0.69	0.21-0.25	17-19
C. minus CCAP 1537/5 (Kudryavtsev 2006)	0.64-0.77	0.58-0.69	0.17-0.18	17-19
C. megatetrastylus (Anderson and Tekle 2013)	0.6-1	0.5-0.7	0.2-0.4 (0.3)	16

Sequence data analysis

SSU rRNA gene

An overview of *Cochliopodium* strains sequenced during this study is presented in Table 3. Sequenced pieces of the SSU rRNA gene in *Cochliopodium plurinucleolum* strains 8 and 86 were identical to each other and had a GC content of 40 %. In other studied strains, cloning of the SSU rDNA amplicons demonstrated slight variation among different molecular clones obtained from the same PCR product. The range of this variation was 0.1-0.3 % in *Cochliopodium minus* CPE (n=5), 0.4-1 % in *C. minutoidum* CCAP 1537/7 (n=4), 1.8-1.9 % in *C. kieliense* and 0.3 % in *C. minus* CCAP 1537/5 (n=2). Among 8 sequenced clones of *C. minus* CCAP 1537/1A 4 clones were identical, the other 4 varied by 0.1-0.3 %.

Cochliopodium minus CCAP 1537/1A was obtained from the culture collection and re-investigated to facilitate a more precise molecular identification of *C. minus* CPE strain studied here, and to evaluate the identity of a sequence JF298257 previously assigned to *Cochliopodium minus* CCAP 1537/1A (Kudryavtsev et al. 2011). Several newly obtained light and electron micrographs of these amoebae are shown in Fig. 26A-G. Our results show that the strain investigated in this study is identical to the one previously studied microscopically (Kudryavtsev 2006; several new micrographs of the scales are shown in Fig. 26F-G). Yet, the SSU rRNA gene sequence of the newly obtained strain is highly divergent from a sequence JF298257 previously designated as *C. minus* CCAP 1537/1A (Kudryavtsev et al. 2011).

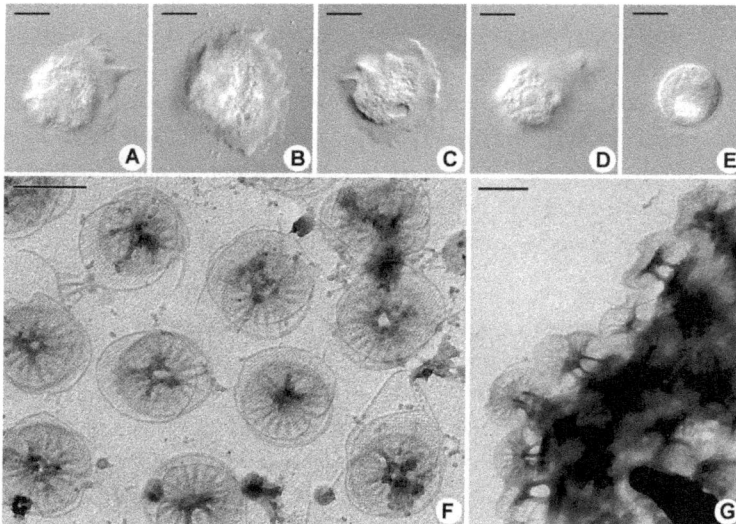

Fig. 26. *Cochliopodium minus* CCAP 1537/1A, newly obtained light (A-E) and electron (F-G) micrographs of re-ordered and sequenced strain. A-C. Locomotive forms on glass surface; C showing scales of the dorsal surface of the cell. D. Nucleus in a living amoeba. E. Cyst. F-G. Scales in a whole-mount TEM preparation shadowed with chromium. Scale bar = 10 μm in A-E, 0.5 μm in F-G.

The use of improved primers and cycling conditions enabled us to update the available sequence database of *Cochliopodium* spp. by obtaining complete SSU rRNA sequences of *C. kieliense* and *C. minus* CCAP 1537/5. Previously, the SSU rRNA gene of both strains was partially sequenced from the amplicons obtained with the primer pair s6-RibB, therefore missing approximately 600 nucleotides from the 5' end (Kudryavtsev et al. 2005; GenBank accession numbers AY785057 and AY785056 respectively), as the full-length amplicons could not be obtained. During this study we obtained and sequenced the full-length amplicons of SSU rRNA genes from the same DNA samples as used in the previous work, and additionally obtained partial Cox1 gene sequences of these species. The newly obtained SSU rRNA gene sequences are largely identical to the previous ones. Differences in several nucleotides are most probably due to a sequencing protocol used during the present study that applied a BigDye Terminator sequencing kit in a single reaction run on a capillary electrophoresis, while the previous one applied 4 separate sequencing reactions and gel electrophoresis (Kudryavtsev et al. 2005) that may have a higher error rate. We therefore substitute the previously published sequences AY785056 and AY785057 with the new ones, to enable the inclusion of more nucleotide positions in phylogenetic analyses.

The phylogenetic trees based on maximum likelihood and Bayesian algorithms robustly placed sequences of *C. plurinucleolum* n. sp. (strains 8 and 86) within the monophyletic genus *Cochliopodium* in the same position. They branched as a sister to a monophyletic clade comprising *C. actinophorum*, *C. kieliense* and a clade consisting of the very shortly branching species *C. minus*, *C. minutoidum*, *C. megatetrastylus* and *C. pentatrifurcatum* (*C. minus*-clade; Fig. 27) with strong support (PP/BS=1.0/92).

In the phylogenetic trees *Cochliopodium minus* CPE branched closely together with *C. minus* CCAP 1537/1A and CCAP 1537/5,

C. pentatrifurcatum ATCC 30935 and *C. megatetrastylus* ATCC 30936 in the *C. minus*-clade. *C. minutoidum* CCAP 1537/7 was always sister to the clade formed by different strains of *C. minus*, and the whole *C. minus*-clade was robustly sister to a clade comprising *C. kieliense* and *C. actinophorum*. Extremely short branches in the *C. minus*-clade were in agreement with very high sequence similarity between these strains (Supplementary Table 1). Yet, distinct nucleotide motifs could be identified in these sequences (Supplementary Table 2), so that each species and clade could be characterized by specific nucleotide signatures. Therefore, a robust topology obtained by phylogenetic analyses was in accordance with gene sequence features.

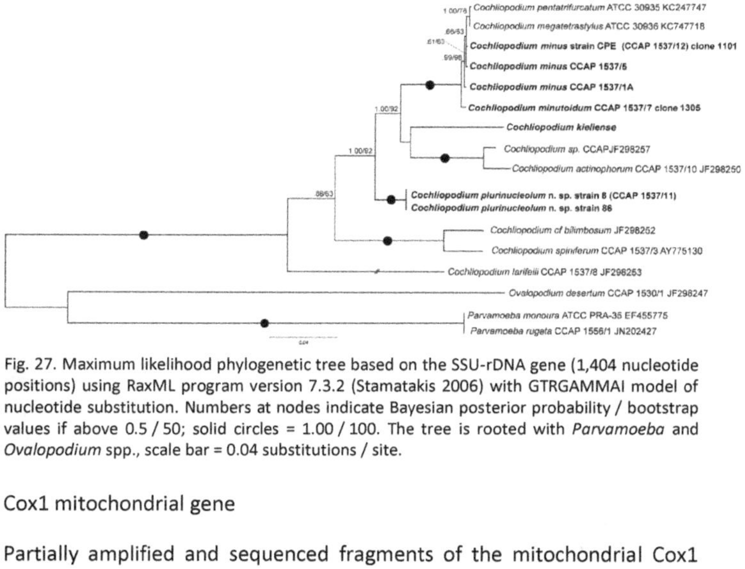

Fig. 27. Maximum likelihood phylogenetic tree based on the SSU-rDNA gene (1,404 nucleotide positions) using RaxML program version 7.3.2 (Stamatakis 2006) with GTRGAMMAI model of nucleotide substitution. Numbers at nodes indicate Bayesian posterior probability / bootstrap values if above 0.5 / 50; solid circles = 1.00 / 100. The tree is rooted with *Parvamoeba* and *Ovalopodium* spp., scale bar = 0.04 substitutions / site.

Cox1 mitochondrial gene

Partially amplified and sequenced fragments of the mitochondrial Cox1 gene of each strain studied were 666 base pairs long not including PCR primers, and did not contain any indels. Cloning of the amplicons has shown identical sequences among different molecular clones in

C. actinophorum, *Cochliopodium minus* CPE and *C. minutoidum* CCAP 1537/7 (5 clones sequenced for each), whereas *C. minus* CCAP 1537/1A, *C. minus* CCAP 1537/5 and *C. kieliense* have shown differences in nucleotide sequences between molecular clones obtained from the same amplicon in the range of 1-3 variable nucleotide positions per sequence (8 clones sequenced). Six out of 8 sequenced molecular clones of *C. minus* CCAP 1537/1A have shown a complete identity to the Cox1 gene sequence of *C. pentatrifurcatum* ATCC 30935 (KC489470; Tekle et al. 2013) at the nucleotide level. Partial Cox1 gene sequences of *C. actinophorum* CCAP 1537/10 and *C. minutoidum* CCAP 1537/7 obtained earlier (Nassonova et al. 2010; GQ354207 and GQ354208, respectively) were completely identical to sequences from the same strains obtained during the present study. The analysis of translated sequences revealed that differences between molecular clones within a single amplicon were synonymous in most cases, but caused variability in 1 to 2 amino acid positions in two molecular clones of *C. minus* CCAP 1537/1A and *C. minus* 1537/5 amplicons, respectively. *C. plurinucleolum* n. sp. differed from other *Cochliopodium* spp. included in this study by an average of 12.9 % at the nucleotide level (11.8-16.4 %), based on 495 nucleotide positions, and by 5.6 % (3.7-11.6 %) at the amino acid level. The identity matrix between all strains of the *C. minus*-clade is shown in Supplementary Table 3. The maximum likelihood phylogenetic analysis of Cox1 sequences of *Cochliopodium* spp. (Fig. 28) basically resulted in the same tree topology and branch support values as in the tree based on the SSU rRNA gene revealing two clades of *Cochliopodium* spp., one comprising *C. actinophorum* and *C. kieliense*, and the second one corresponding to a *C. minus*-clade outlined above. *C. plurinucleolum* n. sp. forms the most basal branch of the *C. minus*-clade, being sister to a monophyletic branch comprised by the other strains. Phylogenetic analysis of amino acid sequences (not shown) resulted in the same topology. However, as most of the substitutions were synonymous in the *C. minus*-clade, the latter did not show any defined topology, while *C. plurinucleolum* n. sp. was sister to it.

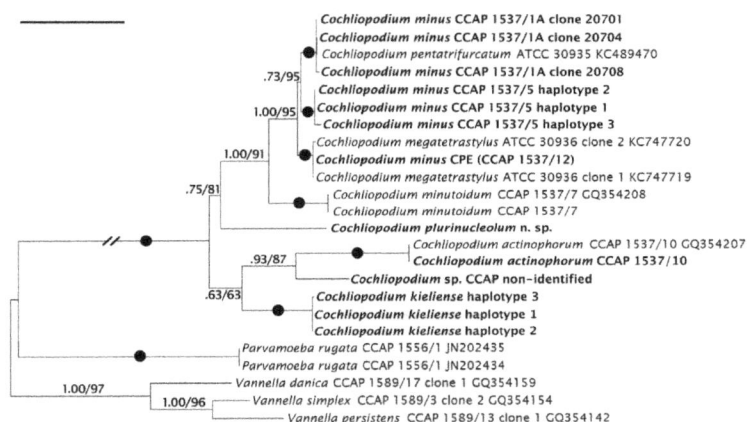

Fig. 28. Maximum-likelihood phylogenetic tree of Cochliopodium spp. based on the partial nucleotide sequences of mitochondrial Cox1 gene (495 nucleotide positions). The tree shown was derived using a RaxML program version 7.3.2 (Stamatakis 2006) with GTRGAMMAI model of nucleotide substitution. Numbers at nodes indicate Bayesian posterior probability / bootstrap values if above 0.5 / 50; solid circles = 1.00 / 100. The tree is rooted with Vannella spp., scale bar = 0.1 substitutions / site.

Discussion

Morphological identification of the studied strains

Cochliopodium plurinucleolum n. sp.

Light and electron microscopic analyses unambiguously identify the studied strains 8 and 86 as a species of the genus *Cochliopodium* due to the presence of a tectum and a characteristic locomotive form. Several morphological characteristics specific for this species distinguish it from all other currently described *Cochliopodium* spp., including *C. barki* that is morphologically most similar to *C. plurinucleolum*, and is one of the few isolates reported so far to occur in soils (Kudryavtsev et al. 2004). Cells of

C. barki are, however, bigger, sometimes produce subpseudopodia and occasionally contain more than one nucleus. The structure of the nucleus is the character that best distinguishes *C. plurinucleolum* from all other known *Cochliopodium* spp. This presence of multiple nucleoli appears to be a reliable character, because no variation of the observed nuclear structure among the studied cells was seen, and the results of the electron microscopic study corresponded to the light microscopic data. The only described species that has a similar nucleus with peripheral nucleoli is *C. clarum* Schaeffer, 1926. However, this species is larger, was isolated from marine habitats and has generally a smoother outline of the cell with several short subpseudopodia (Schaeffer 1926). Other described species of *Cochliopodium* possess a vesicular nucleus containing a large central nucleolus (Dyková et al. 1998, Kudryavtsev 1999, Kudryavtsev 2000, Kudryavtsev 2004, Kudryavtsev et al. 2004, Kudryavtsev 2005, 2006, Kudryavtsev and Smirnov 2006, Anderson and Tekle 2013, Tekle et al. 2013). We therefore recognize *C. plurinucleolum* as a new species. The scale structure of this species is very similar to that of *C. barki*, *C. megatetrastylus*, *C. minus* and *C. minutoidum* (Dyková et al. 1998, Kudryavtsev 2004, 2006, Anderson and Tekle 2013). Size differences between scales in these species are summarized in Table 2. We have to admit that the structural pattern of the scales is identical, and the slight quantitative differences between them may be purely intraclonal, but no statistical data are available to check this at the moment. A study investigating statistical variation of the scale parameters in genetically identical isolates of *Cochliopodium* would be highly appreciated to reliably resolve boundaries between species.

Cochliopodium minus CPE

Based on the light microscopic characters of trophic amoebae and cysts, *Cochliopodium minus* CPE could be identified as most similar to *C. minus* and *C. minutoidum* as described by Kudryavtsev (2006). The most significant characters to identify *C. minus* CPE are oval and almost fan-shaped locomotive forms with a very uneven margin of the frontal hyaline area producing numerous subpseudopodia. This strain differs from *C. minutoidum* by forming cysts, but this character may vary depending on individual clones: for example, Page (1976b) mentioned "encystment present in some strains" for *C. minus*. Scales of *Cochliopodium* sp. CPE strain were most similar to those of *C. minutoidum* as described by Kudryavtsev (2006). Ambiguous identification of this strain based on the microscopic data demands a more precise identification based on the combination of morphology/ultrastructure and gene sequences. We therefore obtained and analysed sequences of SSU rRNA and Cox1 genes from all available strains similar to *C. minus* and *C. minutoidum* in addition to *C. plurinucleolum*.

Molecular phylogeny and species distinction in *Cochliopodium*

Phylogenetic position and relationships of *C. plurinucleolum* n. sp.

Analysis of both SSU rRNA and Cox 1 gene sequences were in accordance with morphological data, unambiguously placing *C. plurinucleolum* n. sp. as a distinct species within *Cochliopodium*. The new species branches deeply in the phylogenetic tree, but trees based on SSU rRNA and Cox1 genes demonstrate conflicting topologies: whereas in the SSU rRNA gene tree *C. plurinucleolum* n. sp. is a sister to the clade comprising *C. minus* and related species, as well as *C. actinophorum* and *C. kieliense*, Cox1 gene analysis places this species as a sister to the *C. minus*-clade, while *C. actinophorum* and *C. kieliense* branch outside this clade (Fig. 27-8). Visual comparison of the aligned nucleotide sequences of Cox1 gene shows that no shared sequence motifs that may serve as molecular signatures to support the topology of either SSU rRNA or Cox1 gene trees can be found in *C. plurinucleolum* n. sp.; there are a number of positions with the motifs shared between *C. plurinucleolum* n. sp. and species of the *C. minus*-clade as well as some positions that are identical in

C. plurinucleolum n. sp. and *C. kieliense*, but differ in other species. Moreover, approximately unbiased (AU) test (Shimodaira 2002) performed on both alternative topologies using SSU rRNA and Cox1 alignments did not give a clear preference to any of the topologies. Several explanations of this incongruence between different markers are currently possible, but none of them can provide sufficient clarifications based on the present data. As the position of *C. plurinucleolum* n. sp. in the tree based on the Cox1 gene was less resolved, as well as several other deep nodes of the tree, it is possible that the tree based on the Cox1 gene may be misleading due to incomplete taxon sampling compared to the tree based on the SSU rRNA gene, and a lower number of nucleotide positions available for phylogenetic analysis (495 nucleotide positions compared to 1404 available for the SSU rRNA gene). Moreover, faster evolution of the Cox1 gene compared to SSU rRNA with unevenly conserved nucleotide positions across the gene (as Cox1 is a coding gene, third codon position is less conserved than the other two) may lead to the saturation of the 3rd codon positions resulting in homoplasies, favouring the evolutionary scenario reconstructed based on the SSU rRNA over Cox1 gene. We also cannot exclude that different mitochondria host different Cox1 gene sequences within single species or incongruence in the evolutionary history between nuclear and mitochondrial genomes in these species, in which case not all markers will yield gene trees congruent to a species tree.

Anyway, the position of *C. plurinucleolum* n. sp. in the phylogenetic tree reconstructed based on both markers is in accordance with scale structure and molecular synapomorphies present in the SSU rRNA gene of this species. The scale structure of these amoebae is identical to that of the whole *C. minus*-clade (scales of Category 1 according to Anderson and Tekle 2013; Fig. 22). Scale structure of *C. actinophorum* falls under the same category as shown by the results of the study on neotype strain CCAP 1537/10 (Kudryavtsev 2014), whereas. initial descriptions of scales of this species were performed by Nagatani et al. (1981) and Yamaoka et al. (1984). The suggestion that this type of scales is a synapomorphy of the clade unifying *C. plurinucleolum* n. sp., *C. actinophorum* and the *C minus*-clade is rather attractive. But as this type of scales is not the only type of scales known among these species with *Cochliopodium kieliense* Kudryavtsev, 2006 and a recently described *C. pentatrifurcatum* Tekle et al., 2013 possessing completely different scale structures, while robustly branching close to *C. actinophorum* and among the *C. minus*-clade, respectively. Moreover, whereas *C. kieliense* forms a relatively long branch that is distantly related to *C. actinophorum* in the trees based on both, SSU rRNA and Cox1 genes, gene sequences of *C. pentatrifurcatum* are completely identical to the majority of molecular clones that we sequenced for the type strain of *C. minus*. Currently, we do not have a clear explanation for these discrepancies on morphological and molecular levels, but an artefact seems unlikely, as the same was independently observed by Y. Tekle (pers. comm.). If this identity is not due to the errors occurred at the stage of data collection caused by erroneous assignment of either the sequences or the scale structure, we have to suggest that this is the first case of sequence identity between distinct morphospecies that considerably differ from each other based on the scale structure. The first explanation for this case may be that both genetic markers sequenced are not capable to distinguish two different species that may differ from each other in other markers. The second possible explanation is that *C. pentatrifurcatum* and *C. minus* are indeed genetically identical, and differences in scale structures may reflect environmental plasticity within a single species, e.g. depending on life stage or environmental differences. Presently there are no data that might favour any of these explanations, and additional studies are necessary to clarify this situation.

The demonstrated position of *C. plurinucleolum* n. sp. in the phylogenetic tree is also in accordance with the molecular synapomorphies revealed within its SSU rRNA gene sequence. A secondary structure pattern of helices 28-30 of the SSU rRNA previously demonstrated for Cochliopodiidae (Kudryavtsev et al. 2011) is also present in this species (Fig. 29A). Comparison of this structure among different species of *Cochliopodium* reveals characters that appear to be synapomorphies of different phylogenetic lineages within this genus. In particular, there are several distinct patterns of nucleotides between helices E29-1 and E29-2. *C. plurinucleolum* n. sp., all species belonging to a clade that comprises *C. actinophorum*, *C. kieliense* and species related to *C. minus* share an 8- to 18-nucleotide AT-rich insertion that forms an extensive loop (Figs. 9A-H). In the same site, *C. spiniferum* and *C.* cf. *bilimbosum* that form *Cochliopodium* clade contain only 2 or 3 adenines (Figs. 9I-J), whereas *C. larifeili* branching separately at the base of the *Cochliopodium* tree possesses an additional helix consisting of 17 nucleotides in this site (Fig. 29K).

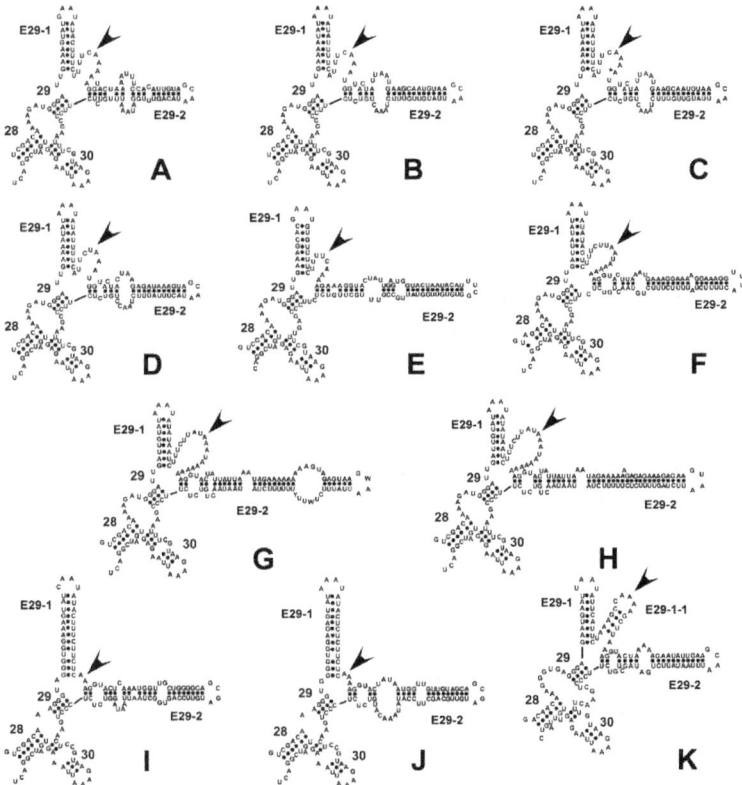

Fig. 29. Secondary structure of the SSU rRNA helices 28 - 30 in *Cochliopodium* spp. A. *C. plurinucleolum* n. sp. B. *C. minus* CCAP 1537/1A and *C. pentatrifurcatum* ATCC 30935 (KC247747). C. *Cochliopodium* sp. CPE, *C. minus* CCAP 1537/5 and *C. megatetrastylus* ATCC 30936 (KC747718). D. *C. minutoidum* CCAP 1537/7. E. *C. kieliense*. F. *C. actinophorum* CCAP 1537/10 (JF298250). G. *Cochliopodium* sp. CCAP non-identified (JF298257). H. *Cochliopodium* sp. from rice (*Oryza sativa* cDNA clone OSIGCRA115O12, CT837767). I. *C. spiniferum* CCAP 1537/3 (AY775130). J. *C.* cf. *bilimbosum* (JF298252). K. *C. larifeili* CCAP 1537/8 (JF298253). Arrowheads mark a loop between helices E29-1 and E29-2 defining major clades of *Cochliopodium*. Asterisk in C marks a single extra nucleotide that distinguishes this structure from B.

Molecular identification of *C. minus* CPE and borders between species related to *C. minus*

Whereas identification and position of *C. plurinucleolum* n. sp. in the phylogenetic tree of *Cochliopodium* are relatively clear apart from incongruence of SSU rRNA and Cox1 phylogenies, identification of *Cochliopodium minus* CPE is more problematic, and relationships between species comprising a *C. minus*-clade are obscure. Our results presented here reveal that the SSU rRNA sequence JF298257 was wrongly attributed to *C. minus* CCAP 1537/1A, and now we provide correct sequences of this strain. This error is due to the source of DNA material used to obtain the previous sequence, which was a genomic DNA sample available in the collection of amoebozoan DNA kept at the University of Geneva and designated as *C. minus* CCAP 1537/1A dating back to early 2000s. SSU rRNA, actin and Cox1 gene sequences were obtained from this DNA sample; sequences of the first two genes were published (Kudryavtsev et al. 2011) and attributed to *C. minus* CCAP 1537/1A as there was no reason to doubt the identity of the source DNA sample. However, accumulation of further sequences of amoebae similar to *C. minus* and *C. actinophorum* raised doubts about this identity. Therefore, a strain of *C. minus* 1537/1A was ordered from CCAP and sequenced again. The sequences obtained differ significantly from the one previously published by Kudryavtsev et al. (2011; JF298257). At the same time, the identity of the presently sequenced strain could be confirmed morphologically, whereas no morphological data that might help to identify the previously used DNA sample are available. In this case we have to conclude that the identity of the previously published sequence was erroneous due to a mislabelled DNA sample. Yet, this sequence branches robustly within *Cochliopodium*, being most closely related to *C. actinophorum* (Fig. 27). We therefore retain the published SSU rRNA gene sequence JF298257 and actin gene sequences JF298270-JF298272 (Kudryavtsev et al. 2011) in the database and designate them as *Cochliopodium* sp. to avoid confusion during subsequent studies. Correct sequences of a type strain of *C. minus* are now made available for analyses as the result of this study.

As mentioned above, all species of the *C. minus*-clade have virtually identical scales (Table 2), with the exception of *C. pentatrifurcatum*, whose scales are significantly different and mostly resemble those of *C. kieliense* (Tekle et al. 2013). Yet this species is virtually identical to *C. minus* based on sequence data. Another striking example is *C. megatetrastylus* that shares an identical scale structure and gene sequences with *C. minus*, while being significantly larger (up to 60 μm) and possessing a smoother outline without pronounced subpseudopodia (Anderson and Tekle 2013). Finally, *C. minutoidum* CCAP 1537/7 has virtually the same scales and only slightly differs from *C. minus* by the continuous peripheral hyaloplasmic sheet in most of the cells and the absence of cysts (Kudryavtsev 2006). *C. minutoidum* comprises the smallest members of this clade, despite the size range of this species largely overlaps with that of the others. These differences may be due to variability between individual strains, yet *C. minutoidum* clearly differs from the remaining species on the molecular level (sequence difference of about 4 % in the SSU rRNA and about 7 % in Cox1, Supplementary Table 1, 3).

In the situation outlined, *Cochliopodium minus* CPE can hardly be assigned to any listed species on the basis of specific characters, as borders between species based on molecular data are only partly congruent with those that can be outlined from morphological and ultrastructural data. On the basis of gene sequences, we have to recognize that *C. megatetrastylus* seems to be the closest relative of *Cochliopodium minus* CPE that contradicts strongly the morphological dissimilarities of these species outlined above. *C. minus* strains CCAP 1537/1A and CCAP 1537/5 are the next closest relatives of *Cochliopodium minus* CPE; morphologically more similar to this strain than *C. megatetrastylus*. We therefore provisionally identify the studied amoeba as another strain of *C. minus*. At the same time our data clearly revealed that species borders in at least some of the phylogenetic lineages in *Cochliopodium* are far from being clear, and additional extensive studies on other closely related strains of the *C. minus*-clade, evaluating their morphological, ultrastructural and genetic diversity are necessary to clarify the taxonomy of this part of the *Cochliopodium* phylogenetic tree. To facilitate further studies, we provide a checklist of all described strains of *C. minus* and species closely related to it, with literature references and gene sequence data accession numbers where available.

The close similarities of the SSU rRNA gene sequences especially in the members of the *C. minus*-clade that can be distinguished based on morphological characteristics, where morphological differences can be clearly distinguished, raise an important issue: the recent advance of high-throughput sequencing (HTS) technologies enabling environmental surveys without the need to cultivate species entirely relies on sequence information, primarily that of the SSU rRNA gene (Urich et al. 2008, Medinger et al. 2010, Stoeck et al. 2014). The data presented here show that a number of species with distinct morphologies may be missed by these approaches because their sequences are nearly identical, and these species may be clustered in a single OTU when the sequencing results are analysed (Creer et al. 2010, Medinger et al. 2010). As only partial sequences can be obtained in HTS studies, those species with identical sequences and even species differing only in certain regions of the SSU rRNA are likely to be missed in HTS studies, since they are impossible to distinguish solely based on sequence information (Bass et al. 2007, Boenigk et al. 2012). This highlights the importance of cultivation based efforts that allow detailed morphological and functional analyses on cultivated species. The data presented here also emphasize that the intragenomic variation of the SSU rRNA may be the second important issue for these studies. Like in previous molecular studies of naked lobose amoebae (e.g. Kudryavtsev et al. 2009, 2011; Nassonova et al. 2010; Smirnov et al. 2007), we confirm that the SSU rRNA gene can show slight sequence variations within a single genome, as demonstrated by comparisons of sequences among different molecular clones of the same amplicon obtained from the total genomic DNA. The degree of this sequence divergence is comparable to the degree of the divergence between strains that show morphological differences from each other. This raises further issues on the evaluation of the diversity of Amoebozoa based on environmental sequencing and calls for a more precise estimation of the borders between different operational taxonomic units (traditionally called species), as outlined above by different taxonomic approaches.

"Golgi attachment" in *Cochliopodium* is a MTOC: electron microscopic evidence

Members of the genus *Cochliopodium* possess a characteristic cytoplasmic structure that looks like an electron-dense bar adjacent to a dictyosome usually referred to as "Golgi attachment" (Yamaoka et al. 1984, Kudryavtsev 2004, Kudryavtsev et al. 2004, Kudryavtsev et al. 2005, Kudryavtsev 2006). Previous results published by Kudryavtsev (2004) and Kudryavtsev et al. (2004) suggested that this structure is in fact a microtubule-organizing centre (MTOC). However, we admit that the data on which this suggestion was based were rather ambiguous. Several published electron micrographs have shown elongated structures that were connected with one tip to the Golgi attachment and continuing into the cytoplasm (see Fig. 13 in Kudryavtsev 2004 and Fig. 8 in Kudryavtsev et al. 2004). These structures were cautiously interpreted as microtubules; therefore the Golgi attachment was suggested to be a MTOC. Here we confirm this initial idea by using a fixation of better quality compared to the earlier works, clearly demonstrating that microtubules are radiating from the Golgi attachment into the cytoplasm (Fig. 23F). Therefore, in addition to our recent data on *Stenamoeba* (Geisen et al. 2014), and earlier published results on Centramoebida (Bowers and Korn 1968), *Stygamoeba* (Smirnov 1996), *Gocevia* (Pussard et al. 1977), *Endostelium* (Bennett 1986), *Pellita* (Kudryavtsev et al. 2014), and *Corallomyxa* and *Stereomyxa* (Benwitz and Grell 1971a, b, Grell and Benwitz 1978), *Cochliopodium* appears to be another lineage of amoebae where the presence of a cytoplasmic MTOC associated with dictyosomes is confirmed.

Diagnosis

Phylum Amoebozoa, Subphylum Lobosa, Class Discosea, Order Himatismenida, Family Cochliopodiidae, Genus *Cochliopodium*.

Cochliopodium plurinucleolum n. sp.

Length in locomotion 8.8 – 16.4 µm (mean 12.0 µm), breadth 10.2 – 14.8 µm (mean 12.0 µm), length: breadth ratio 0.7 – 1.3 (mean 1.0); shape variable, sometimes oval, triangular, fan-shaped, round or crescent; thin, hyaloplasmic veil surrounding the remaining cell smooth or slightly irregular; traversed by stripes with the hyaloplasm between slightly bulging out; hyaloplasm only in fast directed locomotion reduced in the posterior and sometimes replaced by one or more short trailing filaments; no sub-pseudopodia; one single spherical nucleus, 2.2 – 4.0 µm diameter (mean 3.4 µm) with one or more small, often decentralized nucleoli ranging in diameter from 0.6 to 1.2 µm (mean 1.0 µm). Stable double-walled cysts formed in older cultures. Scales made up of a circular, grid-like base plate with the mesh size of 0.015 µm, 4 stalks attached to the base plate converging towards a funnel-shaped apical part consisting of ca. 15 radial spokes with very fine, poorly discernible concentric filaments between them. Diameter of a base plate 0.465-0.76 µm (average

0.627 µm), of the apical part, 0.445-0.577 µm (average, 0.5 µm), height of a scale 0.185-0.3 µm (average, 0.25 µm).

Etymology: The peculiar presence of several nucleoli inside the nucleus was eponymous for designating the species name *"plurinucleolum"*; type material: type culture is deposited with CCAP (UK), accession number CCAP 1537/11; observed habitat: Grassland soil on Sardinia, Italy (40°46′N, 9°10′E); molecular sequence data: KJ569732 (SSU rRNA gene), KJ569731 (Cox1 gene).

Differential diagnosis. Due to its unique nuclear structure, *C. plurinucleolum* is only comparable to *C. clarum* Schaeffer, 1926. However, it differs from this species in being smaller, having a more irregular outline, and inhabiting soils in contrast to marine habitats.

Checklist for *Cochliopodium minus*-similar strains and species

C. minus Page, 1976 CCAP 1537/1A (type strain). References: Dyková et al. 1998, Kudryavtsev 2006, Page 1968 (as *Hyalodiscus actinophorus* var. *minor*), 1976, 1988; GenBank accession numbers: KJ569700- KJ569704 (SSU rRNA), KJ569705- KJ569707 (Cox1).

C. minus CCAP 1537/5 (strain perished in CCAP). References: Kudryavtsev 2006, Kudryavtsev et al. 2005; GenBank accession numbers: KJ569708-KJ569709 (SSU rRNA), KJ569710- KJ569712 (Cox1).

C. minus CPE (CCAP 1537/12). References: this study; GenBank accession numbers: KJ569713- KJ569717 (SSU rRNA), KJ569719- (Cox1).

C. barki Kudryavtsev et al., 2004 CCAP 1537/4 (type strain, perished in CCAP). References: Bark 1973, Kudryavtsev et al. 2004; GenBank accession numbers: not available.

67

C. megatetrastylus Anderson et Tekle, 2013 ATCC 30936 (type strain). References: Anderson and Tekle 2013; GenBank accession numbers: KC747718 (SSU rRNA), KC747719- KC747720 (Cox1).

C. minutoidum Kudryavtsev, 2006 CCAP 1537/7 (type strain). References: Kudryavtsev 2006, this study; GenBank accession numbers: KJ569718, KJ569720-- KJ569722 (SSU rRNA), KJ569723 (Cox1).

C. pentatrifurcatum Tekle et al., 2013 ATCC 30935 (type strain). References: Tekle et al., 2013; GenBank accession numbers: KC247747 (SSU rRNA), KC489470 (Cox1).

Materials and Methods

Establishing cultures

Soil from the upper 20 cm of an intensively managed grassland plot at the Berchidda-Monti long term observatory, managed by the University of Sassari, Italy, 40°46′N, 9°10′E (Bagella et al. 2013) was collected after 2 mm sieving and transferred in thermo isolated containers to the lab. For inoculation, 50 g of soil mixed with 50 ml of sterile distilled water were shaken for 20 min and allowed to settle for 15 min. Ten enrichment cultures were established by transferring 5 µl of the soil suspension in 90 mm Petri dishes filled with soil extract solution (Page 1988). Each dish was carefully examined twice (at days 10-14 and 24-28) with an inverted Nikon Diaphot phase contrast microscope under 100x and 400x magnifications. To establish clonal cultures, amoebae of different morphotypes were individually transferred with a glass pipette to new 60 mm Petri dishes filled with Prescott-James (PJ) medium (Page and Siemensma 1991), enriched with 0.15 % wheat grass (WG) (Weizengras, Sanatur GmbH, D-78224 Singen). Two strains of *C. plurinucleolum* n. sp. designated 8 and 86 were isolated and shown to be identical in gene sequence data and light microscopic characters; therefore, only one (strain 8) was further described in detail. A freshwater strain of *Cochliopodium minus* CPE was donated by Dr. Rolf Michel (Department of Parasitology, Central Institute of the Federal Armed Forces Medical Services, Koblenz, Germany). It was isolated from *Elodea canadensis* purchased from a local pet shop in Neuwied, Germany, and cultured initially on 1.5 % non-nutrient agar prepared with PJ medium. Later the strain was transferred into 0.025 % Cerophyl infusion (roughly equivalent to WG) prepared with PJ. *C. minus* CCAP 1537/1A was obtained from CCAP on NN agar supplemented with *Escherichia coli* and further cultured on Cerophyl infusion.

Light and electron microscopic observations

Several hundred cells of each clonal culture of *C. plurinucleolum* were placed on glass cover slips and observed using a Leica DM2500 microscope with an attached Nikon DS-Fi 1 digital 5-megapixel microscope camera; 30 cells were measured. *C. minus* CPE was observed using a Carl Zeiss Axiovert 200 inverted microscope. Both microscopes were equipped with phase contrast and differential interference contrast optics (DIC). For transmission electron microscopy (TEM), amoebae were briefly (up to 5 min) prefixed at +4 ºC with 0.5 % (w/v) osmium tetroxide followed by fixation with 2.5 % (v/v) glutaraldehyde for 30–40 min and postfixation with 1 % (w/v) osmium tetroxide for 1 h; all fixatives were prepared with 0.05 M cacodylate buffer, pH 7.4. Cells were washed with the same buffer (3x5 min) between fixation steps and before dehydration. Dehydration was conducted in a graded ethanol series, followed by 100 % acetone and embedding in medium hardness epoxy embedding medium (Fluka) or Araldite M (Serva Electrophoresis). Silver to light gold ultrathin sections were cut with Reichert Ultracut E or Leica UC 6 ultramicrotomes, double stained with 2 % (w/v) uranyl acetate in 70 % ethanol for 20–30 min and Reynolds' lead citrate for 10 min and observed with Philips EM208 and Jeol JEM 1400 electron microscopes. Whole mounts of the cells for TEM observations were prepared by rinsing amoebae with glass-distilled water and placing the cell suspension on formvar-coated copper grids. Cells were allowed to settle and fixed for 5 min with osmium vapour. Grids were then air-dried and shadowed with chromium at an angle of ca. 15º using a Jeol JEE-420D vacuum evaporator. Grids were observed with TEM as described above.

Molecular phylogenetic studies

Genomic DNA was isolated from fresh cell cultures using a guanidine isothiocyanate method (Sambrook et al. 1989). In short, liquid medium was replaced with 100 µl of guanidine isothiocyanate after rinsing the dishes twice with sterile WG. Amoebae were scraped off the plate using a disposable cell scraper, and transferred into 2 ml centrifuge tubes. Subsequent stages were performed according to the cited protocol.

The gene encoding the small subunit (SSU) ribosomal RNA (rRNA) was fully amplified from strains 8 and 86 using the universal eukaryotic primers RibA (5'-ACC TGG TTG ATC CTG CCA GT-3') and RibB (5'-TGA TCC ATC TGC AGG TTC ACC TAC-3') (Cavalier-Smith and Chao 1995, Pawlowski 2000). Cycling consisted of initial denaturation at 95 °C for 5 min, followed by 35 cycles of denaturation at 95 °C for 30 s, annealing at 50 °C for 45 s and elongation at 72 °C for 1.5 min with a final elongation at 72 °C for 5 min. 8 µl of the PCR products were enzymatically purified by adding 0.15 µl Endonuclease I (20 U / µl, Fermentas GmbH, St. Leon-Rot, Germany), 0.9 µl Shrimp Alkaline Phosphatase (1 U / µl, Fermentas, Germany) and 1.95 µl H_2O and incubating the mixture for 30 min at 37 °C followed by incubating the samples at 85 °C for 20 min to stop the reaction. Subsequently, purified products were partially sequenced by GATC (Konstanz, Germany) using RibB as a sequencing primer. Lengths of the resulting sequences was 1073 bp and 1145 bp for strains 8 and 86, respectively. In addition, we amplified, cloned and sequenced the full SSU rRNA gene of *Cochliopodium minus* CPE, *C. minutoidum* CCAP 1537/7 (type strain of *C. minutoidum*) and *C. minus* type strain CCAP 1537/1A. To increase the number of nucleotide positions available for phylogenetic analysis, we also updated previously published incomplete SSU rRNA gene sequences of *C. kieliense* and *C. minus* CCAP 1537/5 (AY785057 and AY785056, respectively in Kudryavtsev et al. 2005) by amplification, cloning and sequencing the full-length SSU rDNA from the same DNA samples that were used in the previous study (op. cit.). PCR primers used and procedures for amplification, cloning and sequencing were as described in Kudryavtsev et al. (2009a).

5′ fragment of mitochondrial Cox1 gene was in all cases amplified from the same DNA samples as SSU rDNA. We sequenced this marker for all strains used in this study as well as *C. actinophorum* CCAP 1537/10 and *Cochliopodium* sp. previously identified as *C. minus* (Kudryavtsev et al. 2011). Amplicons of all strains except *Cochliopodium* sp. 8 and 86 were cloned, and 5-8 molecular clones from each amplicon were sequenced in both directions. Primers and protocol for amplification, cloning and sequencing were the same as in Nassonova et al. (2010). A list of all strains sequenced during this study with GenBank accession numbers is given in Table 3.

Table 3. Molecular markers of different *Cochliopodium* strains sequenced during this study and their GenBank accession numbers.

	SSU rRNA	Cox1
C. plurinucleolum n. sp., strains 8 and 86	KJ569732 (both strains)	KJ569731 (both strains)
Cochliopodium sp. CPE	KJ569713- KJ569717	KJ569719
C. minus CCAP 1537/1A	KJ569700- KJ569704	KJ569705- KJ569707
C. minus CCAP 1537/5	KJ569708- KJ569709	KJ569710- KJ569712
C. minutoidum CCAP 1537/7	KJ569718, KJ569720-KJ569722	KJ569723 (identical to GQ354208; Nassonova et al. 2010)
C. actinophorum CCAP 1537/10	JF298248-JF298251 (Kudryavtsev et al. 2011)	KJ173779 (identical to GQ354207; Nassonova et al. 2010
Cochliopodium sp. CCAP non-identified (formerly designated *C. minus* CCAP 1537/1A; Kudryavtsev et al. 2011)	JF298257 (Kudryavtsev et al. 2011)	KJ569724
C. kieliense	KJ569725- KJ569727	KJ569728- KJ569730

Sequences were manually aligned in Seaview 4 (Gouy et al. 2010) using all published sequences within the genus *Cochliopodium* (SSU rRNA: 14 sequences in total with 1403 unambiguously aligned positions, Cox1: 24 sequences in total with 495 aligned nucleotide positions). *Ovalopodium* and *Parvamoeba* were treated as outgroups in the trees based on SSU rRNA gene; *Vannella* spp. were used as outgroup in Cox1 gene trees. Phylogenetic trees were reconstructed using maximum likelihood algorithm with RaxML (Stamatakis 2006). A GTR+γ+I model of evolution with 25 substitution rate categories was applied for the analysis. All model parameters were estimated from the data. The stability of the clades was assessed using a non-parametric bootstrap with 1,000 pseudoreplicates. Bayesian reconstructions of phylogenetic relationships were performed using Mr Bayes Version 3.2.1 with covarion and autocorrelation models for among-site rates (Huelsenbeck and Ronquist 2001). Two runs of four simultaneous Markov chains were performed for 5,000,000 generations (default heating parameters) and sampled every 100 generations; 25 % of the samples were discarded as a burnin. All analyses were run at the University of Oslo Bioportal computer service (http://www.bioportal.uio.no; Kumar et al. 2009) . Approximately unbiased (AU) test of alternative tree topologies resulting from SSU rRNA and Cox1 gene analyses was performed using the Treefinder program, version of June 2008 (G. Jobb, freely distributed at http://www.treefinder.de).

Acknowledgements

SG was supported by research grants involved in the EU-project 'EcoFINDERS' No. 264465. This study was partially supported by the Russian Foundation for Basic Research grant 12-04-01835-a to AK, as well as a grant of St-Petersburg State University, and utilized equipment of core facility centres "Development of Molecular and Cell Technologies" and "Culturing of Microorganisms" of SPSU. We are grateful to Dr. Pier Paulo Roggero for providing the soils used to isolate *C. plurinucleolum*, Dr. Rolf Michel (Department of Parasitology, Central Institute of the Federal Armed Forces Medical Services, Koblenz, Germany) for the strain of *C. minus* CPE and Culture Collection of Algae and Protozoa for donating *Cochliopodium minus* strain CCAP 1537/1A. Part of the data were collected by AK while staying as a postdoc at the Molecular Systematic Group, Department of Genetics and Evolution, University of Geneva (Switzerland); many thanks are due to Prof. Jan Pawlowski, Dr. José Fahrni and Dr. Maria Holzmann for providing facilities and continuous help in the lab.

Supplementary files

Supplementary Table 1. Percentage difference between SSU rRNA gene sequences of different strains within the *Cochliopodium minus* / *C. minutoidum* / *C. megatetrastylus* / *C. pentatrifurcatum* clade; *C. pentat.* = *C. pentatrifurcatum*; *C. megat.* = *C. megatetrastylus*

	C. minus CCAP 1537/1A	*C. minus* CCAP 1537/5	*C. pentat.* ATCC 30935	*Cochliopodium* sp. CPE	*C. megat.* ATCC 30936
C. minus CCAP 1537/5	0.7				
C. pentat. ATCC 30935	0.1	0.8			
Cochliopodium sp. CPE	0.9	0.9	0.9		
C. megat. ATCC 30936	0.9	0.4	0.8	0.1	
C. minutoidum CCAP 1537/7	3.9	4	4	3.9	4

Supplementary Table 3. Average percentage difference between Cox1 gene sequences of different strains within *Cochliopodium minus* / *C. minutoidum* / *C. megatetrastylus* / *C. pentatrifurcatum* clade based on nucleotide and amino acid (bold number in brackets) data; *C. pentat.* = *C. pentatrifurcatum*; *C. megat.* = *C. megatetrastylus*

	C. minus CCAP 1537/1A	*C. minus* CCAP 1537/5	*C. pentat.*	*Cochliopodium* sp. CPE	*C. megat.*
C. minus CCAP 1537/5	2.3 (0.5)				
C. pentat.	Identical sequences	2.3 (**0.4**)			
Cochliopodium sp. CPE	2.8 (**0**)	2.7 (**0.4**)	2.7 (**0**)		
C. megat.	3.1 (**1**)	3 (**1.4**)	3 (**1**)	0.4 (**0.5**)	
C. minutoidum	6.8 (**0.7**)	7.3 (**1.1**)	6.7 (**0.7**)	6.7 (**0.7**)	7 (**1.6**)

Supplementary Table 2. SSU rDNA sequence motifs distinguishing the *Cochliopodium minus /
C. minutoidum / C. megatetrastylus / C. pentatrifurcatum* clade; - : GAP position; *C. minut.* =
C. minutoidum; C. pentat. = C. prentatrifurcatum; C. megat. = C. megatetrastylus

Helix #	Start position (in *C. pentat.*)	Length	Seq. in *C. minut.* CCAP 1537/7	Seq. in *C. minus* CCAP 1537/1A	Seq. in *C. pentat.*	Seq. in *C. megat.*	Seq. in *C.* CPE	Seq. in *C. minus* CCAP 1537/5
9-11	150	1	T	T	T	-	-	-
9-11	215	1	A	G	G	G	G	G
9-11	237	1	T	C	C	C	C	C
9-11	241	5-8	TAACAAAG	TTAAAG	TTAAAG	TTAAG	TTAAG	TTAAG
9-11	251	1	A	A	A	G	G	G
9-11	256	2-3	CTT	CTT	CTT	CT	CT	CTT
16	467	1	G	A	A	A	A	A
16	469	1	A	T	T	T	T	T
17	510	4	GGGA	AGGA	AGGA	AGGG	AGGG	AGGG
17/18	543	1	-	-	-	C	-	-
E23-1-7	775	1	A	A	A	T	T	A
E23-13	895	1	T	G	G	A	A	G
E23-13	906	1	G	C	C	T	T	T
24	1033	1	T	A	T	T	T	T
29	1146	6-7	TCTAAA	TTCAAA	TTCAAA	TTCAAAA	TTCAAAA	TTCAAAA
E43	1489	1	C	T	T	C	C	C
E43	1495	2	AG	AA	AA	GA	GA	GA
E43	1507	1	T	T	T	C	C	C
E43	1513	1	A	G	G	A	A	A
E43	1559	1	T	C	C	T	T	C
E43	1578	2	GA	AA	AA	GG	GG	GG
E43	1588	1	T	C	C	T	T	T
E43	1612	1	T	C	C	T	T	T
43	1624	1	T	T	T	T	T	A
49	1979	1	T	A	A	A	A	G
49	2000	1	T	C	C	T	T	C
49	2007	1	A	A	A	A	G	A

Part 1 – Chapter 3

Expansion of the 'reticulosphere': diversity of novel branching and network-forming amoebae helps to define Variosea (Amoebozoa)

Cédric Berney[1,a,*], Stefan Geisen[2,b], Jeroen Van Wichelen[3], Frank Nitsche[4], Pieter Vanormelingen[3], Michael Bonkowski[2], David Bass[1]

[1] Department of Life Sciences, The Natural History Museum, Cromwell Road, London SW7 5BD, United Kingdom
[2] Department of Terrestrial Ecology, Zoological Institute, University of Cologne, Zülpicher Str. 47b, 50674 Cologne, Germany
[3] Research Unit Protistology and Aquatic Ecology, Biology Department, Ghent University, Krijgslaan 281 (S8), 9000 Gent, Belgium
[4] Department of General Ecology, Zoological Institute, University of Cologne, Zülpicher Str. 47b, 50674 Cologne, Germany
[a] present address: UMR7144, groupe EPEP, CNRS - Station Biologique, Place Georges Teissier, 29680 Roscoff, France
[b] present address: Department of Terrestrial Ecology, Netherlands Institute of Ecology, Droevendaalsesteeg 10, 6708 PB Wageningen, The Netherlands

Abstract

Amoebae able to form cytoplasmic networks or displaying a multiply branching morphology remain very poorly studied. We sequenced the small-subunit ribosomal RNA gene of 15 new amoeboid isolates, 14 of which are branching or network-forming amoebae (BNFA). Phylogenetic analyses showed that these isolates all group within the poorly-known and weakly-defined class Variosea (Amoebozoa). They are resolved into six lineages corresponding to distinct new morphotypes; we describe them as new genera *Angulamoeba* (type species *Angulamoeba microcystivorans* n. gen., n. sp.; and *A. fungorum* n. sp.), *Arboramoeba* (type species *Arboramoeba reticulata* n. gen., n. sp.), *Darbyshirella* (type species *Darbyshirella terrestris* n. gen., n. sp.), *Dictyamoeba* (type species *Dictyamoeba vorax* n. gen., n. sp.), *Heliamoeba* (type species

Heliamoeba mirabilis n. gen., n. sp.), and *Ischnamoeba* (type species *Ischnamoeba montana* n. gen., n. sp.). We also isolated and sequenced four additional variosean strains, one belonging to *Flamella*, one related to *Telaepolella tubasferens*, and two members of the cavosteliid protosteloid lineage. We identified a further 104 putative variosean environmental clone sequences in GenBank, comprising up to 14 lineages that may prove to represent additional novel morphotypes. We show that BNFA are phylogenetically widespread in Variosea and morphologically very variable, both within and between lineages.

Introduction

Naked heterotrophic reticulose and branching amoebae, characterized by more or less thin cytoplasmic extensions from, or as part of, their cell bodies, occur in several places across the eukaryote tree of life. The appearance of the cytoplasmic extensions varies considerably. In some lineages parts of the cells are able to anastomose, connecting their cytoplasm, whereas in others this never happens. While this characteristic is known for some taxa, in general the distinction is difficult to make because very little is known about most lineages. Some highly branching amoebae are not known to form networks (e.g. *Mesofila*; Bass et al. 2009), whereas others appear to form networks under some conditions but perhaps not others. Where networks occur they can be very fine, with or without distinct cell bodies (e.g. *Filoreta* and *Reticulamoeba*; Bass et al. 2009, 2012), or more compact with proportionally smaller lacunae (e.g. *Leptomyxa*, *Protomyxa*, and some vampyrellids; Berney et al. 2013; Goodey 1915b; Hess et al. 2012; Rhumbler 1904; Smirnov et al. 2008), sometimes resembling 'sheets' of cytoplasm within which lacunae may occur, e.g. *Thalassomyxa* (Berney et al. 2013). In these cases lacunae are often formed by cleavage of the cytoplasm into separate streams, so there is no clear distinction between cell body and cytoplasmic network. A large range of variants of the organization of the cell body (hereafter referred to as 'morphotypes') can be found between these main forms.

Despite their morphological distinctiveness, these branching and network-forming amoebae (BNFA) are remarkably poorly known. Ultrastructural and molecular phylogenetic studies have revealed that BNFA can be found in many places across the eukaryotic tree of life within major groups such as Amoebozoa, Rhizaria, and Stramenopiles. However there are also many such amoebae described in the literature for which sequence data are not yet available (e.g. Adl et al. 2012; Lee et al. 2000; Bass et al. 2009, 2012). Table 4 lists formally described BNFA genera from the literature from before the advent of DNA amplification and sequencing (excluding those known to belong to the well characterized and in several ways distinct Mycetozoa and testate Foraminifera, as well as photosynthetic species such as chlorarachniophytes). This list is not exhaustive, but highlights two significant facts: (1) molecular data are still missing for most of these BNFA, and (2) many BNFA isolated and sequenced more recently were morphologically distinct enough from these characterised genera to be described as new taxa, for example *Acramoeba* (Smirnov et al. 2008), *Filoreta* (Bass et al. 2009), and *Telaepolella* (Lahr et al. 2012). Together these observations reinforce the idea that the morphological and molecular diversity of BNFA remains largely unexplored. Supplementary Table 4 provides further morphological and ecological information about different species and isolates of these BNFA where it is available. It also emphasizes the difficulty of species identification of BNFA in absence of molecular data, which can lead to taxonomic confusion and probable misidentifications or incorrect lumping of unrelated organisms in the same genera.

The range of food items utilized by BNFA is strikingly wide, including bacteria, other protists, diatoms, algae, fungi, and even small metazoans. In some cases feeding behaviour has been elusive, further emphasizing the interesting nutritional modes shown by BNFA. It is thought that cytoplasmic networks enable BNFA to be more efficient than many heterotrophic protists in finding and ingesting surface-attached prey, and exploiting food sources in interstitial spaces in particulate sediments (Butler and Rogerson 1997; Rogerson et al. 1996). Networks are often very thin and flat and maximize the cell-surface to volume ratio (i.e. foraging area), making them theoretically more energetically efficient in certain microhabitats in comparison with larger, rounder cells. Per cell volume they have been shown to consume bacteria more quickly than non-reticulose cells, and there is evidence that some lineages can digest bacteria or other prey such as diatoms within the pseudopodia themselves (Rogerson et al. 1996; Grell 1994, 1995).

Table 4. Formally described genera of branched or network-forming amoebae and their taxonomic affiliation.

taxon [1]	reference	habitat [2]	phylogenetic position
Aletium	Trinchese 1884	M	unknown
Arachnula	Cienkowski 1876	F	Vampyrellida (Bass et al. 2009) [3]
Biomyxa	Leidy 1875	F	unknown - possibly Foraminifera
Cichkovia	Valkanov 1931	?	unknown - possibly Variosea
Cinetidomyxa	Chatton & Lwoff 1924	M	unknown - possibly Variosea
Corallomyxa	Grell 1966	M	unknown - probably Amoebozoa [4]
Dictiomyxa	Monticelli 1897	M	unknown
Gephyramoeba	Goodey 1915b	F	unknown - probably Leptomyxida [5]
Gloidium	Penard 1902	F	unknown - possibly Leptomyxida
Gobiella	Cienkowski 1881	F	Vampyrellida (but unsequenced)
Gringa	Frenzel 1892	?	unknown
Gymnophrydium	Dangeard 1891	F	unknown - possibly Vampyrellida
Gymnophrys	Cienkowski 1876	M	unknown - possibly Foraminifera
Haeckelina	Schepotieff 1912	M	unknown - possibly Foraminifera
Lateromyxa	Hülsmann 1993	F	Vampyrellida (but unsequenced)
Leptomyxa	Goodey 1915b	F	Leptomyxida (Amaral Zettler et al. 2000) [6]
Leptophrys	Hertwig & Lesser 1874	F	Vampyrellida (Hess et al. 2012)
Leucodictyon	Grell 1991	M	unknown - probably Cercozoa, Granofilosea
Leukarachnion	Geitler 1942	F	Stramenopiles (Grant et al. 2009)
Megamoebamyxa	Nyholm 1950	M	unknown
Myxastrum	Haeckel 1870	?	unknown
Penardia	Cash 1904	F	Vampyrellida (Berney et al. 2013)
Pontomyxa	Topsent 1892	M	unknown - possibly Foraminifera
Protogenes	Haeckel 1870	M	unknown - possibly Foraminifera
Protomyxa	Haeckel 1870	M	unknown
Reticulamoeba	Grell 1994	M	Cercozoa, Granofilosea (Bass et al. 2013)
Reticulomyxa	Nauss 1949	F	Foraminifera (Pawlowski et al. 1999)
Reticulosphaera	Grell 1989	M	Haptophyta (Cavalier-Smith et al. 1996)
Rhizoplasma	Verworn 1903	M	unknown - possibly Foraminifera
Stereomyxa	Grell 1966	M	unknown - probably Amoebozoa
Synamoeba	Grell 1994	M	unknown - probably Amoebozoa
Thalassomyxa	Grell 1985	M	Vampyrellida (Berney et al. 2013)
Theratromyxa	Zwillenberg 1953	F	Vampyrellida (Parfrey et al. 2010)
Vampyrella	Cienkowski 1865	F	Vampyrellida (Hess et al. 2012)
Vampyrelloides	Schepotieff 1912	M	unknown - probably Vampyrellida

[1] See Supplementary Table 4 for more detailed information on different species and isolates of each taxon (when available) and additional references.

[2] M = marine, F = non-marine (freshwater or terrestrial).

[3] *Arachnula* sp. ATCC 50593 sequenced by Tekle et al. 2008 is *Telaepolella tubasferens*

[4] *Corallomyxa tenera* ATCC 50975 sequenced by Tekle et al. 2007 is a *Filoreta tenera*

[5] *Gephyramoeba* sp. ATCC 50654 sequenced by Amaral Zettler et al. 2000 is *Acramoeba dendroida*

[6] *Leptomyxa reticulata* ATCC 50242 sequenced by Amaral Zettler et al. 2000 could be a *Rhizamoeba* (Smirnov, pers. comm.)

The current state of knowledge of BNFA suggests that, although they can be found across the whole eukaryotic tree of life, there is a relatively small number of lineages with these morphological characteristics (compared to non-BNFA, heterotrophic flagellates, etc.), and that they are present in relatively low numbers in the environment. However, this perception is influenced by some important factors: (1) the cells are easily disrupted and broken by standard sample collection methods, (2) they are often very slow growing in culture or do not thrive in standard laboratory culturing conditions (which usually favour relatively fast-growing bacterivores and ecological generalists), and (3) they are not readily seen or recognized due to lack of dispersed expertise and because they are rarely the focus of experimental work. Rogerson et al. (1996) found that although they were not as numerous as other heterotrophic protists, direct counting methods suggested they were an order of magnitude more abundant in some coastal marine sediments than culturing-based estimates allowed. Feest and Campbell (1986) found BNFA in most soils they investigated, and correlated abundance of BNFA and dictyostelids with low levels of take-all disease. Branching amoeboid morphologies may also confer nutritional advantages for parasites and predators of relatively large organisms: *Grellamoeba* has recently been identified from fish kidney tissue (Dyková et al. 2010a), the stramenopile labyrinthulids include parasites of seaweeds and green plants, *Protogenes* and *Protomyxa* are particularly associated with the seaweeds *Gelidium* and *Bryopsis* respectively (perhaps ectoparasitically), and some vampyrellids enter host algal cells in the process of consuming their contents (Hess et al. 2012). It is perhaps significant that some BNFA taxa are evolutionarily very close to important parasite groups, for example vampyrellids to phytomyxids (Berney et al. 2013) and *Filoreta* and *Gromia* to Ascetosporea (Bass et al. 2009a).

The large physical extent/reach of the cells, the ability to take up and metabolize prey at relatively high rates, and the wide prey range suggest that BNFA may play diverse, distinct, and significant ecological roles. We provide new evidence for this by revealing previously unknown but apparently common variosean (Amoebozoa) BNFA lineages from soils or freshwater habitats that represent novel morphotypes and are phylogenetically very diverse.

Results

Morphology and phylogenetic placement of our new isolates

We isolated and sequenced the small-subunit ribosomal RNA gene (SSU rDNA) from 19 strains of amoebae, all markedly branching (ramose) or network-forming (reticulose), with the exception of FN414 (*Flamella* sp.) and M134 (*Heliamoeba* n. gen.). Information about these new isolates is summarised in Table 5. An appropriate sequence dataset was constructed to refine the phylogenetic placement of our isolates with respect to morphologically characterised amoebozoan taxa. The results of our analyses are shown in Fig. 30 and confirm that all isolates belong to class Variosea (Amoebozoa). The isolates fall into nine main lineages. Some were closely related to previously characterised lineages, congruently with their morphology: FN414 is a *Flamella* sp., Tib190 is related to and resembles the recently described species *Telaepolella tubasferens* (Lahr et al. 2012), F3 is very close to *Schizoplasmodiopsis vulgaris* (EF513180) within the cavosteliid lineage, and M71 branches at the base of that lineage and is likely to represent a novel or previously unsequenced species within or near the genus *Schizoplasmodiopsis*. The 15 other isolates were not closely related to known taxa and form six new variosean lineages with novel, distinct morphotypes. We describe below six new genera for these lineages, with one isolate designated as a type species in each one. One of the novel lineages, *Angulamoeba* n. gen. (Figs 31 & 32), of which we isolated four strains and describe two species, is highly divergent from all described varioseans in terms of SSU rDNA sequence (Fig. 30). Another substantial, but short-branched clade was identified by our isolates, comprising two lineages – one with five reticulose isolates (*Darbyshirella* n. gen.; Fig. 33) and one with three branching but not reticulose isolates (*Ischnamoeba* n. gen.; Fig. 34). The remaining three isolates branch individually as three additional lineages; one of them is filose/ramose (M134, *Heliamoeba* n. gen.; Fig. 35) while the other two are reticulose (WalEn, *Dictyamoeba* n. gen. and Tib182, *Arboramoeba* n. gen.; Figs 36 & 37, respectively).

Table 5. Origin of the 18 new variosean strains isolated in this study.

isolate	taxonomic identity	collection date	geographic origin	sample type	phylogenetic position [1]
Tib190	*Telaepolella* sp.	2011	Tibet	High altitude (4556 m) meadow soil	lineage V07
FN414	*Flamella* sp.	2010	Cologne, Germany	Air sample	lineage V08
M71	*Schizoplasmodiopsis* sp.	2012	Veluwe, the Netherlands	Meadow soil (abandoned field)	lineage V11
F3	undescribed cavosteliid	2012	Frankfurt, Germany	Meadow soil	lineage V11
Tib85	*Ischnamoeba montis*, gen. et sp. nov.	2011	Tibet	High altitude (4556 m) meadow soil	lineage V12
F4	*Ischnamoeba* sp.	2012	Frankfurt, Germany	Meadow soil	lineage V12
FN352	*Ischnamoeba* sp.	2010	Cologne, Germany	Air sample	lineage V12
Tib177	*Darbyshirella terrestris*, gen. et sp. nov.	2011	Tibet	High altitude (4149 m) meadow soil	lineage V13
Esthw	*Darbyshirella* sp.	2007	Oxford, England, UK	Freshwater lake littoral sediment	lineage V13
M68	*Darbyshirella* sp.	2011	Veluwe, the Netherlands	Meadow soil (abandoned field)	lineage V13
JDgam	*Darbyshirella* sp.	2010	Scotland, UK	Garden soil	lineage V13
M77	*Darbyshirella* sp.	2011	Veluwe, the Netherlands	Meadow soil (abandoned field)	lineage V13
WalEn	*Dictyamoeba vorax*, gen. et sp. nov.	2008	Gregynogg, Wales, UK	mixture of soil, moss and lichens from a garden	lineage V14
M134	*Heliamoeba mirabilis*, gen. et sp. nov.	2011	Veluwe, the Netherlands	Meadow soil (abandoned field)	lineage V15
Tib182	*Arboramoeba reticulata*, gen. et sp. nov.	2011	Tibet	High altitude (4556 m) meadow soil	lineage V16
23-6A	*Angulamoeba fungorum*, gen. et sp. nov.	2005	Oxford, England, UK	Freshwater lake littoral sediment	lineage V17
F2	*Angulamoeba* sp.	2012	Frankfurt, Germany	Meadow soil	lineage V17
FN806	undescribed angulamoebid	2010	Cologne, Germany	Air sample	lineage V17

[1] See Fig. 31 and Supplementary Table 5

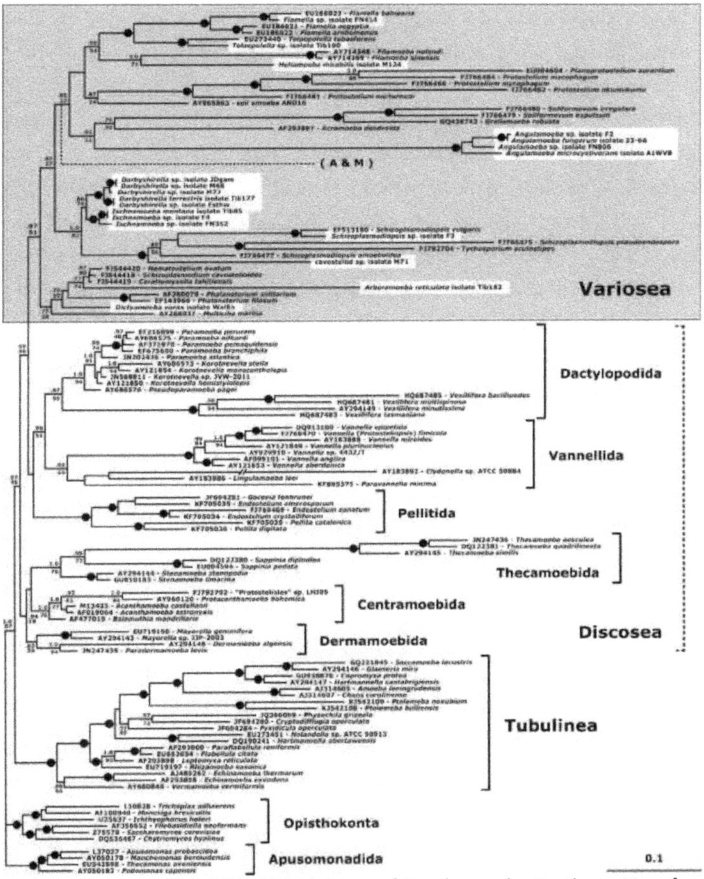

Fig. 30. Maximum likelihood SSU rDNA phylogeny of Amoebozoa, showing the position of our new isolates. Our new isolates are highlighted in white boxes; all belong to Variosea (shaded box). The overall amoebozoan topology is congruent with previous publications using the same genetic marker, supporting the monophyly of Tubulinea and (more weakly) of Variosea, but not of the many orders currently classified in Discosea. Bootstrap support values after 1000 replicates and Bayesian posterior probabilities (see Methods) are indicated at nodes when above 50% and/or 0.75, respectively. Black blobs represent support values at or above 95%/0.99. Opisthokonts and apusomonads were used to root the tree. All branches are drawn to scale, except the one leading to *Clydonella* sp., reduced to half of its length. The branching position of Archamoebae and Macromycetozoa (A & M) when added to this dataset is indicated with a dashed line (see Discussion); Supplementary Fig. 1 shows the alternate position they occupy, separate from Variosea, when some of the most divergent varioseans are excluded.

None of our isolates was seen to produce fruiting bodies of any kind, despite some being maintained in variable culture conditions for up to several years, including dehydration and rehydration of culture vessels and growing on sterilized nutrient agar. However at least two (F3 and M71) clearly belong to a protosteloid lineage (the cavosteliids) where described taxa can produce fruiting bodies. Morphological variation between individual cells within each isolate is typically very high, and can exceed that between genetically different isolates within a lineage or even that between distantly related lineages, especially in younger, smaller cells. Nevertheless, by observing many cells within an isolate it becomes clear that each lineage corresponds to a distinct morphotype and is distinguishable from all others based on the shape, size, and organisation of the cells and/or the networks.

The reticulose genera *Darbyshirella* (Fig. 33; five isolates), *Dictyamoeba* (Fig. 36; isolate WalEn), and *Arboramoeba* (Fig. 37; isolate Tib182) all form indefinitely large networks. In *Darbyshirella* the branches of the network are relatively long and thin and regularly spaced, and the network is formed through what appear to be random anastomoses between branches. The density of the network is relatively regular but with lacunae of various sizes, and often no clear direction of movement can be determined. In cases where such a direction can be observed, the network does appear denser at the anterior edge. In *Dictyamoeba* young reticulose trophozoites have quite thin branches, with one or few non-branching, non-filose posterior endings and a visibly denser branching/reticulose area at the front; they resemble small networks of the genus *Darbyshirella*. As the network grows in presence of suitable food (yeasts seem to be preferred) the branches become markedly thicker and more evenly spaced and the network can then grow in all directions, with many "terminal" areas where denser, fine branching extensions gradually replace the network. *Arboramoeba* exhibits the densest and most impressive networks and in older cultures always looks "tree-like", with one or few non-branching, non-filose posterior endings and an extremely branching and reticulose anterior front forming a wide, non-permeable frontier, significantly denser than the posterior part of the network. These three morphotypes are not directly related according to our phylogenetic analyses.

The branching genera *Angulamoeba* (Figs 31 & 32; four isolates, two of them described as new species *A. microcystivorans* and *A. fungorum*) and *Ischnamoeba* (Fig. 34; three isolates) were never observed to become reticulose; trophozoites of *A. microcystivorans* have a tendency to cluster together tightly within a mucilaginous matrix when feeding on *Microcystis* colonies (Fig, 31I, J), but were not observed to actually fuse. In both genera the branching pattern usually remains quite simple, more so than in the similar genus *Acramoeba* (Smirnov et al. 2008). In *A. fungorum*, the cell body separates into three main branches on average (rarely more than four), which can sometimes (but rarely) be further subdivided. The branches are thicker at the base and quite short, ending with very fine extensions. In *Ischnamoeba*, branches are thinner overall on average, more commonly further subdivided, and more randomly distributed, not mostly originating from a common central point as is usually the case in branching *Angulamoeba* cells. They vary more in size and length but also end in very fine extensions. A flagellated form was observed in *A. microcystivorans* (Fig. 31X); transition between amoeba and flagellate was frequent and happened quite rapidly. Flagellated forms were never observed in *A. fungorum*, although the two species are closely related (Fig. 30). *Heliamoeba mirabilis* (Fig. 35) was never reticulose or clearly branching, but this isolate exhibits fine pseudopodia sometimes exceeding in length the size of the main cell body. Finally, isolate Tib190 displays many of the morphological characteristics of *Telaepolella tubasferens* (ATCC 50593), with young amoebae resembling *Flamella* spp. while older ones become much larger, branching cells, sometimes reticulose with lacunae.

Like many Variosea, our new isolates strongly vary in size, often even within clonal cultures. The number of nuclei per cell varies widely. Directed movement as present in most representatives within the supergroup Amoebozoa and even in the related variosean genera *Filamoeba* and *Flamella* was not easily visible in most of our BNFA isolates because of their extreme slowness; only in a time series of micrographs taken over several hours does it become clear that these amoebae can display directed movement. A shared feature of all isolates was the presence of very fine, filose-like pseudopodia, typical for amoebae in the class Variosea but not generally across Amoebozoa (Lahr et al. 2012). These pseudopodia can be observed all around the cell but are typically denser at the end of branches or at the anterior front of the reticulose genera. Movement could only be observed when the filose-like pseudopodia attached or detached from the substratum, while locomotion of entire trophozoites was very slow and could usually not be observed microscopically. In contrast, intracellular activity was high as granuloplasm movement and vacuolar activity was permanently visible. All isolates formed cysts in older cultures, which often varied strongly in size and shape.

Variosea: phylogeny and lineage diversity

Fig. 38 presents the results of separate phylogenetic analyses focusing on Variosea alone. We used the sequences from all previously known members of Variosea and from all our new isolates as BLASTn search seeds against the NCBI GenBank database and recovered a further 101 environmental clone sequences of variosean origin, 20 of which were identified as chimeric (Supplementary Table 5). Phylogenetic analysis of all these sequences resulted in 26 phylogenetic clusters (that we refer to as lineages hereafter); Supplementary Table 5 lists six other environmental clones forming five additional clusters that are probably also variosean, but were not included in the dataset used for Fig. 38 because the sequences were too short. We defined the 26 lineages listed in Fig. 38 based on (1) a shared distinct morphology, where known, and (2) robust phylogenetic support (Bayesian posterior probability of 1.0 and maximum likelihood bootstrap support above 95%). Accordingly, we identify 17 morphologically described lineages within Variosea (shaded in Fig. 38), six of which are new (darker shading). Those with known branching or network-forming ability are indicated by 'R', while a 'P' denotes lineages known to have a protosteloid life cycle in at least some members.

Five of our new genera represent the first described taxa in lineages previously comprising only environmental sequences from GenBank. A Mexican alkaline lake (Couradeau et al. 2011) had revealed sequences belonging to the genera

Darbyshirella and *Angulamoeba*, while sequences of *Ischnamoeba* sp. and *Arboramoeba* sp. were known from trembling aspen rhizosphere (Lesaulnier et al. 2008). Lake moss pillars in Antarctica (Nakai et al. 2012) were hosts to the genera *Ischnamoeba*, *Dictyamoeba*, and *Angulamoeba*. A sequence of *Arboramoeba* sp. was also retrieved from mixed hardwood soil (O'Brien et al. 2005), and a sequence of *Ischnamoeba* sp. from a water purification plant (Chiellini et al. 2012). An undescribed amoeba from a rice field in Italy (Murase and Frenzel 2008) also corresponds to the new genus *Ischnamoeba*, which is consistent with its morphology (Murase, pers. communication). Finally, other *Angulamoeba* sequences were found before in a hospital water network (Thomas et al. 2006), in *Phythophthora parasitica* biofilm (Galiana et al. 2011), and in an artificial reservoir lake (Lepère et al., unpublished).

Fig. 38 also shows that SSU rDNA alone does not allow resolution of the relationships between variosean lineages. Only two well-supported groups can be inferred from our analyses, one containing the genera *Ischnamoeba* and *Darbyshirella*, and one containing the genera *Telaepolella* and *Flamella*, and environmental lineage V18. All other higher-level relationships recovered in our analyses are weakly supported and should only be seen as working hypotheses for future studies when additional molecular data become available for members of Variosea. These include the possible basal position of *Phalansterium* among Variosea, or the existence of a relatively short-branched group including genera *Dictyamoeba*, *Arboramoeba*, and the schizoplasmodiids. The genera *Ischnamoeba* and *Darbyshirella* may be part of a wider group also including cavosteliids and *Angulamoeba*, while the clade including *Telaepolella*, *Flamella*, and environmental lineage V18 may belong to a wider group of longer-branched varioseans together with genera *Filamoeba* and *Heliamoeba* and environmental lineages V20 and V21.

The length of the SSU rDNA sequences of our new isolates ranged from around 1850 bp to almost 2100 bp. This is above the average length of the SSU rDNA for most eukaryotes (typically in the range 1750-1850 bp), and in all isolates we observed an expansion of helix 43 in the variable region V7. Upon further investigation, we observed that this expansion of helix 43 is associated with the presence of a highly specific secondary structure pattern that we found in all known variosean sequences present in Fig. 38. This specific secondary structure pattern corresponds in part to a conserved motif of eight nucleotides (GGGTGAAG) that was mentioned by Pawlowski and Burki (2009) as a potential signature for Variosea. However, addition of new variosean sequences shows that the motif itself is not conserved across all members of Variosea, but the larger secondary structure pattern to which it belongs is, and we suggest this pattern as a whole is a molecular synapomorphy for Variosea (Supplementary Fig. 2). It is absent in all other known amoebozoan lineages, where variable region V7 is sometimes of the typical length observed in most other eukaryotes (e.g., most members of class Tubulinea), or contains expansions corresponding to a distinct secondary structure pattern from that observed in Variosea (many lineages in the possibly non-monophyletic class Discosea).

All isolates of the genera *Ischnamoeba* and *Darbyshirella* also exhibited expansions in the variable region V4 of the SSU rDNA. This region (that has been defined as a universal "pre-barcode" for protists (Pawlowski et al. 2012) has an average length of 380-400 bp in eukaryotes (and in all of our other isolates), but is about 450 bp long in the *Ischnamoeba* isolates, and 530-570 bp long in the *Darbyshirella* isolates. These shared V4 expansions are congruent with a sister-relationship between the two genera.

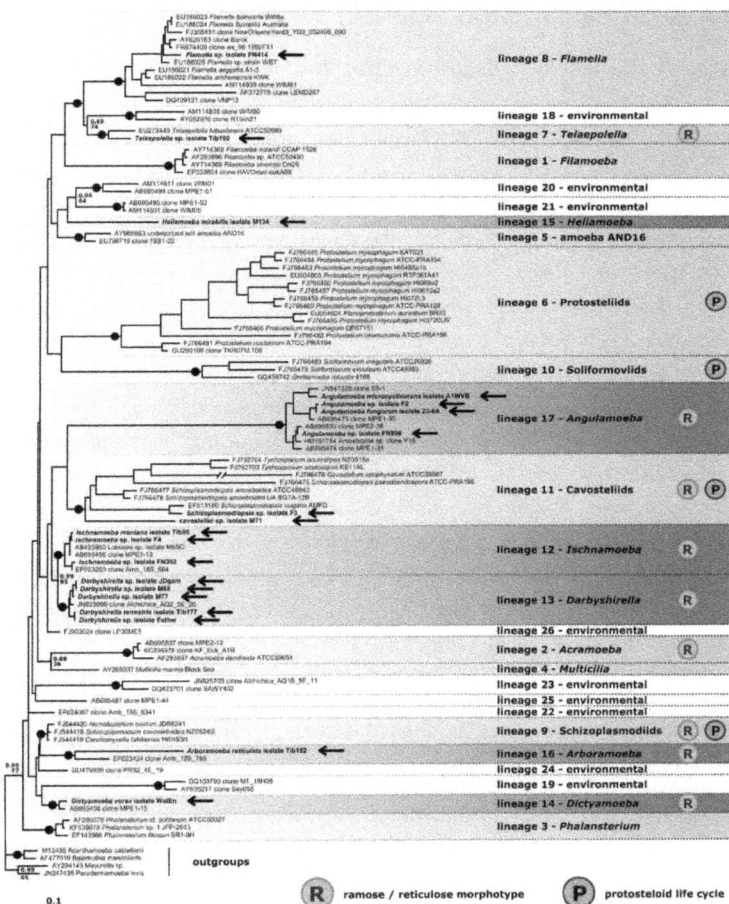

Fig. 38. Maximum likelihood phylogeny of Variosea, highlighting the 26 lineages identified in this study. Bootstrap support values after 1000 replicates and Bayesian posterior probabilities (see Methods) are indicated at nodes when above 50% and/or 0.75, respectively. Black blobs represent support values at or above 95%/0.99. Four discoseans were used to root the tree. Seventeen lineages with morphologically described members are shaded, the nine others are so far known only from environmental clone sequences found in GenBank. Those with known branching or network-forming ability are indicated by a circled R, while the four lineages where at least some members have a protostelid life cycle are indicated by a circled P. Our nineteen new isolates (in bold) are highlighted with arrows; fifteen of them belong to six lineages that were previously unknown and are described as new genera in the present study (darker shaded lineages).

89

Description of the new genera and species

In previously described variosean taxa, the lineages identified in Fig. 38 correspond to a genus (e.g. *Acramoeba, Filamoeba, Phalansterium*), or in the case of the four protosteloid lineages only, to an assemblage of morphologically similar genera. To be taxonomically consistent with this observation, we decided to describe the six new lineages uncovered in this study as new genera, and we describe a new type species in each genus. In three cases, we isolated several genetically distinct strains per lineage, and the genus description reflects what is common to all. For the others, genus diagnoses are based on a single isolate; they are not meant to represent ultimate boarders for further species of the same genus, as we cannot accurately estimate how much these may morphologically depart from each other. In most cases, it is however important to note that genetically distinct isolates (which we consider different species) within a lineage/genus can have completely overlapping morphological characteristics even when many different cells at different stages and of different sizes are observed. Therefore in Variosea species delineation within a genus cannot always be satisfactorily achieved using morphology alone, and for all species descriptions below, the SSU rDNA sequence of the type strain should be considered the main (if not only) distinguishing character.

Angulamoeba Berney, Bass & Geisen, n. gen.

Diagnosis: Uninucleate, branching naked amoebae with slender, pointed and/or filose-like, sometimes branched pseudopodia (Figs 31 & 32). Trophozoites moving too slowly to be seen in light microscopy; main cell body elongated, consisting of up to four main branches often with several smaller lateral branches, never forming a network; numerous fine pseudopodia concentrated mostly at the extremity of the lateral and terminal branches (Fig. 31A), but can be formed anywhere around the cell body; multiple contractile vacuoles. Movement of entire cells mostly too slow to be directly observable. At least one species with flagellate forms. Floating forms observed in at least one species. Cysts of very regular size and shape, round and relatively small.

Type species: *Angulamoeba microcystivorans*, n. sp.

Etym.: Angulus L. angle, corner.

Angulamoeba microcystivorans Van Wichelen & Vanormelingen, n. sp. (Fig. 31)

Diagnosis: Non-marine, main body irregularly shaped, often elongated and branched; individuals often aligned and seemingly connected to each other with filose-like pseudopodia (Fig. 31A, B, K), eventually clustering tightly into a yellowish-brown 'rusty'-coloured mucilage matrix (Fig. 31I, J). Individual size varies considerably. Length (based on 126 individuals in different stages of cultivation) 11–170 µm (mean 55), width 2–32 µm (mean 10). One globular vesicular nucleus with a diameter of about 4 µm, containing one globular nucleolus of about 2 µm. Each amoeba displays up to 20 small, slender, straight, regularly bifurcating pseudopodia that can extend from anywhere around the cell body but mostly originate from several broad cytoplasmic projections. 2–14 contractile vacuoles (mean 6) visible, distributed all over the cell body. Rarely, a floating form is observed (Fig. 31C), more or less globular, 8–17 µm in diameter, either with many (20–40) radiating, small (up to 12 µm), pointed pseudopodia or only a few, broader (up to 2 µm), furcated pseudopodia. A flagellated form is frequently observed, especially shortly after re-inoculation (Fig. 31D-G). They originate from the fast (1–2 minutes) transformation of floating or resting amoebae into a spherical to oval, uni-, bi- to three-flagellated form with a length of 10–20 µm (mean 15) and a width of 5–15 µm (mean 10). First, the amoeba becomes more oval in shape with retraction of most pseudopodia except for one to three that elongate rapidly (up to 50 µm) from proboscis-like cell extensions. In general, each cell extension bears only one flagellum, however up to three flagella can originate from the same extension. Often, these extensions give the flagellates a characteristic L- to Y- shaped outline. After an initial period of slow sigmoidal tapering movements, the flagella start to beat very fast, resulting in a strong spinning movement of the flagellate before it finally takes off. The flagellates' locomotion is rather slow and interrupted. Transformation back from flagellate to amoeba happens just as fast, with detraction of the flagella and proboscis-like cell extensions, gradual expansion of the cell body and eventually formation of pseudopodia. Intermediate forms were observed with on one side a flagellum and on the other side pseudopodia. Spherical to oval cysts with a length and width between 8–25 (mean 13) and 7–20 µm (mean 12), respectively always present in the cultures. They mostly occur clustered in the mucilage matrix (Fig. 31H, L). *A. microcystivorans* feeds voraciously on the chroococcal cyanobacterium *Microcystis aeruginosa* (Fig. 31K). It does not grow with other *Microcystis* (*M. viridis*, *M. wesenbergii*), heterotrophic bacteria or green algae (*Scenedesmus*) as food. *Microcystis* colonies become infested either after contact with the matrix containing many amoebae or by flagellates that readily transform into amoebae after attachment to a colony. Infested colonies are easily recognizable by the change in colour from pale blue-green to yellowish-brown. After inoculation with amoebae, *Microcystis* cultures are completely grazed away in a matter of days.

Type strain: Isolate A1WVB (deposited at the Microbial Culture Collection of the National Institute for Environmental Studies, Tsukuba, Japan, under accession number NIES-XXXX). Type location: a small eutrophic urban pond in Westveldpark (51° 04' 059" N, 3° 46' 427" E), Sint-Amandsberg (Gent, Flanders, Belgium). Isolated on 23/07/2008 from a *Microcystis*-bloom sample. SSU rDNA sequence of the type strain: GenBank accession number KP864083.

Etym.: *vorans* L. devouring, ravenously eating, and *Microcystis* its food.

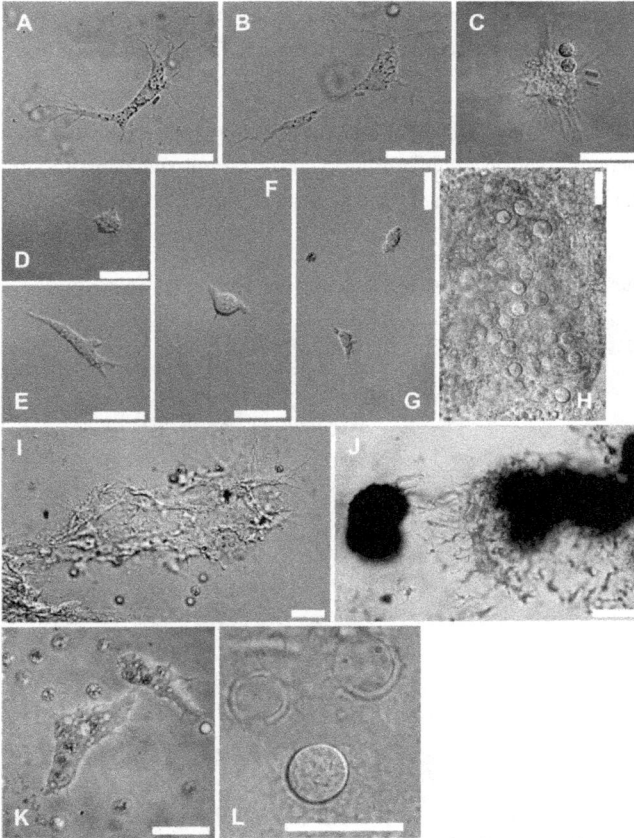

Fig. 31. A-L. Differential interference contrast light micrographs of *Angulamoeba microcystivorans*, n. gen., n. sp. (isolate A1WVB). A, B: trophozoites; C: floating form; D, E, F: typical flagellate; G: early-stage flagellates; H: cysts in mucilage; I, J: trophozoites in mucilage matrix; K: trophozoites with Microcystis in food vacuoles; L: cyst. Scale bars: 20 μm, except J (50 μm).

Angulamoeba fungorum Berney, Bass & Geisen, n. sp. (Fig. 32)

Diagnosis: Non-marine, with morphological characteristics of the genus. Main cell body usually consisting of three (up to four) main branches; length usually between 30–300 μm, cysts between 8–20 μm in diameter. Simple life cycle including trophozoite and cyst stages; flagellated forms or clustering of cells in a mucilage matrix never observed in the culture conditions used. Can survive on bacteria alone, but will attach to and consume fungal filaments when present.

Type strain: Isolate 23-6A (available from the NHM Microbiology Laboratory protist collection). Type location: freshwater lake littoral sediment, Oxford, England, UK. SSU rDNA sequence of the type strain: GenBank accession number KP864084.

Etym.: *Fungus* L. referring to its frequent association with fungal hyphae.

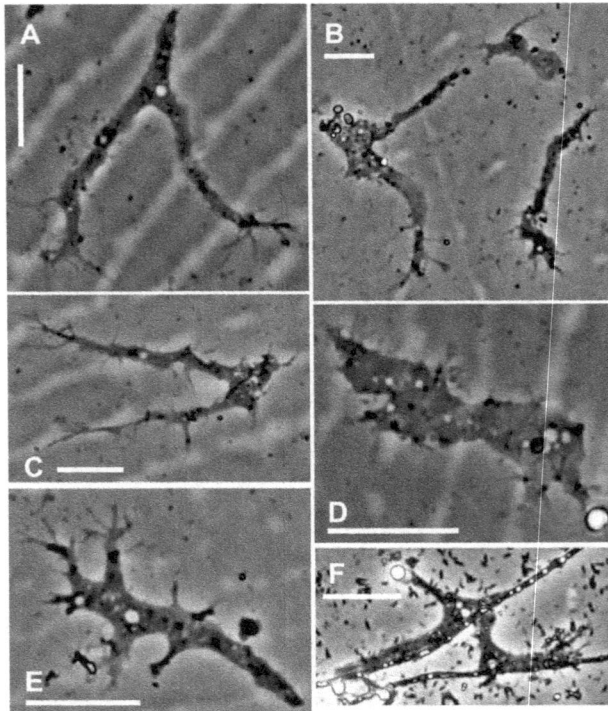

Fig. 32. A-F. Phase contrast light micrographs of *Angulamoeba fungorum*, n. gen., n. sp. (isolate 23-6A). A: most commonly observed shape of the trophozoites, with three main branches; B to E: other possible shapes of the trophozoites; F: trophozoites attached to fungal hyphae; contractile vacuoles visible in A and E, cyst visible in D. Scale bars: 50 μm.

94

Other isolates in the *Angulamoeba* lineage:

- *Angulamoeba* sp. isolate F2 from a meadow soil, Frankfurt, Germany; SSU rDNA sequence (KP864085) very similar to that of *A. fungorum* (96.2% identity in the V4 region).

- *Angulamoeba* sp. isolate FN806 from an air sample, Cologne, Germany; SSU rDNA sequence of this isolate (KP864086) represents a clearly distinct type with only 73.5% identity in the V4 region compared to the type species.

- Trophozoites similar to *A. microcystivorans* were observed on *Microcystis* colonies from bloom samples taken from a recreational lake (Zilverstrand, Mol, Belgium) on 04/08/2014 and from a fishpond (Waarloos, Belgium) on 21/08/2014, but monoclonal cultures were not established from these samples.

Darbyshirella Berney, Bass & Geisen, n. gen.

Diagnosis: Multinucleate, highly branching and reticulate naked amoebae with slender, pointed, sometimes branched pseudopodia (Fig. 33A-K). The whole cell body is strongly branching and of narrow width (e.g. Fig. 33A), especially in the most extended parts, while more condensed parts are wider (Fig. 33F). Pseudopodia and branches are usually formed when cells condense in the anterior extending regions, resulting in up to three new branches (Fig. 33H). Posterior end usually pointed with no or few pseudopodia and no branching. Anastomoses occur randomly between branching parts of the cell body that come into contact, leading to networks of varying complexity (Fig. 33D). Very elongated amoebae are often less reticulate than more condensed ones (Fig. 33K). Many contractile vacuoles present in the entire cell. Movement of entire cells too slow to be directly observable. Cysts present, varying both in size and shape.

Type species: *Darbyshirella terrestris*, n. sp.

Etym.: Named after Dr John Darbyshire, a key Fig. in the study of soil protists and particularly amoebae, who isolated a strain of this amoeba.

Darbyshirella terrestris Berney, Bass & Geisen, n. sp. (Fig. 33)

Diagnosis: Non-marine, with morphological characteristics of the genus. Usually above 600 μm in length. Cell body often narrower than 2 μm, more condensed parts up to 30 μm. Cysts with two clearly separate walls, spherical, oval or bean-shaped, varying between 10–40 μm in diameter. Mainly bacterivorous, but some cells may ingest small eukaryotes.

Type strain: Isolate Tib177 (deposited at the Culture Collection of Algae and Protozoa, Argyll, Scotland, UK, under accession number CCAP 2511/1). Type location: high altitude (4149 m) meadow soil, Mila Mountain East slope (29° 52′ N, 92° 34′ E), Tibet. SSU rDNA sequence of the type strain: GenBank accession number KP864087.

Etym.: *Terrestris* L. terrestrial; by or on land.

Other isolates in the *Darbyshirella* lineage:

- *Darbyshirella* sp. isolate Esthw from freshwater lake littoral sediment, Esthwaite Water, Cumbria, England, UK; SSU rDNA sequence (KP864088) very similar to that of the type species (97.8% identity in the V4 region).

- *Darbyshirella* sp. isolate JDgam from a garden soil, Scotland, UK, and *Darbyshirella* sp. isolate M68 from a meadow soil (abandoned field), Veluwe, Netherlands; these two isolates have very similar SSU rDNA sequences (97.2% identity in the V4 region) but represent a clearly distinct SSU rDNA type from the type species with only 83.3% identity in the V4 region (GenBank accession numbers KP864089 and KP864090, respectively).

- *Darbyshirella* sp. isolate M77 from a meadow soil (abandoned field), Veluwe, Netherlands; SSU rDNA sequence of this isolate (KP864091) represents a third clearly distinct type with only 80.4% identity in the V4 region compared to the type species.

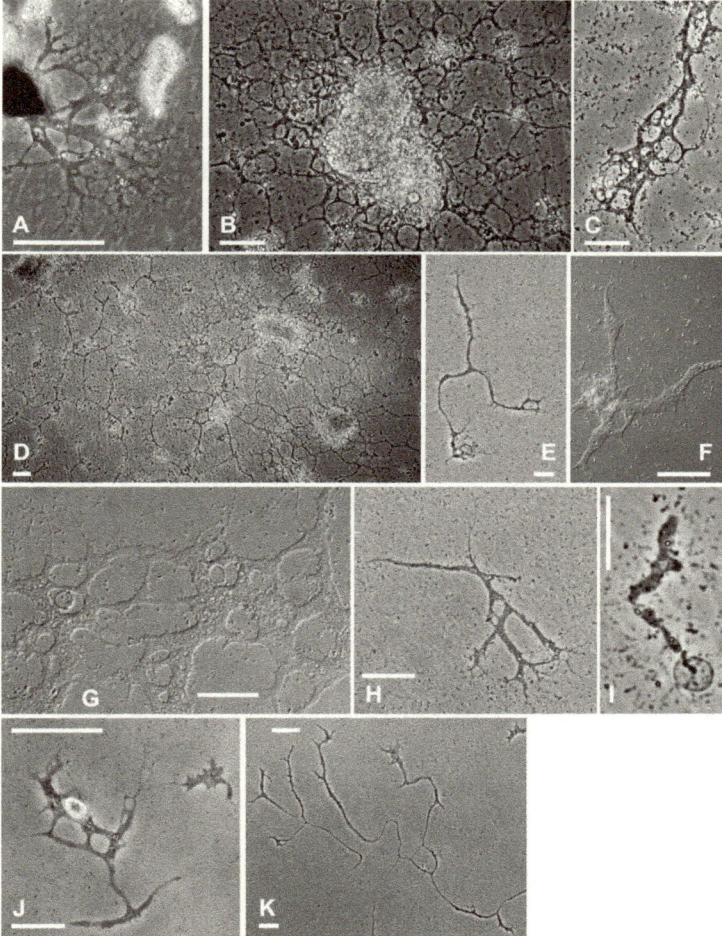

Fig. 33. A-K. Phase contrast light micrographs of *Darbyshirella terrestris*, n. gen., n. sp. (isolate Tib177). A: relatively compact form, with denser pseudopodia at the anterior extending front (right); B-D: networks of varying complexity; E, K: elongated amoebae, less reticulate than the more condensed ones; F, G: fragment of a network showing varying width of the branches; H-J: younger forms shortly after exiting cysts; many contractile vacuoles visible in H, cysts visible in D and J. Scale bars: 100 μm.

Ischnamoeba **Geisen, Bass & Berney, n. gen.**

Diagnosis: Uninucleate, branching naked amoebae (Fig. 34A-H). Cells usually thin, extended and flat, showing no well-defined cell body, except often a slight broadening in the area containing the nucleus. Never reticulate, with whole cells often bent, but not very branched; branching more pronounced in condensed cells or in condensed parts of individual cells. Very thin pseudopodia produced almost exclusively at distal parts of cells and more pronounced in condensed organisms, often branching. Movement of entire cells too slow to be directly observable. Cysts round or sometimes oval.

Type species: *Ischnamoeba montana*, n. sp.

Etym.: *Ischn* Gr. thin, lean.

Geisen, Bass & Berney, n. sp. (Fig. 34)

Diagnosis: Non-marine, with morphological characteristics of the genus. Studied organisms between 35–140 µm in size. Branches of narrow width of often only 2 µm (up to 12 µm in rarely found, non-elongated amoebae, e.g. Fig. 34E). Exclusively bacterivorous. Cysts round, sometimes oval, 6–16 µm in diameter (Fig. 34F).

Type strain: Isolate Tib85 (deposited at the Culture Collection of Algae and Protozoa, Argyll, Scotland, UK, under accession number CCAP 2533/1). Type location: high altitude (4556 m) meadow soil, Sejila Mountain (29° 37' N, 92° 39' E), Tibet. SSU rDNA sequence of the type strain: GenBank accession number KP864092.

Etym.: *montana* L. mountain or hill dweller.

Other isolates in the *Ischnamoeba* lineage:

- *Ischnamoeba* sp. isolate F4 from a meadow soil, Frankfurt, Germany; SSU rDNA sequence (KP864093) almost identical to that of the type species (99.6% identity in the V4 region).

- *Ischnamoeba* sp. isolate FN352 from an air sample, Cologne, Germany; SSU rDNA sequence of this isolate (KP864094) represents a clearly distinct type with only 80.4% identity in the V4 region compared to the type species.

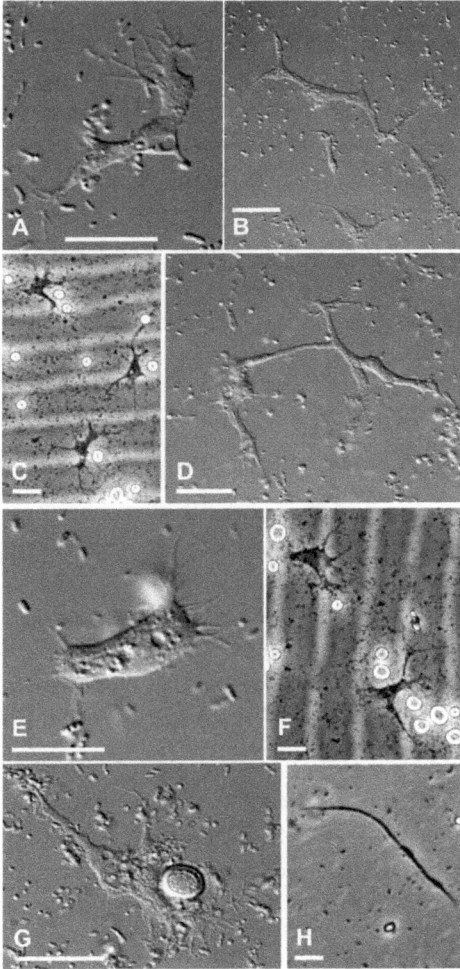

Fig. 34. A-H. Phase contrast light micrographs of *Ischnamoeba montana*, n. gen., n. sp. (isolate Tib85). A-D, F, G: typical trophozoites, never highly branching, with the very fine pseudopodia produced mostly in the distal parts of the cells; E: example of a particularly condensed cell; H: more elongated cells often bent but not branched; cysts visible in C and F. Scale bars: 50 μm, except F, H (30 μm).

99

Heliamoeba Geisen, Bass & Berney, n. gen.

Diagnosis: Binucleate, naked amoebae with filose-like pseudopodia (Fig. 35A-F). A clearly distinct cell body is always present, the pronounced pseudopodia making up most of the total cell dimension. Very rarely branching and if so only slightly, at edges of the cell body; never reticulate (Fig. 35B). Pseudopodia are often branching and present mostly in the anterior and posterior parts of fully extended cells, or all around the cell body in more condensed forms. When disturbed under a coverslip, cells often condense and eventually produce short, round, lobose-like extensions all around the cell body. Cell movement slow but observable under the microscope. Movement occurs with cell bodies narrowing at the posterior end and widening at the anterior end, where new pseudopodia are formed. Cysts of regular round shape.

Type species: *Heliamoeba mirabilis*, n. sp.

Etym.: *Helios* Gr. personification of the sun in Greek mythology; referring to the pseudopodia radiating from the central cell body.

Heliamoeba mirabilis Geisen, Bass & Berney, n. sp. (Fig. 35)

Diagnosis: Non-marine, with morphological characteristics of the genus. Reaching up to 170 μm in size, cell body up to 60 μm in length, width 10 μm. Cysts between 10–15 μm in diameter, with two clearly separate walls. Feeds on bacteria, fungi and small flagellates. If detached from the substratum, a distinct floating form is produced with a spherically condensed cell body and numerous radially extending filose pseudopodia (similar to that of *Angulamoeba microcystivorans*; see Fig. 31C). Amoebae rapidly resume their normal shape and movement when re-attaching to the substratum.

Type strain: Isolate M134 (available from the NHM Microbiology Laboratory protist collection). Type location: meadow soil (abandoned field), Veluwe, Netherlands. SSU rDNA sequence of the type strain: GenBank accession number KP864095.

Etym.: *mirabilis* L. wonderful, marvellous, extraordinary.

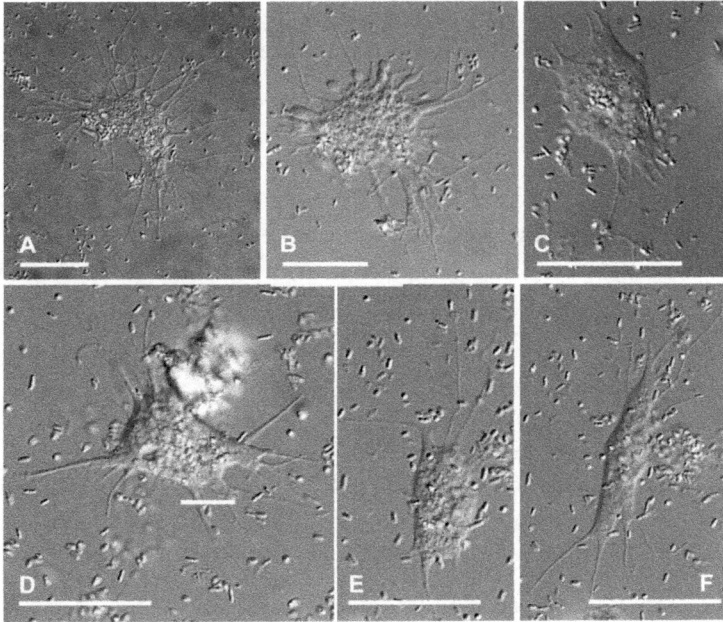

Fig. 35. A-F. Phase contrast light micrographs of *Heliamoeba mirabilis*, n. gen., n. sp. (isolate M134). A-B: trophozoites at rest, with the fine, branching pseudopodia (which can exceed the cell body in length) formed all around the cell; C-F: range of possible shapes of moving trophozoites, with the pseudopodia present mostly in the anterior and posterior parts of the cells. Scale bars: 50 μm.

Dictyamoeba Berney, Bass & Geisen, n. gen.

Diagnosis: Multinucleate, highly branching and reticulate naked amoebae with slender, pointed, sometimes branched pseudopodia (Fig. 36A-D). Trophozoites moving too slowly to be seen in light microscopy. The main cell body is multiply branched and anastomosing even in its simplest form, and can grow into gigantic networks (up to several mm) with intersecting segments of varying width and numerous terminal branching areas (Fig. 36A). Abundant fine pseudopodia are concentrated mostly at the extremity of lateral and terminal branches, especially in complex networks, but can be formed anywhere around the cell body in simpler forms. Numerous contractile vacuoles observed everywhere in the network, but more commonly at intersections of the main

101

branches. Movement of entire cells too slow to be directly observable. Cysts of varying sizes and shapes, the simplest ones rounded, the larger ones irregular in shape.

Type species: *Dictyamoeba vorax*, n. sp.

Etym.: *Dictyon* Gr. net.

Dictyamoeba vorax Berney, Bass & Geisen, n. sp. (Fig. 36)

Diagnosis: Non-marine, with morphological characteristics of the genus. Size across between 30–300 μm in young trophozoites (Fig. 36C). Can survive on bacteria alone, but in that case retains a relatively simple form and medium size as in Fig. 36C. Grows into a gigantic network in presence of yeasts, which are consumed very rapidly; in that case yeasts tend to be collected in clumps at nodes of the network before being consumed (Fig. 36A). When all food has been consumed, the whole network contracts simultaneously in different areas, leading to the formation of very many cysts of all possible sizes and shapes, the position of which mirrors the original layout of the network.

Type strain: Isolate WalEn (available from the NHM Microbiology Laboratory protist collection). Type location: garden soil sample with moss and lichens, Gregynog, Wales, UK. SSU rDNA sequence of the type strain: GenBank accession number KP864096.

Etym.: *vorax* L. ravenous, insatiable, devouring.

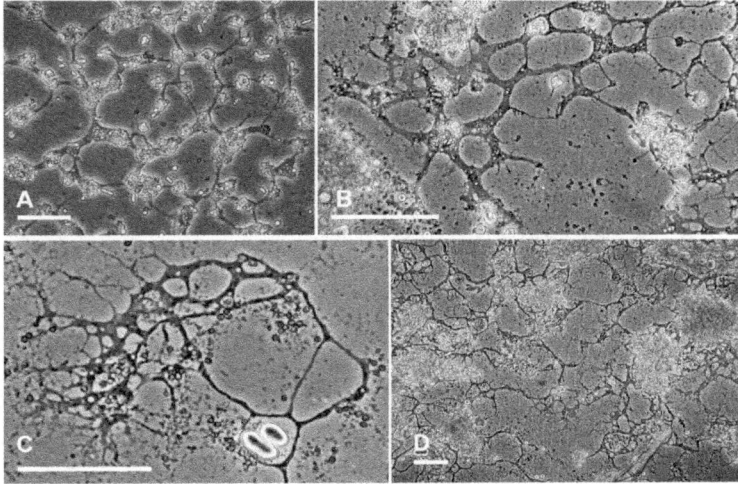

Fig. 36. A-D. Phase contrast light micrographs of *Dictyamoeba vorax*, n. gen., n. sp. (isolate WalEn). A: very extensive network characteristic of cultures supplemented with yeasts as food, with the yeasts collected in clumps at nodes of the network; B: medium-size network with intersecting segments of varying width; C: young trophozoite, multiply branched and anastomosing even shortly after exiting cyst - when feeding on bacteria alone, cells do not grow much larger than this; D: part of an extensive network, not long before encystment, when most yeasts have been consumed. Scale bars: 100 μm.

Arboramoeba Geisen, Bass & Berney, n. gen.

Diagnosis: Multinucleate, highly branching and reticulate naked amoebae (Fig. 37A-E). Cell body not really distinct; nuclei and other cell contents are distributed across the whole network. The network is significantly more complex at the anterior front, forming a wide, very densely reticulate, non-permeable frontier where prey phagocytosis occurs, leaving behind a nearly sterile zone as the cell moves forward. The posterior part of the cells is much less reticulate and branching, with one or few non-branching, non-filose posterior endings (Fig. 37A). Branching, filose-like, pseudopodia are mostly present at the anterior front of the cell, but can also be found in other parts of the network, regularly fusing with other parts of the cell. Very strong vacuolar activity across the whole network (Fig. 37B). Movement of entire cells too slow to be directly observable. No distinct floating form was observed; cells retain their complex shape when carefully detached from the substratum. Cysts with two clearly separate walls varying both in size and shape (never perfectly round, often ovoid). Type species: *Arboramoeba reticulata*, n. sp.

Etym.: *Arbor* L. tree.

Arboramoeba reticulata Geisen, Bass & Berney, n. sp. (Fig. 37)

Diagnosis: Non-marine, with morphological characteristics of the genus. Size usually larger than 600 µm, cysts between 10–20 µm in diameter. Feeds on bacteria, fungi and small flagellates.

Type strain: Isolate Tib182 (available from the NHM Microbiology Laboratory protist collection). Type location: high altitude (4556 m) meadow soil, Sejila Mountain (29° 37′ N, 92° 39′ E), Tibet. SSU rDNA sequence of the type strain: GenBank accession number KP864097.

Etym.: *Reticulum* L. net.

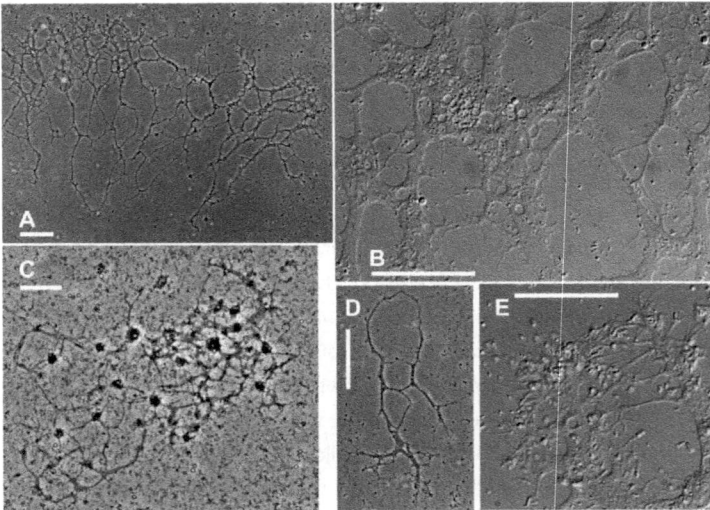

Fig. 37. A-E. Phase contrast light micrographs of *Arboramoeba reticulata*, n. gen., n. sp. (isolate Tib182). A, C: typical shape of networks, significantly more complex at the anterior front (up in A, right in C); B: part of a dense area of a network with strong vacuolar activity; D: younger form, not yet very reticulate; E: detail of the area at the anterior front of a network where fine, branching pseudopodia are produced. Scale bars: 100 µm, except in E (50 µm).

Discussion

Expanding our understanding of Variosea

Class Variosea is a morphologically puzzling assemblage of mostly amoeboid but sometimes flagellated organisms within Amoebozoa. It was never recognised as a potential taxon before the advent of molecular phylogenies because of the highly disparate morphologies observed among its members. Variosea provide an excellent example of the many challenges that modern protistologists have to face when defining higher-level clades based primarily on molecular data, in absence of an obvious morphological identity. In the case of Variosea, that task was made all the more difficult for two reasons. Firstly, amoeboid organisms can be particularly difficult to distinguish and easily misidentified because more or less convergent morphotypes are found across the eukaryote tree of life. Until very recently, their taxonomy has suffered greatly from a striking underestimation of their true diversity; as a result new morphotypes would often more likely be mistaken for members of a few long-known genera instead of being correctly described as new ones. Secondly, the molecular marker used to define the clade (here the SSU rDNA) is unfortunately highly prone to phylogenetic reconstruction artefacts in Amoebozoa, because of the unusually wide range of evolutionary rates observed between the various sequenced lineages. As a result, the history of the recognition of class Variosea has been plagued with several issues that obscured the evolutionary and ecological significance of the group for more than a decade. Classically, the various genera and/or morphotypes that we now know belong to Variosea fell in one of the three following categories: (1) described genera that were previously classified with non-Variosea amoebae based on convergent characters such as general pseudopod morphology, for instance the genera *Filamoeba* and *Flamella* (see below); (2) described genera that had been left *incertae sedis* in classification systems of eukaryotes, such as *Phalansterium* and *Multicilia*; and (3) organisms that represent amoeboid morphotypes that had never been properly described before, and when observed were generally ignored as unidentifiable filose or reticulose amoebae - this is true of most of the new isolates described in the present study.

105

The first molecular evidence for the existence of Variosea was published when Amaral Zettler et al. (2000) showed a relationship between the genus *Filamoeba* and an ATCC strain (50654) that was then believed to belong to the leptomyxid genus *Gephyramoeba*. This result was surprising (see, e.g., Smirnov et al. 2005) because *Filamoeba* was believed to be closely related to the tubulinean genus *Echinamoeba* based on seemingly shared pointed subpseudopodia, since then revealed as the result of convergent evolution (category 1 above). In addition, later re-investigation of the ATCC strain 50654 by Smirnov et al. (2008) conclusively showed that it is actually not the leptomyxid described by Goodey (1915b) but corresponds to a new genus that they named *Acramoeba*. Class Variosea was introduced by Cavalier-Smith et al. (2004) when a third member, the flagellate *Phalansterium* was sequenced. Later Nikolaev et al. (2006) showed that the mysterious multiflagellated taxon *Multicilia* also belongs to this assemblage; both genera are typical examples of taxa belonging to category 2 above. Since then, Kudryavtsev et al. (2009b) added the long known genus *Flamella* to the list of taxa belonging to Variosea, showing that it is not related to flabellinids as traditionally believed. Another ATCC strain (50593) has been shown to belong to Variosea (Tekle et al. 2008); initially it was tentatively identified as a potential member of the genus *Arachnula* but was since described as a new genus *Telaepolella* (Lahr et al. 2012). Finally, several lineages of amoebae with a protosteloid life cycle have been shown to be possibly related to or part of Variosea (Brown et al. 2007, Shadwick et al. 2009).

In the present study, the combined effect of increased taxon sampling and use of secondary structure information in the variable region V7 of the SSU rDNA allows us to indisputably define which of the Amoebozoa morphotypes sequenced to date are part of Variosea, and how many clones from environmental SSU rDNA libraries in GenBank represent additional lineages of yet unknown morphology that also belong to that clade. We identify 26 lineages that clearly belong to Variosea based on both phylogenetic evidence and a specific secondary structure pattern of the variable region V7 of the SSU rDNA (see Fig. 38, Supplementary Fig. 2, and Supplementary Table 5). Of these, 17 have a known morphotype, six of which are novel and described here for the first time as new genera. The other nine variosean lineages presented in Fig. 38 and Supplementary Table 5 are so far known only from environmental clone libraries. Supplementary Table 5 lists five additional environmental lineages that may also belong to Variosea, but their sequences either lack the variable region V7 or were too short to be included in our phylogenetic analyses.

Members of the Archamoebae (pelobionts, mastigamoebids, and entamoebids) and the Macromycetozoa (dictyosteliid and myxogastrid slime moulds) have very divergent SSU rDNA sequences. In phylogenetic studies containing a good taxon sampling for Amoebozoa, they often appear to branch within class Variosea, either together or separately, usually next to some of the fast evolving protosteloid lineages within that class (see, e.g., Nikolaev et al. 2006, Tekle et al. 2008, Shadwick et al. 2009, Pombert et al. 2013) . This apparent paraphyly of Variosea with respect to Archamoebae and Macromycetozoa (A & M) could be genuine, especially for Macromycetozoa, in agreement with the idea that cellular and acellular slime moulds evolved from a protosteloid ancestor (e.g. Olive 1970). However it could also be the result of long-branch attraction artefacts. Because SSU rDNA alone cannot be expected to resolve the position of A & M within Amoebozoa, we decided to exclude them from the analyses shown in Fig. 30, to minimise artefacts when inferring internal relationships within Variosea. Our results show that even in absence of the highly divergent A & M, the range of evolutionary rates observed in SSU rDNA sequences among Variosea lineages precludes resolution of their relationships based on that gene alone (see also Fig. 38). Other molecular markers will evidently be needed to resolve variosean relationships.

Among the lineages of protosteloid amoebae sequenced by Shadwick et al. (2009), five were branching in the Variosea + A & M part of the amoebozoan tree in their analyses. However, only four of them (the protosteliids, the schizoplasmodiids, the soliformoviids, and the cavosteliids) possess the V7 secondary structure signature of the Variosea and are regarded as variosean lineages in the present study (see Fig. 38). The protosporangiids (genera *Clastostelium* and *Protosporangium*) lack this V7 secondary structure signature, and importantly, so do all members of A & M, suggesting they are separate from Variosea (although this could also be a result of the divergent nature of their SSU rDNA sequences). To address this issue we performed phylogenetic analyses on two additional datasets, including the two protosporangiids and representatives of A & M, with or without some of the fastest evolving variosean taxa. In all analyses, protosporangiids robustly grouped within Macromycetozoa, as a sister taxon to *Ceratiomyxa*, a result also obtained recently by Zadrobilkova et al. (2015). In presence of the fastest evolving variosean taxa, A & M branched among Variosea as discussed above (in the position indicated by a dashed line in Fig. 30). However when fast-evolving varioseans were excluded A & M branched separately from Variosea (Supplementary Fig. 1), in agreement with observed V7 secondary structure patterns. These results together strengthen the case for A & M being indeed separate from Variosea, and suggest that protosporangiids belong to Macromycetozoa and are with *Ceratiomyxa* the closest relatives of dictyosteliid and myxogastrid slime moulds among sequenced protosteloid amoebae. Multigene datasets including protein-coding genes will be necessary to confirm these hypotheses.

Current definitions of Variosea (Cavalier-Smith et al. 2004, Smirnov et al. 2011b) are morphology-based, but in absence of any defined synapomorphy for the clade, merely consist of an enumeration of the possible morphologies observed among known members of the clade. Such definitions are impractical, because they are quite subjective and make amendments necessary every time new lineages with new morphotypes are shown to also belong to Variosea based on molecular data. However, we believe there is satisfactory molecular evidence to indicate the existence of a clade containing (for now) the 17 morphologically identified lineages presented in Fig. 38 and additional lineages known only from environmental clone sequences (see also Supplementary Table 5). Here we propose the following phylogenetic, node-based definition of Variosea to supplement its morphological diagnosis: it is the clade containing the last common ancestor of the 17 morphologically identified lineages presented in Fig. 38 and all its descendants. It is supported by a molecular synapomorphy (the specific V7 secondary structure pattern in the SSU rDNA;

Supplementary Fig. 2) that is present in all lineages listed in Supplementary Table 5 and absent in any other sequenced amoebozoan lineage that we know of. Other taxa such as Archamoebae and Macromycetozoa may or may not later be shown to belong to Variosea based on multigene data, in which case the defining molecular character mentioned above would have been secondarily lost in some lineages (while the node-based definition would stand).

Variosean BNFA: ecology, abundance, and food preference

As well as improving the resolution and limits of variosean phylogeny, our results also provide the first morphological data for lineages previously only represented by environmental sequences: five of the new genera described here provide previously unknown morphotypes for available environmental clone sequences. In addition, our new data also demonstrate that other sequences available in GenBank belong to Variosea, sometimes representing additional lineages of yet unknown morphology. In total, we identified 81 environmental clones belonging to Variosea, plus 20 chimeric ones of partial variosean origin (Supplementary Table 5). These clones were rarely annotated as variosean, but either (in most cases) as unidentified eukaryotes, or sometimes as different organisms altogether, e.g. in the case of sequences from Lesaulnier et al. (2008) labelled as a basidiomycete (EF023260; in fact *Ischnamoeba* sp.), or as *Acanthamoeba* sp. (EF023424; in fact *Arboramoeba* sp.), or as an apicomplexan (EF024977; in fact *Flamella* sp.). This highlights a recurring problem when taxonomically annotating environmental clones: not only is a reference database needed (which is often at best incomplete), but also a phylogenetic or signature-based approach should ideally be preferred over a BLAST / sequence similarity approach mostly because of the widely heterogeneous rates of substitutions between eukaryotic lineages.

It is striking that most of our new isolates require describing as new genera; we found no corresponding taxa previously formally described in the literature. In some cases, unnamed and unsequenced organisms were apparently very similar to one or more of our new taxa. For example, Reynolds et al. (1981) mentioned the rapid deterioration of natural *Microcystis* bloom samples (from

Blelham Tarn, Cumbria, UK) under laboratory conditions upon amoebal infestations. Their description of a fast disappearance of *Microcystis* cells and the striking colour adopted by colonies after infection due to the appearance of brown, ferruginous deposits in the mucilage matrix, suggests an infection with *A. microcystivorans*. Likewise, Canter-Lund and Lund (1995) observed a 'big freshwater plasmodium' feeding on *Volvox* colonies that is reminiscent of *Darbyshirella* or *Dictyamoeba*. Recently, sequence data has been published for several non-sporocarp-forming branched and reticulate amoebae that were previously only represented by (sometimes basic) morphological descriptions, e.g. the rhizarians *Thalassomyxa*, *Penardia*, *Vampyrella*, *Leptophrys* (Hess et al. 2012, Berney et al. 2013), *Reticulamoeba* (Berney et al. 2013), *Arachnula* (Bass et al. 2009a), the stramenopile *Leukarachnion* (Grant et al. 2009), and many amoebozoans mentioned previously. Others, superficially similar to those described in this paper remain unsequenced, e.g. *Asterocaelum*, *Branchipocola*, *Chichkovia*, *Cinetidomyxa*, *Enteromyxa*, *Gymnophrydium*, *Protomyxa*, *Synamoeba* (from Lee et al. 2000). However, none of these corresponded to our new isolates.

Somewhat conversely, the morphological conservation and/or convergence in variosean BNFA present taxonomic challenges. For example *Darbyshirella* is similar in form to reticulate amoeboid stages of *Schizoplasmodiopsis pseudoendospora* and other cavosteliids, yet it is phylogenetically distinct from them and apparently lacks fruiting stages. The *Darbyshirella* and *Ischnamoeba* lineages appear to be closely related, yet multiple independent isolates confirm sufficient morphological differences for them to be assigned to separate genera (Figs 33 & 34). Similarly, the amoeba-flagellate transition of *Cavostelium* (Olive 1967) is strongly reminiscent of that seen in *Angulamoeba microcystivorans*, but again these two genera are phylogenetically distinct. The two *Angulamoeba* species we describe are very closely related, yet an amoeba-flagellate transition is present in one and apparently absent in the other. These observations underline the fact that evolutionary distances in terms of morphological traits and SSU rDNA sequences are highly uncoupled in Variosea (as is true of some other protist groups). We suggest that this may partly be explained by evolutionary divergence from a common ancestor with amoeba and flagellate stages and the ability to produce fruiting bodies, with subsequent simplification of the life cycle and/or loss of the flagellate stage in different lineages.

The high rate of lineage discovery as new BNFA isolates are sequenced is particularly interesting in light of the fact that BNFA were encountered in almost all soil samples analysed in a recent study by Geisen et al. (2014a), suggesting that they are numerically abundant in many soils, as well as being phylogenetically highly diverse. Geisen et al. (2014a) used a modified version of the liquid aliquot culturing method (LAM; Butler and Rogerson 1995). They did not detect known fruiting body-forming protosteliids or BNFA from other groups, for example the rhizarian vampyrellids. Some Mycetozoa were isolated, but at low frequency – fewer than 1000 per gram of soil, which is at the lower limit of detection using LAM. In contrast, BNFA numbered up to 5000 per gram of soil, and represented 5-20% of all amoebae isolated per sample. Every culturing/cell isolation method has biases, so we cannot use our results to say that variosean BNFA are necessarily more abundant in soils than other BNFA taxa, although this may well be the case. The liquid aliquot method is likely biased towards bacterivores that can survive the physical disruption of this approach, or survive as resistant encysted stages until the culture conditions favour them. Nonetheless our results show that BNFA are likely to be key ecological players in soil habitats, and raises the hypothesis that non-sporocarp-forming variosean amoebae are the dominant BNFA taxa in soils. There are several reasons why their diversity and prevalence has not previously been recognised using culturing methods: (1) they take longer to appear and grow in culture than protists normally detected by these methods; (2) at least in their early stages they are morphologically similar, which without molecular analyses would give no indication of their genetic diversity, and means they could even be misidentified as smaller amoebae such as *Filamoeba*; and (3) their unusual morphotypes are often not recognised, or confused with other organisms, e.g. fungi. Even molecular sequencing studies have drastically underestimated them; their SSU rDNA sequences are often divergent in sequence and length, leading to biases against them in general eukaryote

111

surveys, while misidentifications due to wrongly labelled sequences as detailed above lead to further confusion.

The physical nature of BNFA also points to their exerting a significant and distinctive ecological impact in soils. The networks/branches comprise a great range of cell body and pseudopodial dimensions, of extreme morphological plasticity, allowing the amoeba to probe interstitial spaces of many dimensions, range of relatively large areas, cross and both exploit and survive chemical and ecological microgradients, and potentially acquire and re-locate large amounts of resource around the cells. Therefore it is viable for them to exploit unusually small (or difficult to access) and large food resources, from bacteria to small metazoans. The fact that the BNFA morphology has evolved several times across the eukaryotic tree of life, often followed by significant radiations (in many amoebozoan and rhizarian lineages, stramenopiles, etc.) demonstrates the success of this foraging strategy. In culture their ability to densely cover large areas can rapidly virtually eliminate bacterial and eukaryote prey populations. Their eukaryote prey range is large, including fungi, diatoms, green algae, and possibly other heterotrophic protists and small metazoans. Some lineages, in particular *Darbyshirella* and *Dictyamoeba*, develop particularly dense and extensive networks in the presence of eukaryotic prey, but will also grow in the presence of bacteria alone, while *Heliamoeba* and *Arboramoeba* appear to require other eukaryotes as food. Their flat, extensive cells are often closely associated with soil particles, which as well as allowing highly efficient food acquisition may also protect them from being eaten themselves; in any case they are multinucleate and individuals can be fragmented by physical disruption and presumably also be attacked by other organisms, and then re-grow.

In conclusion, the new genera and species we describe in this paper possibly represent a much larger diversity of branching or network-forming variosean Amoebozoa that are abundant in many soil/sediment types, where they occupy a very distinctive niche, as well as in freshwater habitats where they can also have a profound effect on the foodweb (Van Wichelen et al. 2010). We infer from their morphology and behaviour in culture that they are significant

112

grazers of bacteria and predators of a wide range of other eukaryotes. They are therefore likely to have a profound impact on soil ecology, which if so has been largely overlooked in ecological studies. Other BNFA groups have been better studied in soils, for example amoebae feeding on pathogenic fungi and plant parasitic nematodes (Sayre 1973, Old and Darbyshire 1980, Old and Oros 1980), but those are either in different groups (*Arachnula* and *Theratromyxa* are vampyrellids) or do not correspond to our isolates. Furthermore it will be necessary to ensure that these often highly genetically divergent lineages are accounted for in molecular soil ecology studies, by revising existing primers, designing new ones, and/or by RNA-based and metagenomic approaches that should be less biased against such lineages.

Materials and Methods

Sample collection, culture isolation, microscopy, and DNA extraction

Localities and dates of collection of samples for isolation are given in Table 5. In the culture lab, samples were placed in Volvic® mineral water (Danone, Paris, France), Prescott-James medium (Page 1991) enriched with 0.15 % wheat grass (Weizengras, Sanatur GmbH, Singen, Germany), or freshwater WC-medium (Guillard and Lorenzen 1972), and grown at room temperature and/or in a 14ºC incubator in the dark. Dishes were checked regularly for the presence of BNFA. Amoebae of interest were subcultured and when possible purified by serial dilution until devoid of other eukaryotic organisms. For some observed morphotypes, a subset of dishes was supplemented with distinct selected food sources (heterotrophic bacteria alone, various *Microcystis* cyanobacteria, baker's yeasts, the green alga *Scenedesmus*, or the freshwater diatom *Achnanthes*) to see if it had any influence on the growth rate and morphology of the amoeba's active trophic stage. For differential interference contrast (DIC) light micrographs, we used either a Nikon (Tokyo, Japan) Eclipse 80i microscope with x40 (NA 0.6) and x60 (NA 1.0) DIC water immersion lenses, a Nikon Eclipse 90i with x63 (NA1.0) DIC oil immersion lenses, or a Leitz Diaplan microscope with x40 (NA 0.7) PL Fluotar lenses. Phase contrast micrographs were taken from a Leica (Wetzlar, Germany) DM IRB microscope or a Nikon Eclipse TS100 microscope. Images were recorded on a Sony (Tokyo, Japan) HD HDR-XR155 camcorder, a HDV 1080i Handycam®, or Nikon DIGITAL SIGHT DS-fi1 camera. Frames were captured using the software PMB version 5.2 (Sony) or Final Cut Express HD 3.5.1 (Apple Inc., Cupertino, CA, USA). DNA was extracted either from mixed cultures where a single BNFA morphotype was observed and had reached a reasonable density or, whenever possible, from purified clonal cultures. For mixed cultures, this was done by picking individual cells (20 to 50) with a micropipette. DNA extraction was then performed either by using the UltraClean™ Soil DNA Isolation Kit (MoBio Laboratories, Carlsbad, CA, USA) following the Maximum Yield Protocol, or by placing the cells in guanidine thiocyanate buffer and using a protocol described in Tkach and Pawlowski (1999). Where pure, total DNA extractions were performed on the whole culture. Most of the culture medium was decanted off, and using a sterile scraper, cells were collected from the bottom of the culture dish. Total DNA was then extracted from the pellet of organic material using the UltraClean™ Soil DNA Isolation Kit as above.

Ethics statement: No specific permission or permits were required for the described field studies. The sites were not privately owned or protected in any way and were fully open to public access. No endangered or protected species were involved in this study.

Amplification and sequencing of the SSU rDNA

The complete or nearly complete SSU rDNA was amplified from all isolates except F2. For the latter we only amplified a 644 bp fragment spanning the variable regions V4 and V5, which turned out to be more than 98% identical to the sequence of isolate 23-6A. SSU rDNA sequences were obtained in a single or up to three overlapping fragments, using universal (eukaryotic) primers and, when necessary, a 'primer-walking' approach with lineage-specific primers (details available on request from the authors). PCR amplifications were done in a total volume of 30 µl with an amplification profile typically consisting of 35 cycles with 30 s. at 95ºC, 30 s. at 56ºC, and 90 s. at 72ºC, followed by 5 min. at 72ºC for the final extension. PCR products of clonal strains or individually picked cells were quality controlled for the expected length of amplification product on a 1.5% TAE agarose gel and subsequently purified by adding 0.15 µl Endonuclease I (20 U/µl, Fermentas, St. Leon-Rot, Germany), 0.9 µl Shrimp Alkaline Phosphatase (1 U/µl, Fermentas, St. Leon-Rot, Germany) and 1.95 µl H_2O. The resulting mixture was incubated for 30 min at 37°C, followed by 20 min at 85°C. In non-clonal cultures, bands of the appropriate length were excised, and cleaned following the protocol of the QIAquick® Gel Extraction Kit (Qiagen, Hilden, Germany). Purified PCR amplicons were sequenced directly, or when necessary cloned into StrataClone™ SoloPack® Competent Cells using the StrataClone™ PCR Cloning Kit (Stratagene, Agilent Technologies, Santa Clara, CA, USA). White colonies were screened using the primers M13for (5' - CGT TGT AAA ACG ACG GCC AGT - 3') and M13rev (5' - CAC AGG AAA CAG CTA TGA CCA - 3'). Positive PCR products were cleaned using a polyethylene glycol (PEG) protocol: for 20 µl PCR reactions, 20µl of a 20% PEG / 2.5M NaCl mixture was added to each tube. The tubes were mixed by vortexing and incubated for 30 min at 37°C, then centrifuged at 3000 rpm for 30 min to pellet the PCR products. Supernatant was discarded by pulse-spinning the inverted tubes at 600 rpm. The pellet was then washed with ice-cold 75% ethanol, spun for ten minutes at 3000 rpm, again inverted and pulse-spun to remove the supernatant. The ethanol wash was repeated; the PCR pellet was re-suspended in de-ionised water, and stored at −20°C. Sequencing was performed with the Big Dye Terminator v1.1 Cycle Sequencing Kit, and analysed with an ABI-3730xl DNA sequencer (Applied Biosystems, Life Technologies, Carlsbad, CA, USA).

BLASTn searches and construction of the sequence datasets

The new SSU rDNA sequences reported in this paper have been deposited in the NCBI GenBank database under accession numbers KP864083 to KP864101. They were edited and aligned manually using the BioEdit software (Hall 1999), following the secondary structure model proposed by Wuyts et al. (2000). Visualization of the secondary structure of variable region V7 in selected species (see Supplementary Fig. 2) was achieved with the program RnaViz2 (De Rijk et al. 2003). Visual screening of the sequences in search of sequence signatures, BLASTn searches (Altschul et al. 1990) against the NCBI GenBank nr/nt database using default parameters, and preliminary phylogenetic analyses performed on a large dataset including a wide range of eukaryotes congruently suggested that our isolates all belong to class Variosea within phylum Amoebozoa. A first dataset was constructed to confirm this phylogenetic placement of our isolates; it contains 1350 unambiguously aligned positions and 121 taxa, and includes all our isolates, representatives of all morphologically identified lineages within class Variosea and of most other morphologically defined higher-order taxa of lobose amoebae, and ten outgroup sequences (belonging to opisthokonts and apusomonads). GenBank accession numbers of all sequences used in this dataset are given in Fig. 30. The divergent sequences of Archamoebae and Macromycetozoa were excluded from the main analyses but included in two variants of the first dataset, with or without some of the fastest evolving variosean taxa (see Discussion and Supplementary Fig. 1).

Exhaustive BLASTn searches (default parameters) were then performed using both our new and existing Variosea SSU rDNA sequences to seed searches to identify all Variosea clone sequences from environmental libraries present in the NCBI GenBank nr/nt database (Supplementary Table 5). Further BLASTn searches were performed using manually truncated sequences (both at the 5' and 3' ends) as queries. This allowed retrieval of shorter clone sequences that would escape identification when using complete sequences as queries because of lower overall similarity scores compared to more distantly related but full-length sequences. All identified Variosea sequences were used as additional queries in further rounds of BLASTn searches until no more new unambiguous Variosea sequences could be found. Clones shorter than 500 bp were not considered. The presence of sequence chimeras was assessed by visual screening of the alignment in search for contradictory sequence signatures, and potential chimeric clones were confirmed as such by distance analyses based on different subsets of unambiguously aligned regions (as described in Berney et al. (2004). In total, we identified 20 environmental clones as chimeric out of a total of 114 (Supplementary Table 5). A second dataset was then constructed to refine phylogenetic relationships and highlight lineage diversity within Variosea, containing 1375 unambiguously aligned positions and 100 taxa, including all our isolates, most morphologically identified members of class Variosea, a selection of identified variosean environmental clones, and four outgroup sequences from other amoebozoan lineages. Only environmental clones longer than 1000 bp and spanning both the V4 and V7 variable regions of the SSU rDNA were included. Whenever multiple highly similar clones from a same environmental library had been identified, only one was kept to avoid over-populating the figure (but all identified clones excluded from the dataset are listed in Supplementary Table 5). Missing data in partial environmental sequences were encoded as such (Ns). The non-Variosea part of any chimeric sequence included in this dataset was also encoded as missing data.

Maximum likelihood and Bayesian phylogenetic analyses

All phylogenetic analyses for both datasets were performed on the CIPRES Science Gateway (Miller et al. 2010). Maximum likelihood (ML) analyses (Felsenstein 1981) were performed with the program RaxML version 7.6.6 (Stamatakis 2006), using the GTRGAMMA model with 8 rate categories. All necessary parameters were estimated from the datasets. The best ML topology was selected from 100 inferences with distinct maximum parsimony starting trees. The reliability of internal branches was assessed with the bootstrap method (Felsenstein 1985) using 1000 replicates. In addition, Bayesian analyses were performed with MrBayes version 3.2.3 (Huelsenbeck and Ronquist 2001, Ronquist and Huelsenbeck 2003) using a GTR + gamma model with 8 rate categories. Two runs of four simultaneous chains were run for 3,000,000 generations (heat parameters set to default), and trees were sampled every 100 generations. For each run 30,000 trees were sampled, 10,000 of which were discarded as the burn-in. Posterior probabilities of the branching pattern were estimated from the 40,000 remaining trees and mapped onto the ML tree when present. For both datasets the Bayesian posterior probability 50% majority-rule consensus tree was fully compatible with the corresponding ML tree. Sequence alignments for Figs 30 and 38 are available on request from the authors.

Acknowledgements

This work was enabled by a NERC Standard Research Grant (NE/H009426/1) supporting CB and DB. SG and MB were supported by EcoFINDERS (EU Reference No. 264465). JVW was supported by the BELSPO (Belgian Science Policy) project B-BLOOMS2. PV is a postdoctoral research fellow with the Research Foundation – Flanders (FWO).

Supplementary files

Supplementary Table 4: Described genera of branched or network-forming amoebae and their taxonomic affiliation, with detailed information on different species and isolates of each taxon (when available) and additional comments and references (provided below). Ref = Reference;

Taxon	Ref	type locality	type habitat	known food(s)	morphological observations	phylogenetic position and other remarks
Aletium pyriforme	Trinchese 1884	Naples, Italy	M / on *Chaetomorpha*	?	bipolar plasmode; elongate cell body (up to 3-4 mm) with only 2 opposite pseudopodia	Eukaryotes *incertae sedis*
Aletium pyriforme	Schepotieff 1912b	Naples, Italy	M / on algae	algae?	bipolar plasmode very similar to Trinchese's description	Eukaryotes *incertae sedis* / resembles *Gymnophrys cometa* according to Schepotieff
"*Amoeba*" *dumetosa*	Penard 1904	Geneva, Switzerland	F / lake sample	?	branched / reticulate; 200-250 um, highly variable shape, temporary anastomoses, multinucleate	Eukaryotes *incertae sedis* / could be a leptomyxid
Arachnula impatiens	Cienkowski 1876	Odessa, Ukraine / Germany	F / ponds and brackish water	not mentioned by Cienkowski	branched; Cienkowski "never saw pseudopodia so quickly formed in any other protozoa"	Rhizaria, Cercozoa, Vampyrellida / an isolate identified as *Arachnula* sp. was sequenced by Bass et al. 2009
Arachnula impatiens	Dobell 1913	River Granta, Cambridge, UK	F / surface mud with cyanobacteria and diatoms	diatoms, cyanobacteria (*Oscillatoria*), flagellates, desmids, ciliates, bacteria	branched; up to 350 um, presence of digestive + reproductive cysts	Rhizaria, Cercozoa, Vampyrellida / *Arachnula* sp. sequenced by Tekle et al. 2008 is *Telaepolella*
Arachnula impatiens	Old & Darbyshire 1980	Europe and North America	F / soil	conidia of *Cochliobolus* + bacteria, cyanobacteria, diatoms, nematodes, fungi	branched / sometimes reticulate; up to several mm, presence of digestive + resting cysts	Rhizaria, Cercozoa, Vampyrellida / could be *Platyreta*

Species	Author	Location	Habitat	Food	Description	Classification
Biomyxa vagans	Leidy 1875	NewJersey & Pennsylvania, USA	F / *Sphagnum* swamp	never seen eating	reticulate (through fusion and lacunae); highly variable size and shape, bi-directional movement	Eukaryotes *incertae sedis* / probably a fragment of the reticulopodial network of "*Gromia terricolo*" (actually a foraminiferan)
"*Biomyxa vagans*"	Leidy 1879	?	F / among desmids	never seen eating	described as a "nucleated form" of Biomyxa (Plate XLVIII, Fig. 21 to 25)	Eukaryotes *incertae sedis* / probably a vampyrellid
"*Biomyxa vagans*"	Anderson & Hoeffler	Sargasso Sea	M / *Sargassum* sp. from surface	bacteria only	presence of resting cysts	Eukaryotes *incertae sedis* / possibly a marine vampyrellid or a naked foraminiferan
"*Biomyxa*" *merdaria*	Hollande 1942	Paris, France	F / hippo dung	never seen eating	mostly filose; 10-100 um, filopodia with granules, never reticulate, uninucleate	Eukaryotes *incertae sedis* / possibly a variosean, looks like *Filamoeba* / *Grellamoeba*
Cichkovia reticulata	Valkanov 1931	Northern Europe	?	?	reference not seen	Eukaryotes *incertae sedis* / could be a variosean or a vampyrellid
Cinetidomyxa nucleoflagellata	Chatton & Lwoff 1924	Roscoff, France	M / ?	*Nitzschiella longissima*	filose, uni- or plurinucleate amoebae able to fuse to form a plasmodial network	Eukaryotes *incertae sedis* / could be a variosean
Cinetidomyxa chattoni	Cachon & Cachon-	?	?	?	reference not seen	Eukaryotes *incertae sedis* / corresponds to *Corallomyxa chattoni* according to Grell & Benwitz 1978
Corallomyxa mutabilis	Grell 1966	Madagascar	M / dead pieces of coral reef	pennate diatoms	meroplasmodial network; 200 um to several mm, multinucleate with multinucleate floating buds	Eukaryotes *incertae sedis* / probably Amoebozoa
Corallomyxa chattoni	Grell & Benwitz 1978	Villefranche-sur-Mer, France / Cornwall, UK	M / mixed algae	*Cryptomonas* sp.	meroplasmodial network; smaller than *C. mutabilis*, same morphology, uninucleate floating buds	Eukaryotes *incertae sedis* / probably Amoebozoa; *Corallomyxa tenera* sequenced by Tekle et al. 2007 is a *Filoreta*
Corallomyxa multipara	Grell 1988	Queensland, Australia	M / Grell did not remember	*Amphiprora* sp. (slow) / *Cryptomonas* sp. (better)	meroplasmodial network; multiple uninucleate floating buds after condensation of network	Eukaryotes *incertae sedis* / probably Amoebozoa
Corallomyxa japonica	Grell 1991	Shimoda Marine Research Centre, Japan	M / sand grains, tide pool	*Amphiprora* sp.	meroplasmodial network; uninucleate floating buds from constrictions in network arms	Eukaryotes *incertae sedis* / probably Amoebozoa
Dictiomyxa trinchesei	Schepotieff 1912a	Naples, Italy	M / on *Chaetomorpha*	algae?	filose / reticulose; large body mass with filo/reticulopodial network, possible heliozoan stage	Eukaryotes *incertae sedis* / pseudopodia insufficiently described, first described by Monticelli 1897

"freshwater plasmode"	Cienkowski 1876	?	F / among Tetraspora	algae?	reticulate (through fusion and lacunae); very slow moving	Eukaryotes *incertae sedis* / Dangeard 1891 refers to it as *Gymnophrydium cienkowskii*
Gephyramoeba delicatula	Goodey 1915b	Harpenden, Hertford, UK	F / cucumber house soil	bacteria, small amoebae	branched; 60-250 um, never reticulate, uninucleate, presence of cysts	Amoebozoa, Tubulinea, Leptomyxida / remains unsequenced; *Gephyramoeba* sp. sequenced by Amaral Zettler et al. 2000 is *Acramoeba*
Gephyramoeba delicatula	Pussard & Pons 1976b	Abbaye de Citeaux, France	F / prairie	same	branched; 35-300um, same	Amoebozoa, Tubulinea, Leptomyxida / remains unsequenced
"*Gloidium inquinatum*"	Penard 1902	Lossy swamp, Geneva, Switzerland	F / swamp sample	?	branched; nucleus not visible	Eukaryotes *incertae sedis* / could be a leptomyxid
Gobiella borealis	Cienkowski 1881	?	F / ?.	*Oedogonium*	mostly filose, *Vampyrella*-like	Rhizaria, Cercozoa, Vampyrellida / remains unsequenced
Gobiella closterii	Poisson & Mangenot	Brittany, France	F / peat bog	*Closterium intermedium*	mostly filose, *Vampyrella*-like	Rhizaria, Cercozoa, Vampyrellida / remains unsequenced; described as *Vampyrella closterii* by Poisson & Mangenot 1933
Gringa sp.	Frenzel 1892/7?	?	?.	?	reference not seen	Eukaryotes *incertae sedis*
Gymnophrydium hyalinum	Dangeard (PA) 1891	Caen, France	F / associated with *Oscillatoria*	*Nitella* (and *Oscillatoria*?)	reticulate (through fusion and lacunae); fast moving, with granules and cysts	Eukaryotes *incertae sedis* / probably a vampyrellid
"*Gymnophrydium hyalinum*"	Dangeard (PA) 1910	Poitiers, France	F / mixed culture with desmids	bacteria	reticulate (through fusion and lacunae); very slow moving, no granules	Eukaryotes *incertae sedis* / clearly distinct from the original description; could be a variosean
Gymnophrydium marinum	Dangeard (P) 1968	Bordeaux, France?	M / mixed culture	*Navicula* + cyanobacteria, green, red and brown algae	reticulate (through fusion and lacunae); multinucleate with uninucleate floating buds	Eukaryotes *incertae sedis* / resembles *Corallomyxa*
"*Gymnophrydium marinum*"	Raghu Kumar 1980	North Sea	M / water bathed with pine pollen	bacteria	filose amoebae fusing to form colonial networks	Eukaryotes *incertae sedis* / clearly distinct from the original description; resembles *Filoreta*
Gymnophrys cometa	Cienkowski 1876	Naples, Italy / Charkow, Russia	M / tide pool - F / swamp	never seen eating	reticulate (through fusion and lacunae); foraminiferan-like, but bi-directional movement not explicitly mentioned	Eukaryotes *incertae sedis*

Haeckelina (=*Protogenes*) *primordialis*	Schepotieff 1912b	Naples, Italy	M / ?	?	reticulose; bi-directional movement not mentioned	Eukaryotes *incertae sedis* / moved from *Protogenes* to *Haeckelina* by Schepotieff
Lateromyxa gallica	Hülsmann 1993	crater lakes, Auvergne, France	F / *Oedogonium* filaments	*Oedogonium*	mostly filose; 25-35 um, multinucleate syncytia of up to 800 um, modified vampyrellid life-cycle	Rhizaria, Cercozoa, Vampyrellida / remains unsequenced; see also Röpstorf, Hülsmann & Hausmann 1993
Leptomyxa reticulata	Goodey 1915b	Harpenden, Hertford, UK	F / cucumber house soil	bacteria, small flagellates and amoebae	reticulate (through fusion and lacunae); 50 um to 1 mm, plasmodial and multinucleate, cysts observed	Amoebozoa, Tubulinea, Leptomyxida / see also McLennan 1931, Singh 1948a, 1948b, Cann 1984
Leptomyxa reticulata	Pussard & Pons 1976a	Bourron-Marlotte, France	F / compost	bacteria, small amoebae	same; 50 um to 1 cm	Amoebozoa, Tubulinea, Leptomyxida
Leptophrys sp.	Hertwig & Lesser 1874	not indicated	F / no details given	diatoms, flagellates and algae in vacuoles	branched	Rhizaria, Cercozoa, Vampyrellida / sometimes considered a synonym of Vampyrella vorax; sequenced by Hess et al. 2012
Leucodictyon marinum	Grell 1991	Shimoda Marine Research Centre, Japan	M / sand grains, tide pool	*Amphiprora* sp.	meroplasmodial network; cells enclosed in a lorica, bi-directional movement, flagellated swarmers	Rhizaria, Cercozoa, Granofilosea / remains unsequenced; Grell & Schueller 1991 see it as a loricate *Reticulamoeba* relative
Leukarachnion batrachospermi	Geitler 1942	?	F / stream	?	meroplasmodial network; up to 2 mm, multinucleate	Stramenopiles, Synchromophyceae / based on an ATCC strain identified as Leukarachnion sp. (Grant et al. 2009)
"*Leukarachnion*-like plasmode"	Geitler 1959	Lunz, Austria	F / lake stream, among diatoms &	diatoms & other algae?	reticulate; fixed material only	Eukaryotes *incertae sedis* / could be a variosean
Megamoebamyxa argillobia	Nyholm 1950	Northern Europe	M / organically enriched sediments	?	reference not seen	Eukaryotes incertae sedis / could be a naked foraminiferan
Myxastrum radians	Haeckel 1870	?	?	?	reference not seen	Eukaryotes *incertae sedis*
Penardia mutabilis	Cash 1904	Epping Forest, Essex, UK	F / *Sphagnum* from a swamp	sometimes rotifers	mostly filose; 90-100 um up to 300-400 um, fast moving, anastomosing filopodia, green/yellow	Rhizaria, Cercozoa, Vampyrellida / based on an isolate identified as Penardia sp. (Berney et al. 2013); see also Cash & Hopkinson 1905
Pontomyxa flava	Topsent 1892	Mediterranea & Atlantic, France	M / ?	?	reticulose; body mass surrounded by a reticulopodial network with bi-directional movement	Eukaryotes *incertae sedis* / probably a naked foraminiferan

121

Pontomyxa flava	Schepotieff 1912a	Naples, Italy	M / on rocks with algae	?	reticulose?; reticulopodia not described, possibly presence of cysts	Eukaryotes *incertae sedis* / looks different from Topsent's original drawing
Protogenes primordialis	Haeckel 1870	Northern Europe	M / ?	?	reference not seen; reticulose; body up to 1 mm / network up to 4 mm	Eukaryotes *incertae sedis* / could be a naked foraminiferan
Protogenes roseus	Trinchese 1884	Naples, Italy	M / on the red alga *Gelidium*	?	reticulose; 5-15 mm according to Trinchese	Eukaryotes *incertae sedis* / see *Vampyrelloides*
Protomyxa aurantiaca	Haeckel 1870	Canary Islands	M / ?	major parasite of *Bryopsis*?	reticulose; anastomosing pseudopodia, presence of heliozoan swarmers and cysts	Eukaryotes *incertae sedis*
Protomyxa aurantiaca	Schepotieff 1912b	Naples, Italy	M / tide pool	?	reticulose; same	Eukaryotes *incertae sedis* / looks different from Haeckel's original drawings
Reticulamoeba gemmipara	Grell 1994	Santorin Island, Greece	M / volcanic stones from the littoral zone	*Amphiprora* sp.	network by fusion of cells; reticulopodia of different cells fusing only when food becomes scarce	Rhizaria, Cercozoa, Granofilosea / based on an isolate identified as *Reticulamoeba gemmipara* (Bass et al. 2013)
Reticulamoeba minor	Grell 1995	Alanya, Turkey	M / little stones from a tide pool	*Amphiprora* sp.	network by fusion of cells	Rhizaria, Cercozoa, Granofilosea / based on an isolate identified as *Reticulamoeba minor* (Bass et al. 2013)
Reticulomyxa filosa	Nauss 1949	New York City, USA	F / decaying leaves, stream	pulverized wheat germs	reticulate (through fusion and lacunae); up to 6 cm, multinucleate, with bi-directional movement	Rhizaria, Retaria, Foraminifera / sequenced by Pawlowski et al. 1999; see also Hülsmann 1984, Ostwald 1988
Reticulosphaera socialis	Grell 1989a	Caribbean coast	M / sandy beach	*Amphiprora* sp.	meroplasmodial network; with amoeboid and heliozoan dispersal stages	Haptophyta / based on an isolate identified as *Reticulosphaera socialis* (Cavalier-Smith et al. 1996)
Reticulosphaera japonensis	Grell 1990	Shimoda Marine Research Centre, Japan	M / sand grains and mollusc shells, tide / *Amphiprora* sp.	mostly photosynthetic	meroplasmodial network; grows into smaller plasmodia than *R. socialis*, more prone to photosynthesis only	Haptophyta / same; see also Grell 1989b, Grell, Heini & Schueller 1990
Rhizoplasma kaiseri	Verworn 1903	Red Sea	M / on coral reefs	?	reticulose; body mass surrounded by a reticulopodial network with bi-directional movement	Eukaryotes *incertae sedis* / probably a naked foraminiferan
Rhizoplasma kaiseri	Schepotieff 1912a	Naples, Italy	M / with sponges in the station's aquariums	diatoms?	reticulose; body 200-500 um / network 2-5 mm / bi-directional movement not mentioned	Eukaryotes *incertae sedis* / could be a naked foraminiferan

Taxon	Reference	Location	Habitat	Food	Description	Classification / Notes
Stereomyxa ramosa	Grell 1966	Madagascar	M / dead pieces of coral reef	pennate diatoms	branched; up to 1 mm, uninucleate, slow, never really reticulate	Eukaryotes *incertae sedis* / probably Amoebozoa; see also Benwitz & Grell 1971a
Stereomyxa angulosa	Grell 1966	Madagascar	M / dead pieces of coral reef	pennate diatoms	branched; up to 300 um, uninucleate	Eukaryotes *incertae sedis* / probably Amoebozoa; see also Benwitz & Grell 1971b
Synamoeba arenaria	Grell 1994	Tenerife, Canary Islands	M / sandy beach	*Amphiprora* sp.	meroplasmodial network; 100-500 um, presence of floating forms, network not permanent	Eukaryotes *incertae sedis* / probably Amoebozoa
Thalassomyxa australis	Grell 1985	Rottnest Island near Perth, Australia	M / sand sample	*Amphiprora* sp. + *Dunaliella* sp. & *Chlorella* sp. (less good)	reticulate (through fusion and lacunae); multinucleate plasmodia with digestive cysts and floating buds	Rhizaria, Cercozoa, Vampyrellida / based on an isolate identified as *Thalassomyxa* sp. (Berney et al. 2013)
Thalassomyxa jamaicensis	Grell 1992a	North coast of Jamaica	M / pieces of dead coral in tidal pools	*Amphiprora, Dunaliella, Pyrenomonas* (leading to differences in life-cycle)	reticulate (through fusion and lacunae); multinucleate plasmodia with digestive cysts and floating buds	Rhizaria, Cercozoa, Vampyrellida / same; see also Grell 1992b
Thalassomyxa canariensis	Grell 1994	Tenerife, Canary Islands	M / little stones from a tide pool	*Amphiprora* sp., *Pyrenomonas* sp.	reticulate (through fusion and lacunae); very similar to *T. jamaicensis*	Rhizaria, Cercozoa, Vampyrellida / same
Theratromyxa weberi	Zwillenberg 1953	Netherlands	F / sandy soil	nematodes	branched; 40-300 um, *Arachnula*-like but without groupings of filopodia	Rhizaria, Cercozoa, Vampyrellida / based on an isolate identified as *Theratromyxa* sp. (Parfrey et al. 2010); see also Sayre 1973
Vampyrella lateritia	Cienkowski 1865	?	F / ?	?	mostly filose	Rhizaria, Cercozoa, Vampyrellida / first described by Fresenius 1856; see also Lloyd 1929, Hoogenrand & De Groot 1942; sequenced by Hess et al. 2012
Vampyrelloides (=Protogenes) roseus	Schepotieff 1912b	Naples, Italy	M / on *Chaetomorpha*	sucks contents of algal cells	branched, filose, *Arachnula*-looking with a pinkish tinge	Rhizaria, Cercozoa, Vampyrellida / remains unsequenced; not clear if Schepotieff observed the same organism as Trinchese's *Protogenes roseus*

Supplementary Table 5: List of all the variosean taxa for which SSU rDNA sequences are available, and of all the environmental clone sequences from the NCBI GenBank database that we identified as variosean, with information on their origin (additional references are given below). Sequences highlighted in blue are from the 45 described taxa included in Fig. 30, and sequences highlighted in green correspond to additional sequences present in Fig. 38. Sequences highlighted in orange were excluded because they are too short and/or of chimeric origin. For habitat, MA = marine, FW = freshwater or terrestrial, BR = brackish. Columns F, G and H indicate the length of the sequence and whether it includes variable regions V4 and V7, respectively; lin = lineage; # = number; len = length

lin	name	Accession #	isolate / clone	habitat	len	V4	V7	reference
V01	Filamoeba	AF293896	Filamoeba sp. ("nolandi") ATCC 50430	MA - USA, offshore sediment core, North Carolina	1846	Y	Y	Amaral Zettler et al. 2000
		AY714369	Filamoeba sinensis CH26 (type)	FW - China, gills of Carassius gibelio	1839	Y	Y	Dyková et al. 2005
		AY714368	Filamoeba nolardi CCAP 1526/1 (type)	FW - USA, shore of Little Deer Lake, Minnesota	1839	Y	Y	Dyková et al. 2005
		AB425944	Filamoeba sp. H9a_6E	FW - Italy, rice field soil	1802	Y	Y	Murase & Frenzel 2008
		AB425947	Filamoeba sp. I4_5E	FW - Italy, rice field soil	1802	Y	Y	Murase & Frenzel 2008
		EF032804	clone HAVOmat-eukA09	FW - Hawaii, lava cave cyanobacterial mat	778	N	Y	Brown et al. unpublished (2006)
		GU320603	Filamoeba sp. COHH87	BR - USA, shore of Mt Hope Bay, Massachusetts	1784	Y	Y	Gast et al. unpublished (2011)
		GU320604	Filamoeba sp. COHH88	BR - USA, shore of Mt Hope Bay, Massachusetts	1139	Y	N	Gast et al. unpublished (2011)
		GU320578	Filamoeba sp. COHH101	BR - USA, shore of Mt Hope Bay, Massachusetts	1845	Y	Y	Gast et al. unpublished (2011)
		GQ371176	Filamoeba sp. JIH56	FW - Czech Republic, hot-water piping system	1840	Y	Y	Peckova et al. unpublished (2012)
V02	Acramoeba	AF293897	Acramoeba dendroida ATCC 50654 (type)	FW - USA, pond in Grand Haven, Michigan	1859	Y	Y	Amaral Zettler et al. 2000; Smirnov et al. 2008
		AB695507	clone MPE2-12	FW - Antarctica, lake moss pillars	1817	Y	Y	Nakai et al. 2012
		AB695457	clone MPE1-14	FW - Antarctica, lake moss pillars	1817	Y	Y	Nakai et al. 2012
		KC306576	clone KF_Euk_A1R	FW - Germany, groundwater of a karstic aquifer	1186	Y	N	Risse-Buhl et al. 2013
		KC306569	clone KF_Euk_F7	FW - Germany, groundwater of a karstic aquifer	1157	Y	N	Risse-Buhl et al. 2013
		KC306572	clone KF_Euk_A9R	FW - Germany, groundwater of a karstic aquifer	1162	Y	N	Risse-Buhl et al. 2013
		KC306575	clone KF_Euk_F7R	FW - Germany, groundwater of a karstic aquifer	1157	Y	N	Risse-Buhl et al. 2013
		KC306581	clone KF_Euk_A1	FW - Germany, groundwater of a karstic aquifer	1186	Y	N	Risse-Buhl et al. 2013
		KC306594	clone KF_Euk_A9	FW - Germany, groundwater of a karstic aquifer	1162	Y	N	Risse-Buhl et al. 2013
		KC306611	clone KF_Euk_D12R	FW - Germany, groundwater of a karstic aquifer	1186	Y	N	Risse-Buhl et al. 2013
		KC306613	clone KF_Euk_D12R	FW - Germany, groundwater of a karstic aquifer	1186	Y	N	Risse-Buhl et al. 2013
V03	Phalansterium	AF280078	Phalansterium cf. solitarium ATCC 50327	FW - USA, outdoor bath at hot springs area, West Virginia	1876	Y	Y	Cavalier-Smith et al. 2004
		EF143966	Phalansterium filosum Thailand (type)	FW - Thailand, forest soil near flooded stream	1858	Y	Y	Smirnov et al. 2011
		FK539978	Phalansterium sp. 1 JFP-2013	FW - contaminant in Monomorphina aenigmatica UTEX 1284	1873	Y	Y	Pombert et al. 2013

V04	*Multicilia*	AY268037	*Multicilia marina*	MA - Ukraine, surface of brown algae, Black Sea	2746	Y	Y	Nikolaev et al. 2006
V05	AND16	AY965863	undescribed amoeba AND16	FW - Spain, silty clay soil, Andújar	1831	Y	Y	Lara et al. 2007
		EU798715	clone 18S1-22	FW - China, lignocellulose decomposing forest soil	1024	Y	N	Huang et al. unpublished (2008)
V06	Protosteliids	EU004603	*Protostelium mycophagum* RTF06-1A-4-1	FW - habitat where strain was isolated not provided	1790	Y	Y	Brown et al. 2007
		EU004604	*Planoprotostelium aurantium* BR33	FW - habitat where strain was isolated not provided	1865	Y	Y	Brown et al. 2007
		FJ766481	*Protostelium nocturnum* ATCC PRA-194	FW - habitat where strain was isolated not provided	1800	Y	Y	Shadwick et al. 2009
		FJ766482	*Protostelium okumukumu* ATCC PRA-156 (type)	FW - USA, Hawaii, on plants	1813	Y	Y	Shadwick et al. 2009
		FJ766483	*Protostelium mycophagum* HI04-85a-1b	FW - habitat where strain was isolated not provided	1819	Y	Y	Shadwick et al. 2009
		FJ766484	*Protostelium mycophagum* ATCC PRA-154 (type)	FW - USA, on reed dead inflorescence, New Jersey	1809	Y	Y	Shadwick et al. 2009
		FJ766440	*Protostelium mycophagum* HI07-6L-3	FW - habitat where strain was isolated not provided	1808	Y	Y	Shadwick et al. unpublished (2010)
		FJ766441	*Protostelium mycophagum* HI07-22L-4	FW - habitat where strain was isolated not provided	1808	Y	Y	Shadwick et al. unpublished (2010)
		FJ766442	*Protostelium mycophagum* TB-A2-1	FW - habitat where strain was isolated not provided	1808	Y	Y	Shadwick et al. unpublished (2010)
		FJ766443	*Protostelium mycophagum* PBR-A3-1	FW - habitat where strain was isolated not provided	1808	Y	Y	Shadwick et al. unpublished (2010)
		FJ766444	*Protostelium mycophagum* BRNFDF05-11A-3	FW - habitat where strain was isolated not provided	1811	Y	Y	Shadwick et al. unpublished (2010)
		FJ766445	*Protostelium mycophagum* KA-T02-1	FW - habitat where strain was isolated not provided	1812	Y	Y	Shadwick et al. unpublished (2010)
		FJ766446	*Protostelium mycophagum* KEN-5A-2	FW - habitat where strain was isolated not provided	1811	Y	Y	Shadwick et al. unpublished (2010)
		FJ766447	*Protostelium mycophagum* HI06-7A-1	FW - habitat where strain was isolated not provided	1814	Y	Y	Shadwick et al. unpublished (2010)
		FJ766448	*Protostelium mycophagum* HI07-4a-III	FW - habitat where strain was isolated not provided	1830	Y	Y	Shadwick et al. unpublished (2010)
		FJ766449	*Protostelium mycophagum* WWL06-3a-1	FW - habitat where strain was isolated not provided	1828	Y	Y	Shadwick et al. unpublished (2010)
		FJ766450	*Protostelium mycophagum* HI06-9a-2	FW - habitat where strain was isolated not provided	1812	Y	Y	Shadwick et al. unpublished (2010)

		FJ766451	*Protostelium mycophagum* KA-T15-1	FW - habitat where strain was isolated not provided	1800	Y	Y	Shadwick et al. unpublished (2010)
		FJ766452	*Protostelium mycophagum* KA-T15-3	FW - habitat where strain was isolated not provided	1811	Y	Y	Shadwick et al. unpublished (2010)
		FJ766453	*Protostelium mycophagum* OMBS04-1-1	FW - habitat where strain was isolated not provided	1800	Y	Y	Shadwick et al. unpublished (2010)
		FJ766454	*Protostelium mycophagum* SI04-01A-1	FW - habitat where strain was isolated not provided	1791	Y	Y	Shadwick et al. unpublished (2010)
		FJ766455	*Protostelium mycophagum* MFB-b	FW - habitat where strain was isolated not provided	1789	Y	Y	Shadwick et al. unpublished (2010)
		FJ766456	*Protostelium mycophagum* BRNFDF05-11A-1	FW - habitat where strain was isolated not provided	1789	Y	Y	Shadwick et al. unpublished (2010)
		FJ766457	*Protostelium mycophagum* HI06-10a-2	FW - habitat where strain was isolated not provided	1810	Y	Y	Shadwick et al. unpublished (2010)
		FJ766458	*Protostelium mycophagum* HI07-2L-3	FW - habitat where strain was isolated not provided	1819	Y	Y	Shadwick et al. unpublished (2010)
		FJ766459	*Protostelium mycophagum* HI07-2a-2	FW - habitat where strain was isolated not provided	1819	Y	Y	Shadwick et al. unpublished (2010)
		FJ766460	*Protostelium mycophagum* ATCC PRA-128	FW - USA, plants on lake shore near Seattle, Washington	1791	Y	Y	Shadwick et al. unpublished (2010)
		FJ766461	*Protostelium mycophagum* BRNFDF05-7C	FW - habitat where strain was isolated not provided	1794	Y	Y	Shadwick et al. unpublished (2010)
		FJ766462	*Planoprotostelium aurantium* LEE06-21-3-1	FW - habitat where strain was isolated not provided	1815	Y	Y	Shadwick et al. unpublished (2010)
		FJ766463	*Protostelium mycophagum* MFA-c	FW - habitat where strain was isolated not provided	1794	Y	Y	Shadwick et al. unpublished (2010)
		FJ766464	*Protostelium mycophagum* KA-T15-2	FW - habitat where strain was isolated not provided	1803	Y	Y	Shadwick et al. unpublished (2010)
		FJ766465	*Protostelium mycophagum* HI07-20L-IV	FW - habitat where strain was isolated not provided	1812	Y	Y	Shadwick et al. unpublished (2010)
		FJ766466	*Protostelium mycophagum* QEC7-15-1	FW - habitat where strain was isolated not provided	1814	Y	Y	Shadwick et al. unpublished (2010)
		FJ766467	*Protostelium nocturnum* MCWTKF06-10L-2-1	FW - habitat where strain was isolated not provided	1810	Y	Y	Shadwick et al. unpublished (2010)
		FJ766468	*Protostelium nocturnum* BADL04-A	FW - habitat where strain was isolated not provided	1821	Y	Y	Shadwick et al. unpublished (2010)
		GU290108	clone TKR07M.106	FW - Lake Tanganyika, metalimnion	1845	Y	Y	Tarbe et al. unpublished (2010)
V07	*Telaepolella*	EU273440	*Telaepolella tubasferens* ATCC 50593	FW - habitat where strain was isolated not provided	1934	Y	Y	Tekle et al. 2008; Lahr et al. 2012
		KP864099	*Telaepolella*	this study	1847	Y	Y	this study

			sp. isolate Tib190					
V08	*Flamella*	EU186021	*Flamella aegyptia* A1-3	FW - Egypt, river Nile near Assuan	1858	Y	Y	Kudryavtsev et al. 2009
		EU186022	*Flamella arnhemensis* KWK - CCAP 1525/2	FW - Netherlands, cooling water circuit, Arnhem	1834	Y	Y	Kudryavtsev et al. 2009
		EU186023	*Flamella balnearia* Will8a - CCAP 1525/3	FW - Germany, physiotherapy hospital pool, Wildbad	1832	Y	Y	Kudryavtsev et al. 2009
		EU186024	*Flamella fluviatilis*	FW - Australia, soil in the floodplain of the Murray river	1824	Y	Y	Kudryavtsev et al. 2009
		EU186025	*Flamella* sp. strain WBT	FW - Germany, drinking water treatment plant near Bonn	1845	Y	Y	Kudryavtsev et al. 2009
		KP864098	*Flamella* sp. isolate FN414	this study	1763	Y	Y	this study
		AF372778	clone LEMD267	FW - USA, anoxic lake sediment	1460	Y	Y	Dawson & Pace 2002
		AM114809	clone WIM81	FW - Netherlands, 1975 agricultural field sample	1869	Y	Y	Moon-van der Staay et al. 2006
		AY626163	clone Borok	FW - Russia, waste treatment plant, Borok	1824	Y	Y	Nikolaev et al. 2006
		DQ409131	clone VNP13	FW - France, hyper-eutrophic lake picoplankton	983	Y	N	Lepère et al. 2007
		FJ355431	clone NewOrleansYard3_YD3_03 2406_090	FW - USA, New Orleans yard, Louisiana	1135	Y	Y	Amaral-Zettler et al. 2008
		EF023373	clone Amb_18S_704 (chimera)	FW - USA, trembling aspen rhizosphere	1865 / 1393	Y	Y	Lesaulnier et al. 2008
		EF023606	clone Amb_18S_844 (chimera)	FW - USA, trembling aspen rhizosphere	1865 / 1393	Y	Y	Lesaulnier et al. 2008
		EF023634	clone Amb_18S_879 (chimera)	FW - USA, trembling aspen rhizosphere	1865 / 1393	Y	Y	Lesaulnier et al. 2008
		EF023656	clone Amb_18S_908 (chimera)	FW - USA, trembling aspen rhizosphere	1868 / 1393	Y	Y	Lesaulnier et al. 2008
		EF023685	clone Amb_18S_948 (chimera)	FW - USA, trembling aspen rhizosphere	1865 / 1393	Y	Y	Lesaulnier et al. 2008
		EF023712	clone Amb_18S_991 (chimera)	FW - USA, trembling aspen rhizosphere	1865 / 1393	Y	Y	Lesaulnier et al. 2008
		EF023441	clone Amb_18S_1015 (chimera)	FW - USA, trembling aspen rhizosphere	1865 / 1393	Y	Y	Lesaulnier et al. 2008
		EF023478	clone Amb_18S_1064 (chimera)	FW - USA, trembling aspen rhizosphere	1865 / 1393	Y	Y	Lesaulnier et al. 2008
		EF023545	clone Amb_18S_1147 (chimera)	FW - USA, trembling aspen rhizosphere	1864 / 1392	Y	Y	Lesaulnier et al. 2008
		EF023826	clone Amb_18S_1275 (chimera)	FW - USA, trembling aspen rhizosphere	1865 / 1393	Y	Y	Lesaulnier et al. 2008
		EF023984	clone	FW - USA, trembling aspen	1865	Y	Y	Lesaulnier et al.

			Amb_135_14 55 (chimera) clone	rhizosphere	/ 1393			2008
		EF024977	Elev_18S_152 5	FW - USA, trembling aspen rhizosphere	1873	Y	Y	Lesaulnier et al. 2008
		AB425943	Flamella sp. H9a_3E clone	FW - Italy, rice field soil	1829	Y	Y	Murase & Frenzel 2008
		JF826388	North_Pole_S W170_108 (chimera) clone ws_96,	MA - North Pole, 170m-deep water under sea ice	1699 / 1457	Y	Y	Bachy et al. 2011
		FR874405	clone 1802F11	MA - Norway, fjord coastal water, Bergen	1872	Y	Y	Newbold et al. 2012
		FJ153641	clone GoC1_G12	MA - Baltic Sea, anoxic water sample, Gotland Deep	1280	Y	Y	Stock et al. unpublished (2008)
V09	Schizoplasmodiids	FJ544418	Schizoplasmodium cavostelioides ATCC PRA-197	FW - habitat where strain was isolated not provided	1920	Y	Y	Shadwick et al. 2009
		FJ544419	Ceratiomyxella tahitiensis HI04-93L-1	FW - habitat where strain was isolated not provided	1883	Y	Y	Shadwick et al. 2009
		FJ544420	Nemctosteliu m ovatum JDS 6241	FW - habitat where strain was isolated not provided	1918	Y	Y	Shadwick et al. 2009
V10	Soliformoviids	FJ766479	Soliformovum expulsum ATCC 48083 (type)	FW - habitat where strain was isolated not provided	1878	Y	Y	Shadwick et al. 2009
		FJ766480	Soliformovum irregulare ATCC 26826 (type)	FW - Mexico, on purple bean pods	1881	Y	Y	Shadwick et al. 2009
		GQ438740	Grellamoeba robusta 4168 clone 111	FW - Czech Republic, gills of Sander lucioperca	1880	Y	Y	Dyková et al. 2010
		GQ438741	Grellamoeba robusta 4168 clone 122	FW - Czech Republic, gills of Sander lucioperca	1880	Y	Y	Dyková et al. 2010
		GQ438742	Grellamoeba robusta 4168 clone 841	FW - Czech Republic, gills of Sander lucioperca	1880	Y	Y	Dyková et al. 2010
		EF513181	Soliformovum irregulare ATCC 26826 (type)	FW - Mexico, on purple bean pods	1848	Y	Y	Fiore-Donno et al. 2010
		HE614594	Soliformovum irregulare ATCC 26826 (type)	FW - Mexico, on purple bean pods	1785	Y	Y	Nandipati et al. unpublished (2012)
V11	Cavosteliids	FJ766475	Schizoplasmodiopsis pseudoendospora ATCC PRA-195	FW - habitat where strain was isolated not provided	1786	Y	Y	Shadwick et al. 2009
		FJ766476	Cavostelium apophysatum ATCC 38567 (type)	FW - Granada, on plants	1777	Y	Y	Shadwick et al. 2009
		FJ766477	Schizoplasmo	FW - Cook Islands, on pigeon	1927	Y	Y	Shadwick et al.

		diopsis amoeboidea ATCC 46943 (type)	pea pods				2009
	FJ766478	*Schizoplasmodiopsis amoeboidea* BG7A-12B	FW - habitat where strain was isolated not provided	1876	Y	Y	Shadwick et al. 2009
	FJ792703	*Tychosporium acutostipes* KEA-11A-L	FW - Kenya, more details not provided	1818	Y	Y	Shadwick et al. 2009
	FJ792704	*Tychosporium acutostipes* NZ05-15a-2	FW - New Zealand, more details not provided	1818	Y	Y	Shadwick et al. 2009
	EF513172	*Cavostelium apophysatum* ATCC 38567 (type)	FW - Granada, on plants	1732	Y	Y	Fiore-Donno et al. 2010
	EF513179	*Schizoplasmodiopsis amoeboidea* ATCC 46943 (type)	FW - Cook Islands, on pigeon pea pods	1909	Y	Y	Fiore-Donno et al. 2010
	EF513180	*Schizoplasmodiopsis vulgaris*	FW - Switzerland, on dead grass leaves near Geneva	1812	Y	Y	Fiore-Donno et al. 2010
	KP864100	*Schizoplasmodiopsis* sp. isolate F3	this study	1745	Y	Y	this study
	KP864101	undetermined cavosteliid sp. isolate M71	this study	1578	Y	Y	this study
V12	*Ischnamoeb a* n. gen.	KP864092	*Ischnamoeba montana* isolate Tib85 (type)	this study	1889	Y	Y	this study
	KP864093	*Ischnamoeba* sp. isolate F4	this study	1878	Y	Y	this study
	KP864094	*Ischnamoeba* sp. isolate FN352	this study	1798	Y	Y	this study
	AB425950	undetermined amoeba Mb_5C	FW - Italy, rice field soil	1891	Y	Y	Murase & Frenzel 2008
	EF023260	clone Amb_18S_564 (chimera)	FW - USA, trembling aspen rhizosphere	2226 / 789	N	Y	Lesaulnier et al. 2008
	EF023267	clone Amb_18S_572 (chimera)	FW - USA, trembling aspen rhizosphere	1897 / 789	N	Y	Lesaulnier et al. 2008
	EF023369	clone Amb_18S_699 (chimera)	FW - USA, trembling aspen rhizosphere	1900 / 789	N	Y	Lesaulnier et al. 2008
	HE575399	uncultured amoeba	FW - Italy, water purification plant	1900	Y	Y	Chiellini et al. 2012
	AB695456	clone MPE1-13	FW - Antarctica, lake moss pillars	1923	Y	Y	Nakai et al. 2012
	AB695508	clone MPE2-13	FW - Antarctica, lake moss pillars	1923	Y	Y	Nakai et al. 2012
V13	*Darbyshirell a* n. gen.	KP864087	*Darbyshirella terrestris* isolate Tib177 (type)	this study	2005	Y	Y	this study

Group	Accession	Name	Source	Location	Length			Reference
	KP864088	*Darbyshirella* sp. isolate Esthw	this study		2092	Y	Y	this study
	KP864089	*Darbyshirella* sp. isolate JDgar	this study		2079	Y	Y	this study
	KP864090	*Darbyshirella* sp. isolate M68	this study		1988	Y	Y	this study
	KP864091	*Darbyshirella* sp. isolate M77	this study		2001	Y	Y	this study
	JN825696	clone Alch chica_AQ 2_5E_20	FW - Mexico, alcaline lake microbialite		1272	Y	N	Couradeau et al. 2011
V14 *Dictyamoeba* n. gen.	KP864096	*Dictyamoeba vorex* isolate WalEn (type)	this study		1850	Y	Y	this study
	AB695458	clone MPE1-15	FW - Antarctica, lake moss pillars		1860	Y	Y	Nakai et al. 2012
V15 *Heliamoeba* n. gen.	KP864095	*Heliamoeba mirabilis* isolate M134 (type)	this study		1660	Y	Y	this study
V16 *Arboramoeba* n. gen.	KP864097	*Arboramoeba reticulata* isolate Tib182 (type)	this study		1773	Y	Y	this study
	AY969212	clone dfmo4344.099	FW - USA, mixed hardwood soil, North Carolina		734	Y	N	O'Brien et al. 2005
	EF023424	clone Amb_18S_765	FW - USA, trembling aspen rhizosphere		1862	Y	Y	Lesaulnier et al. 2008
V17 *Angulamoeba* n. gen.	KP864083	*Angulamoeba microcystivorans* isolate A1WVB (type)	this study		1714	Y	Y	this study
	KP864084	*Angulamoeba fungorum* isolate 23-6A	this study		1847	Y	Y	this study
	KP864085	*Angulamoeba* sp. isolate F2	this study		644	Y	N	this study
	KP864086	*Angulamoeba* sp. isolate FN806	this study		1763	Y	Y	this study
	DQ123626	undetermined amoeba CRIB-09	FW - Switzerland, hospital water network		568	Y	N	Thomas et al. 2006
	HM017144	uncultured amoeba clone A2WVB	FW - Belgium, hypertrophic urban pond		1475	Y	Y	Van Wichelen et al. 2010
	JN825704	clone Alchichica_AQ 2w_5E_66	FW - Mexico, alcaline lake microbialite		1001	Y	N	Couradeau et al. 2011
	HM161754	"Cyanidioschyzon sp." Y16	FW - France, Phytophthora parasitica biofilm		1870	Y	Y	Galiana et al. 2011
	AB695473	clone MPE1-30	FW - Antarctica, lake moss pillars		1804	Y	Y	Nakai et al. 2012

		AB695474	clone MPE1-31	FW - Antarctica, lake moss pillars	1850	Y	Y	Nakai et al. 2012
		AB695529	clone MPE2-35	FW - Antarctica, lake moss pillars	1803	Y	Y	Nakai et al. 2012
		AB695530	clone MPE2-36	FW - Antarctica, lake moss pillars	1841	Y	Y	Nakai et al. 2012
		JN547325	clone S9-1	FW - France, Sep reservoir artificial lake	1576	Y	Y	Lepère et al. unpublished (2012)
V18	"RT5iin21"	AY082976	clone RT5iin21	FW - Spain, green biofilm in the acidic and iron-rich Rio Tinto	1912	Y	Y	Amaral Zettler et al. 2002
		AY082989	clone RT5iin44	FW - Spain, green biofilm in the acidic and iron-rich Rio Tinto	1914	Y	Y	Amaral Zettler et al. 2002
		AM114808	clone WIM80	FW - Netherlands, 1975 agricultural field sample	1961	Y	Y	Moon-van der Staay et al. 2006
		EF441963	clone RT07C_2E_32	FW - Spain, endolithic community in the acidic Rio Tinto basin	1738	Y	Y	Lopez-Garcia et al. unpublished (2012)
		EF441972	clone RT07C_2E_9	FW - Spain, endolithic community in the acidic Rio Tinto basin	1732	Y	Y	Lopez-Garcia et al. unpublished (2012)
		EF441973	clone RT07C_2E_43	FW - Spain, endolithic community in the acidic Rio Tinto basin	1731	Y	Y	Lopez-Garcia et al. unpublished (2012)
V19	"Sey088"	AY605217	clone Sey088	FW - Switzerland, river sediment near Geneva	883	N	Y	Berney et al. 2004
		DQ103790	clone M1_18H06	MA - Denmark, anoxic Mariager Fjord	1656	Y	Y	Zuendorf et al. 2006
		DQ103816	clone M1_18G11	MA - Denmark, anoxic Mariager Fjord	1654	Y	Y	Zuendorf et al. 2006
		EF526901	clone SA1_4A12	MA - Norway, anoxic Framvaren Fjord	1367	Y	N	Behnke et al. 2010
		HQ868115	clone SHAO448 (chimera)	MA - Canada, micro-oxic water column near Vancouver	908 / 595	Y	N	Orsi et al. 2012
V20	"WIM1"	AM114811	clone WIM1	FW - Netherlands, 1975 agricultural field sample	1893	Y	Y	Moon-van der Staay et al. 2006
		AB695494	clone MPE1-51	FW - Antarctica, lake moss pillars	1839	Y	Y	Nakai et al. 2012
V21	"WIM5"	AM114801	clone WIM5	FW - Netherlands, 1975 agricultural field sample	1871	Y	Y	Moon-van der Staay et al. 2006
		AB695495	clone MPE1-52	FW - Antarctica, lake moss pillars	1833	Y	Y	Nakai et al. 2012
		AB695538	clone MPE2-44	FW - Antarctica, lake moss pillars	1833	Y	Y	Nakai et al. 2012
V22	"Amb_18S_6341"	EF024087	clone Amb_18S_6341 (chimera)	FW - USA, trembling aspen rhizosphere	1877 / 1668	Y	Y	Lesaulnier et al. 2008
V23	"SAWY402"	DQ423701	clone SAWY402	FW - USA, acid mine drainage biofilm, California	1163	Y	Y	Baker et al. 2009
		DQ423693	clone SAWY394	FW - USA, acid mine drainage biofilm, California	1146	Y	Y	Baker et al. 2009
		DQ423694	clone SAWY395	FW - USA, acid mine drainage biofilm, California	1155	Y	Y	Baker et al. 2009
		DQ423699	clone SAWY400	FW - USA, acid mine drainage biofilm, California	1146	Y	Y	Baker et al. 2009

	DQ423700	clone SAWY401	FW - USA, acid mine drainage biofilm, California	1157	Y	Y	Baker et al. 2009
	DQ423702	clone SAWY403	FW - USA, acid mine drainage biofilm, California	1089	Y	Y	Baker et al. 2009
	DQ423705	clone SAWY406	FW - USA, acid mine drainage biofilm, California	1155	Y	Y	Baker et al. 2009
	DQ423706	clone SAWY407	FW - USA, acid mine drainage biofilm, California	1172	Y	Y	Baker et al. 2009
	DQ423709	clone SAWY410	FW - USA, acid mine drainage biofilm, California	1160	Y	Y	Baker et al. 2009
	DQ423710	clone SAWY411	FW - USA, acid mine drainage biofilm, California	1022	Y	Y	Baker et al. 2009
	DQ423711	clone SAWY412	FW - USA, acid mine drainage biofilm, California	1097	Y	Y	Baker et al. 2009
	DQ423716	clone SAWY417	FW - USA, acid mine drainage biofilm, California	1151	Y	Y	Baker et al. 2009
	DQ423721	clone SAWY422	FW - USA, acid mine drainage biofilm, California	1150	Y	Y	Baker et al. 2009
	DQ423725	clone SAWY426	FW - USA, acid mine drainage biofilm, California	1091	Y	Y	Baker et al. 2009
	DQ423728	clone SAWY429	FW - USA, acid mine drainage biofilm, California	1028	Y	Y	Baker et al. 2009
	DQ423735	clone SAWY437	FW - USA, acid mine drainage biofilm, California	1164	Y	Y	Baker et al. 2009
	DQ423737	clone SAWY440	FW - USA, acid mine drainage biofilm, California	1180	Y	Y	Baker et al. 2009
	DQ423740	clone SAWY444	FW - USA, acid mine drainage biofilm, California	1168	Y	Y	Baker et al. 2009
	DQ423742	clone SAWY446	FW - USA, acid mine drainage biofilm, California	1074	Y	Y	Baker et al. 2009
	DQ423745	clone SAWY449	FW - USA, acid mine drainage biofilm, California	1158	Y	Y	Baker et al. 2009
	DQ423748	clone SAWY452	FW - USA, acid mine drainage biofilm, California	1130	Y	Y	Baker et al. 2009
	DQ423762	clone SAWY466	FW - USA, acid mine drainage biofilm, California	1135	Y	Y	Baker et al. 2009
	DQ423764	clone SAWY468	FW - USA, acid mine drainage biofilm, California	1161	Y	Y	Baker et al. 2009
	DQ423765	clone SAWY469	FW - USA, acid mine drainage biofilm, California	1148	Y	Y	Baker et al. 2009
	DQ423774	clone SAWY479	FW - USA, acid mine drainage biofilm, California	1040	Y	Y	Baker et al. 2009
	JN825703	clone Alchichica_AQ 1B_5E_11	FW - Mexico, alcaline lake microbialite	1118	Y	N	Couradeau et al. 2011
V24	"PRS2_4E_1 9" GU479959	clone PRS2_4E_19	FW - Switzerland, peat bog in the Jura mountains	1539	Y	Y	Lara et al. 2011
V25	"MPE1-44" AB695487	clone MPE1-44	FW - Antarctica, lake moss pillars	1874	Y	Y	Nakai et al. 2012
	AB695493	clone MPE1-50 (chimera)	FW - Antarctica, lake moss pillars	1790 / 1217	Y	N	Nakai et al. 2012
V26	"LP30ME5" FJ903024	clone LP30ME5	FW - Mexico, phreatic limestone sinkhole	857	Y	Y	Sahl et al. unpublished (2009)
V27	"Elev_18S_6 03" EF024236	clone Elev_18S_603 (chimera)	FW - USA, trembling aspen rhizosphere	1957 / 643	N	Y	Lesaulnier et al. 2008
V28	"9_69" EU087280	clone 9_69	MA - Korea, 9cm deep core	924	Y	N	Park et al. 2008

				sediment in the East Sea				
V29	"9_174"	EU545726	clone 9_174	MA - Korea, 9cm deep core sediment in the East Sea	989	Y	N	Park et al. 2008
V30	"Alchichica_ AQ2_5E_31 "	JN825689	clone Alchichica_AQ 2_5E_31	FW - Mexico, alcaline lake microbialite	1038	Y	N	Couradeau et al. 2011
V31	"Alchichica_ AQ1_5E_65 "	JN825698	clone Alchichica_AQ 1_5E_65	FW - Mexico, alcaline lake microbialite	1119	Y	N	Couradeau et al. 2011
		JN825697	clone Alchichica_AQ 1_5E_60 (chimera)	FW - Mexico, alcaline lake microbialite	1032 / 487	N	N	Couradeau et al. 2011

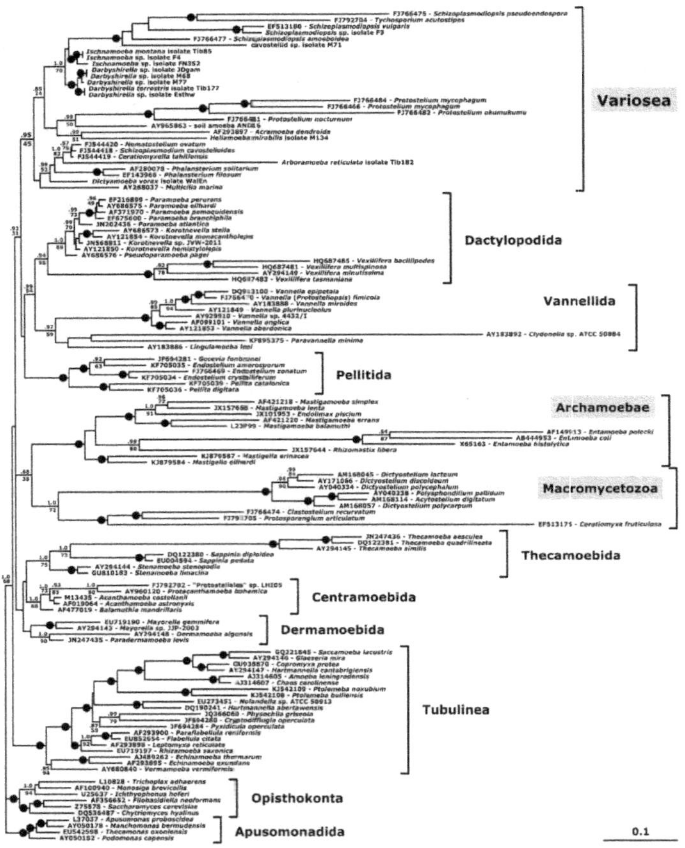

Supplementary Fig. 1: Maximum likelihood SSU rDNA phylogeny of Amoebozoa, showing the position of Archamoebae and Macromycetozoa (A & M) when some of the most divergent varioseans are excluded. A & M are separate from Variosea and occupy an unresolved deep position within Amoebozoa. The overall amoebozoan topology is congruent with previous publications using the same genetic marker. Bootstrap support values after 1000 replicates and Bayesian posterior probabilities (see Methods) are indicated at nodes when above 50% and/or 0.75, respectively. Black blobs represent support values at or above 95%/0.99. Opisthokonts and apusomonads were used to root the tree. See the dashed line in Fig. 30 for the alternate, potentially artefactual position of A & M when all varioseans are included (see Discussion).

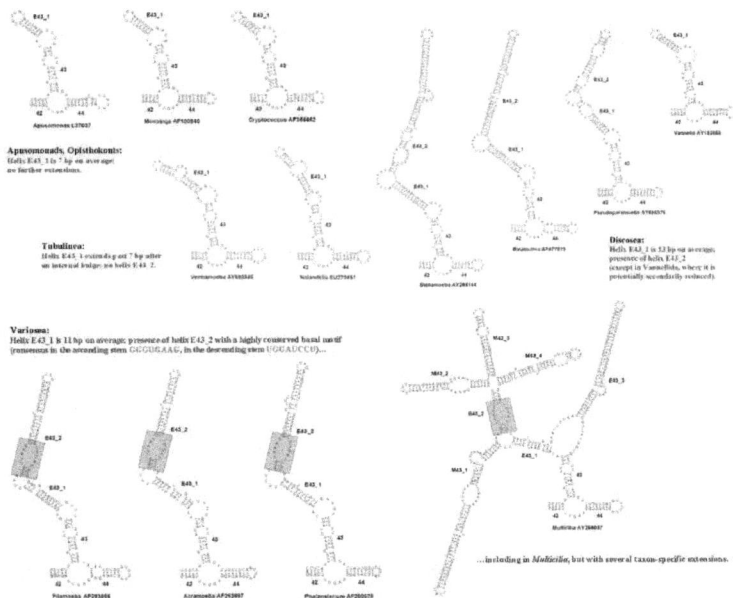

Supplementary Fig. 2: Graphical representation of the secondary structure of variable region V7 of the SSU rDNA of selected amoebozoan taxa, highlighting a specific secondary structure pattern found only in Variosea (helix numbering: see Wuyts et al. 2000). In Variosea helix E43_1 has a conserved length of 11 bp on average, and the base of helix E43_2 contains a highly conserved motif with the consensus GGGUGAAG in the ascending stem, and UGGAUCCU in the descending stem.

Part 1 – Chapter 4

Heterogeneity in the genus *Allovahlkampfia* and the description of the new genus *Parafumarolamoeba* (Vahlkampfiidae; Heterolobosea)

Geisen Stefan[1], Bonkowski Michael[1], Zhang Junling[2], De Jonckheere Johan F.[3,4]

[1] Cologne Biocentre, Zoological Institute, University of Cologne, 50674 Cologne, Germany
[2] College of Resources and Environmental Sciences, China Agricultural University, 100193 Beijing, P R China
[3] de Duve Institute, 1200 Brussels, Belgium
[4] Scientific Institute of Public Health, 1050 Brussels, Belgium

Abstract

Heterolobosean amoebae are common and diverse members of soil protist communities. In this study we isolated seven strains of amoebae from soil samples taken in Tibet (at high altitude), Sardinia and the Netherlands, all resembling to belong to a similar heterolobosean morphospecies. However, sequences of the small subunit (SSU) rDNA and internal transcribed spacers, including the 5.8S rDNA, revealed a high heterogeneity in the genus *Allovahlkampfia* to which six of the isolates belong. Some unnamed strains, of which the sequences had been published before, are also included within the genus *Allovahlkampfia*. One *Allovahlkampfia* isolated in the Netherlands harbours a twin-ribozyme, containing a His-Cys box, similar to the one found in strain BA of *Allovahlkampfia*. The other SSU rDNA sequence grouped in phylogenetic analyses with sequences obtained in environmental sequencing studies as sister to the genus *Fumarolamoeba*. This phylogenetic placement was supported by analyses of the 5.8S rDNA leading us to describe it as a new genus *Parafumarolamoeba*.

Introduction

The name "Heterolobosea" was coined by Page and Blanton (1985) for amoebae with eruptive locomotion, often a differentiated flagellate stage and distinct ultrastructural features. Despite morphological similarities to organisms within the supergroup Amoebozoa, phylogenetic analyses place the Heterolobosea distantly in the supergroup Excavata as a sister taxon to the morphologically different Euglenozoa (Adl et al. 2012). Identification of heterolobosean amoebae based on morphology is difficult, if not impossible. The flagellate stage and cyst structure have been used for species description (Page 1988), but due to the high morphological similarity amongst the Heterolobosea, only molecular tools have made a breakthrough by detecting cryptic species and genera within morphospecies. Within Heterolobosea the genus *Naegleria* (Vahlkampfiidae) has been investigated in most detail as *Naegleria fowleri* causes a fatal disease in humans, i.e. primary amoebic meningoencephalitis (De Jonckheere 2002, Visvesvara et al. 2007, De Jonckheere 2012). Many species within the genus *Naegleria* have been described based on the sequence of the ribosomal internal transcribed spacers (ITS) (De Jonckheere 2004). In a similar way, other Vahlkampfiidae have been separated into different genera based on the small subunit (SSU) rDNA (Brown and De Jonckheere 1999) and ITS, including the 5.8S rDNA, sequences (De Jonckheere and Brown 2005). Vahlkampfiid amoebae without a flagellate stage are morphologically very similar and were traditionally placed all together into a single genus *Vahlkampfia* (Page 1988). Only with the advent of DNA sequencing, this genus was found to be paraphyletic and species formerly placed in the genus *Vahlkampfia* were transferred to other genera and renamed as *Tetramitus*, *Vahlkampfia*, *Neovahlkampfia*, and *Paravahlkampfia* (Brown and De Jonckheere 1999).

The diversity within the Heterolobosea is enormous despite that only about 150 species have so far been described; it contains organisms that either only adopt an amoeboid or a flagellate stage, others alter between both, while even multicellular forms exist (Page and Blanton 1985, Brown et al. 2012). Heterolobosea seem to have a ubiquitous distribution, also occupying extreme environments (Amaral-Zettler et al. 2002, De Jonckheere 2006, De Jonckheere et al. 2011a, Park and Simpson 2011, Pánek et al. 2012).

We have investigated 7 new Vahlkampfiidae strains from diverse environments, four of which were cultivated from extremely high altitudes (> 4100 meters). As these strains are nearly indistinguishable morphologically, the work is focused primarily on molecular tools, i.e. the ITS, including the 5.8S rDNA, and the SSU rDNA sequences. Six strains from highly diverse soils are closely related and branch in phylogenetic analyses together with *Solumitrus palustris* (Anderson et al. 2011) and *Allovahlkampfia spelaea* (Walochnik and Mulec 2009). Furthermore, we describe a new genus for a strain which clusters in phylogenetic trees with several environmental sequences, of which the closest described relative is *Fumarolamoeba ceborucoi* (De Jonckheere et al. 2011b).

Materials and Methods

Isolation and cultivation of amoebae

The top 10 cm of soils were sampled from a pasture soil in the Netherlands, from high altitudes in Tibet and the Berchidda-Monti long term observatory, managed by the University of Sassari, Italy (Table 6).

50 g of each soil was suspended in 200 ml sterile ddH$_2$O and incubated on a shaker at room temperature for 10 min. 200 µl of each suspension was subsequently transferred to ten 90 mm Petri dishes filled with autoclaved liquid wheat grass medium (WG), made by adding vacuum-dry wheat grass powder (Weizengras, Sanatur, Singen, Germany) to PJ medium (Prescott and James 1955b) to the weight concentration 0.15 % to stimulate bacterial growth. Individual dishes were investigated with an inverted microscope (Nikon Eclipse TS100) after 14 and 28 days and individual amoebae were transferred and maintained in 60 mm Petri dishes containing fresh WG. Detailed light-microscopic analyses of each clonal culture and measurements on ~50 cells were performed in hanging drop preparations by observing amoebae attached to glass cover slips immediately after transfer using a Nikon Eclipse 90i with

differential interference contrast (DIC) equipped with a Nikon DIGITAL SIGHT DS-fi1. The strains were tested for growth at temperatures of 30 °C and 37 °C. Formation of flagellate stages was tested by flooding amoebae grown in a Petri dish with distilled water. Formation of fruiting forms were tested by adding autoclave-sterilized *Quercus alba* bark soaked in a sterile slurry of *Rhodotorula mucilaginosa* in ddH$_2$0 as described in Brown et al. (2012). The isolates were feeding on the bacteria that grew with them upon isolation, but they were also tested to see whether they could grow when fed with yeast, by transferring isolates into a culture of the ascomycete *Saccharomyces cerevisiae* and the basidiomycete *R. mucilaginosa*. We subsequently investigated microscopically, whether amoebae actively take up yeast and use it as a food source.

Table 6. List of strains with their origin and strain-specific characteristics; †: could not be tested, culture extinct; °: Max temp (°C) = maximum temperature tolerance in °C; NI = The Netherlands; Loc. = Locomotive; T. = Trophozoite; Cyst dia. (µm) = Cyst diameter (µm)

Strain	Location	Location (Label)	GPS	Elevation (m)	Loc. form	Max temp (°C)	T. length (µm)	T. width (µm)	Cyst dia. (µm)
Sar9	Italy	Sardinia	N 40°46' E 9°10'	181	Mostly limax	< 37	18.2 –	3.6 –	5.6 –
							28.8	5.2	10.8
							23.6 ±	4.6 ±	7.4 ±
							3.4	0.6	1.2
Sar37	Italy	Sardinia	N 40°46' E 9°10'	181	Limax	< 37	20.8 –	3.4 –	5.4 –
							30.0	6.0	8.4
							23.8 ±	4.6 ±	6.6 ±
							2.4	0.6	0.8
Nl64	NI	Veluwe	N 52°06' E 6°00'	57	Mostly flabellate	< 30	21.8 –	3.6 –	5.0 –
							50.8	8.6	7.8
							33.2 ±	5.8 ±	6.2 ±
							10.0	1.4	0.8
Paraf. alta Tib23	Tibet	Mila East Slope	N 29°52' E 92°33'	4149	Limax	†	14.8 –	3.9 –	5.3 –
							31.1	13.4	6.2
							20.9 ±	7.9 ±	5.7 ±
							4.8	3.1	0.4
Tib32	Tibet	Mila Mountaintop	N 29°49' E 92°20'	5033	Mostly flabellate	< 37	21.2 –	5.6 –	5.0 –
							31.2	9.2	8.2
							25.2 ±	6.6 ±	6.6 ±
							4.0	1.2	0.8
Tib50	Tibet	Mila West Slope	N 29°42' E 92°10'	4149	Limax or flabellate	< 37	27.6 –	5.0 –	5.8 –
							36.0	8.8	9.2
							30.0 ±	7.4 ±	8.0 ±
							2.8	1.4	1.0
Tib19 1	Tibet	Mila East Slope	N 29°52' E 92°33'	4149	Limax	< 37	19.4 –	4.2 –	6.4 –
							26.4	7.6	8.6
							23.0 ±	6.2 ±	7.6 ±
							2.2	1.0	0.6

DNA extraction, amplification and sequencing

Petri dishes with dense growth of amoebae were washed 3 times with fresh medium. Subsequently, the medium was replaced with 100 µl of guanidine isothiocyanate buffer. Cells were then scraped off the bottom using a sterile metal cell scraper and the buffer containing amoebae was transferred to an Eppendorf tube. After vortexing, the solution was heated to 72 °C for 10 min, cooled down to room temperature, 200 µl isopropanol was added and left at -20 °C overnight. After vortexing and centrifugation for 15 min at 15,000 rpm the supernatant was removed and replaced by 100 µl 70 % EtOH. After another centrifugation step for 10 min at 15,000 rpm, the supernatant was removed and DNA was resuspended in 50 µl ddH$_2$O. DNA was stored at -20 °C until further use (Maniatis et al. 1982a).

Small subunit (SSU) ribosomal DNA was amplified using the universal eukaryotic primers EukA and EukB (Medlin et al. 1988), while the primers JITS-F and JITS-R were used to amplify and sequence the ITS region, including the 5.8S rDNA (De Jonckheere and Brown 2005). PCR reactions were run in 30 µl volume consisting of 0.6 µl of each primer (10 µM), 0.6 µl nucleotides (10 mM), 2 µl template DNA, 24.5 µl H$_2$O, 3 µl GreenTaq Buffer and 0.15 µl GreenTaq polymerase (5 U * µl^{-1}) (Fermentas, St. Leon-Rot, Germany). The cycling conditions included a 5 min initial denaturation at 95 °C, followed by 30 cycles of 95 °C for 30 s, 50 °C for 60 s, and 72 °C for 120 s, and a final extension at 72 °C for 5 min. 8 µl of the PCR products were enzymatically purified by adding 0.15 µl Endonuclease I (20 U * µl^{-1}, Fermentas, St. Leon-Rot, Germany), 0.9 µl Shrimp Alkaline Phosphatase (1 U * µl^{-1}, Fermentas, St. Leon-Rot, Germany) and 1.95 µl H$_2$O. The resulting mixture was incubated for 30' at 37 °C, followed by 20' at 85 °C. Cleaned PCR products were subsequently sequenced (GATC, Konstanz, Germany) using all amplification end primers and internal ones to obtain full length sequences of the SSU rDNA.

Phylogenetic analyses

Sequences obtained were subjected to BLAST searches to establish taxonomic affiliations. All sequences were aligned using Clustal Omega directly

140

implemented in SEAVIEW v. 4.4.2 (Gouy et al. 2010) together with all closest BLAST hits of cultivated taxa, a subset of sequences of uncultivated taxa and sequences of other heterolobosean genera in order to obtain better resolved phylogenetic affinities of our strains. Two datasets including sequences of the SSU rDNA and the 5.8S rDNA were analysed. 43 sequences were aligned in the SSU rDNA dataset resulting in 1,527 unambiguously aligned positions, while 23 sequences were aligned in the 5.8S rDNA dataset resulting in 145 unambiguously aligned positions. ITS1 and ITS2 could not reliably be aligned and were consequently excluded from phylogenetic analyses of all vahlkampfiids. To explain species differentiation of clones closely resembling *A. spelaea* we aligned the entire ITS region of *A. spelaea*, '*S. palustris*' and six new strains obtained in this study, increasing unambiguously aligned positions to 340.

A general time-reversible model with a proportion of invariable sites and γ distribution (GTR + I + γ) as proposed by jModeltest v. 2.1.3 under the Akaike Information Criterion (Darriba et al. 2012) was used for phylogeny inferences. These were performed directly in SEAVIEW with 5 random starting trees using nearest-neighbour interchange and subtree pruning and regrafting algorithms for tree searching (Guindon and Gascuel 2003, Hordijk and Gascuel 2005). The stability of the clades was assessed using a non-parametric bootstrap with 100 pseudoreplicates using unambiguously aligned positions. Further, a Bayesian analysis was performed using Mr Bayes Version 3.1.2 (Huelsenbeck and Ronquist 2001) with the GTR + I + γ model of substitution. Two runs of four simultaneous Markov chain Monte-Carlo analyses starting from different random trees were performed for 5,000,000 generations (default heating parameters), sampled every 100 generations. Convergence (average deviation of split frequencies < 0.01) was reached after 210,000 generations (SSU dataset) and 300,000 generations (5.8S dataset). All trees before convergence was reached were discarded as burn in and a consensus tree was built from the remaining trees.

Another focused phylogenetic analysis was performed on a clade containing *A. spelaea* to decipher species relations within that clade. Those sequences were aligned using Clustal Omega and subsequently modified manually. Most of the ITS regions could still not reliably be aligned and only the remaining reliable positions shared by all *Allovahlkampfia* spp. were used for subsequent phylogenetic analyses. Maximum likelihood and Bayesian analyses were performed as described above, but only running 2,000,000 generations in the Bayesian analysis as the runs converged after 70,000 generations, leaving sufficient numbers of trees after burn in to construct a consensus tree. This focused alignment is uploaded to TreeBASE (treebase.org), submission ID 15118.

New nucleotide sequences from all Heterolobosea described in this study are available in GenBank under the accession numbers KF547907, KF547908, KF547909, KF547910, KF547911, KF547912 and KF547913 for the SSU-rDNA; KF547914, KF547915, KF547916, KF547917, KF547918, KF547919 and KF547920 for the ITS region including the 5.8S rDNA. Type cultures are deposited at the Culture Collection of Algae and Protozoa (CCAP) under the accession numbers CCAP 2502/1 to CCAP 2502/6.

Results

Cultivation and morphology

All isolated amoebae grew well at room temperature, as well as 30 °C, except for strain NI64 (Table 6). Growth performance of Tib23 could not be tested as the strain went extinct before temperature tests (and any growth tests) could be completed (Table 6).

None of all isolates actively took up and fed on the yeast species tested, i.e. *S. cerevisiae* and *R. mucilaginosa*. Sorocarp formation was not observed under any condition tested. Most of the isolates showed the typical morphology of the Vahlkampfiidae including eruptive lobopodia and an elongated "limax" form during locomotion (Fig. 39 and Fig. 41), although flabellate forms were regularly seen in some. Further, all were uninucleate, possessed a contractile vacuole located at the posterior end, and often produced uroidal structures. These uroids usually attached the amoebae to the substratum, and varied highly in shapes and lengths even in the same amoeba, sometimes branching into finer subpseudopodia with lengths occasionally reaching twice that of amoebae (e.g. Fig. 39e). However, some isolates seemed to differ in their predominantly adopted locomotive form and could be classified into four groups; group I included organisms that always adopted an elongated "limax" shape under directed locomotion (strains Sar37 [Fig. 39B] and Tib191 [Fig. 39F]), group II rarely moved in a typical elongated form most often (~ 70 % of amoebae in directed locomotion under the studied conditions) adopting a more elongated flabellate shape (strains Nl64 [Fig. 39C]and Tib32 [Fig. 39D]), group III almost never formed elongated-limax forms, but usually (~ 90 % of amoebae in directed locomotion under the studied conditions) adopted condensed flabellated forms (strains Sar9 [Fig. 39A] and Tib50 [Fig. 39E]) and group IV adopting a limax shape but morphologically differentiating it from all previously described strains by a more eruptive formation of pseudopodia and the presence of only a small, centrally located nucleolus (Tib23, Fig. 41). The nucleolus of Tib23 reached only about 1/3 of the diameter of the nucleus, while that of the other strains reached about half. Amoebae in stationary or non-directed movement were condensed and formed eruptive pseudopods in all directions, with high intra-strain morphological plasticity that was even present within the same amoeba over time. Amoebae often changed the direction of locomotion after few seconds sometimes in different directions resulting in a strongly elongated and highly variable shape. All morphological characters were similar when observed in Petri dishes or on glass slides.

143

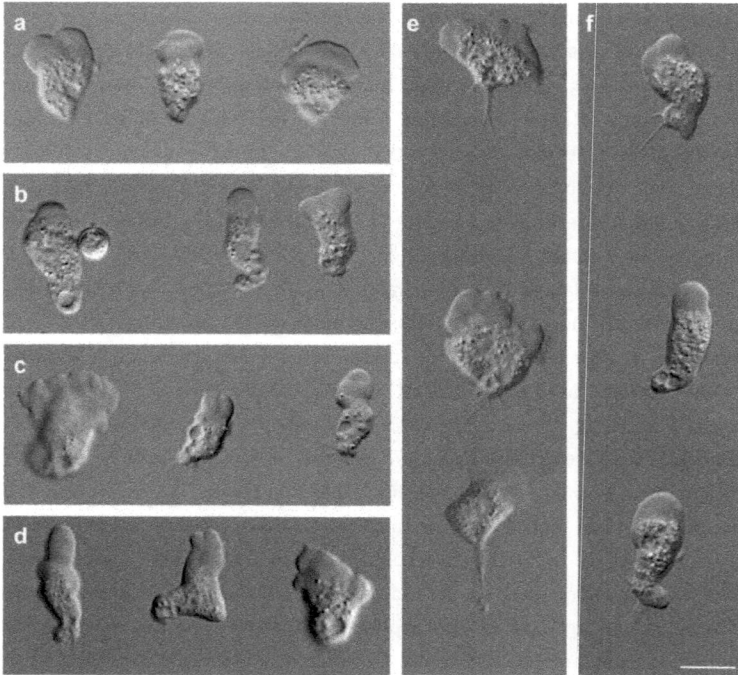

Fig. 39. Differential interference contrast (DIC) pictures showing trophozoites in locomotive form of all new *Allovahlkampfia* strains cescribed in this study, drawn at the same scale. a. Sar9. b. Sar37. Note size difference between the trophozoite and cyst. c. Nl64. d. Tib32. e. Tib50. Note pronounced length of uroidal filaments. f. Tib191. Scale bar: 10 μm.

All strains produced cysts (Fig. 40), with Tib23 again differing strongly in cyst morphology by forming dimorphic cysts, but always of a similar size (Fig. 41C). None of the cultures formed a flagellate stage. After detachment from substratum, floating amoebae immediately settled and continued locomotion without adopting a pronounced floating form. Size differences even within strains were usually profound so size differences between strains were negligible (Table 6).

144

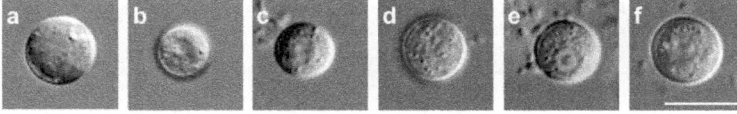

Fig. 40. Differential interference contrast (DIC) pictures showing cysts of all new *Allovahlkampfia* strains described in this study, drawn at the same scale. a. Sar9. b. Sar37. c. NI64. d. Tib32. e. Tib50. f. Tib191. Scale bar: 10 μm.

Despite some morphological and locomotive differences the lack of reliable uniform and constant characteristics, which is typical for Vahlkampfiidae, necessitate an identification based on molecular analyses (Brown and De Jonckheere 1999, De Jonckheere and Brown 2005).

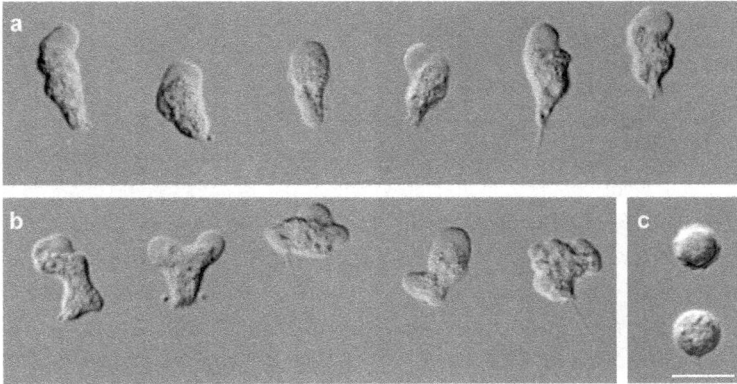

Fig. 41. Differential interference contrast (DIC) pictures showing *Parafumarolamoeba alta* Tib23. a. Locomotive forms. Note pronounced uroidal filaments. b. Amoeba in indirect locomotion forming eruptive pseudopodia everywhere at cell body. c. Irregular cysts. Upper cyst wrinkled, lower cyst round. Scale bar: 10 μm.

Phylogenetic analyses

Full length sequences of the ITS region, including the 5.8S rDNA, were obtained for all strains. The sizes of ITS1 and ITS2 differed strongly between most strains, while the 5.8S rDNA were always 161 bp long, with the exception of Sar37 and Tib191, where it was 2 bp longer (Table 7).

145

Table 7. Length (bp) of the SSU rDNA available, presence and length of SSU group I intron, and length of ITS1, 5.8S and ITS2; -: absent; NA: information not available

Strain	SSU rDNA	Group I intron	ITS1	5.8S	ITS2	Origin
Sar9	2038	-	163	161	128	Italy
Sar37	2026	-	117	163	130	Italy
NI64	2060	1358	155	161	127	The Netherlands
Tib32	1962	-	154	161	127	Tibet
Tib50	2032	-	153	161	125	Tibet
Tib191	2010	-	129	163	162	Tibet
'S. palustris'	671	-	339	161	152	USA
Soil amoeba AND12	2178	-	NA	NA	NA	Spain
Allovahlkampfia BA	2125	1313	NA	NA	NA	Canada
Allovahlkampfia OSA	2129	-	NA	NA	NA	Canada
A. spelaea	2138	-	166	161	108	Slovenia
Paraf. alta Tib23	1738	-	163	157	162	Tibet
F. ceborucoi	2371	-	133	154	125	Mexico
Parav. ustiana	1903	-	129	158	316	Czech Republic
Parav. lenta	NA	-	120	159	397	UK
Parav. francinae	1880	-	129	158	270	USA
Parav. sp. A1PW2	1921	-	NA	NA	NA	Germany
Parav. sp. LA	1909	-	NA	NA	NA	Canada
Parav. sp. Ii3	1876	-	NA	NA	NA	USA

Nearly full length sequences of the SSU rDNA, except for the very beginning and end, were obtained. Interestingly, one of our isolates (NI64) contained a group I intron, which is only present in one other strain investigated in this group (Heterolobosea BA). The intron in strain NI64 was located at the same position as in Heterolobosea BA, but sequence dissimilarity exists not only in the intron region (21.1 % bp dissimilarity), but also in the remaining sequence of the SSU rDNA (1.2 % bp dissimilarity) (Table 8). The group I intron in both strains are twin-ribozymes (Tang et al. 2014) and contain an open reading frame, coding for a homing endonuclease with a His-Cys box (Wikmark et al. 2006) with 38 % dissimilarity in the AA sequence.

Table 8. Percentage identity matrix obtained with manually modified Clustal Omega alignments of the SSU rDNA (below the diagonal) and 5.8S rDNA (above the diagonal) in the Allovahlkampfia clade; NA = no comparison possible due to lack of a reference sequence

	Sar 9	Sar 37	NI 64	Tib 32	Tib 50	Tib 191	'S. palustris'	AND12	BA	OSA	A. spelaea
Sar9	100	95.5	99.4	100	97.5	96.3	93.8	NA	NA	NA	99.4
Sar37	95.1	100	94.6	95.0	94.0	98.8	93.4	NA	NA	NA	95.4
NI64	98.4	94.9	100	99.4	98.1	95.7	93.2	NA	NA	NA	98.8
Tib32	98.4	94.9	99.9	100	97.5	96.3	93.8	NA	NA	NA	98.4
Tib50	96.0	96.0	95.6	95.7	100	95.0	92.5	NA	NA	NA	98.1
Tib191	95.4	98.3	95.0	94.9	96.0	100	93.2	NA	NA	NA	96.9
'S. palustris'	89.0	90.3	87.8	87.8	90.0	90.4	100	NA	NA	NA	94.3
Soil amoeba AND12	90.2	90.2	89.9	89.2	91.2	90.6	83.7	100	NA	NA	NA
A. BA	99.1	95.5	88.8	89.1	95.8	95.1	97.2	90.3	100	NA	NA
A. OSA	96.0	96.0	95.5	95.6	99.8	96.0	89.8	90.6	95.9	100	NA
A. spelaea	96.1	96.6	95.3	95.3	97.0	96.9	89.2	90.3	95.9	97.1	100

Blast searches based on the SSU rDNA indicated closest affinities of strains Sar9, Sar37, NI64, Tib32, Tib50 and Tib191 with *Allovahlkampfia spelaea* (Walochnik and Mulec, 2009), unnamed Heterolobosea BA and OSA (Shut and Gray, unpublished), and *Solumitrus palustris* (Anderson et al. 2011). Blast searches based on the entire ITS region including the 5.8S rDNA similarly detected closest affinities of Sar9, Sar37, NI64, Tib32, Tib50 and Tib191 with *A. spelaea and 'S. palustris'*. Blast hits of Tib23 based on the SSU rDNA matched best with several uncultured eukaryotes, followed by *Fumarolamoeba ceborucoi* (De Jonckheere et al. 2011b). The latter, *F. ceborucoi,* represented the best hit when a Blast search was based on the ITS region.

Maximum likelihood and Bayesian phylogenetic analyses of the SSU rDNA alignments (Fig. 42) confirmed a strongly supported group of our six isolates with *A. spelaea, 'S. palustris'* and Heterolobosea BA and OSA. The *Allovahlkampfia* clade formed a strongly supported sister clade to species of the genus *Acrasis* with the sequence of the uncultivated taxon "soil amoeba AND12" (Lara et al. 2007a) branching outside the *Allovahlkampfia* spp.

Strain Tib23 formed a clade with several uncultured sequences (Valster et al. 2009, Valster et al. 2010, Valster et al. 2011, Farhat et al. 2012), with *Fumarolamoeba* (De Jonckheere et al. 2011b) as the most closely related genus, while *Paravahlkampfia* formed a separate branch.

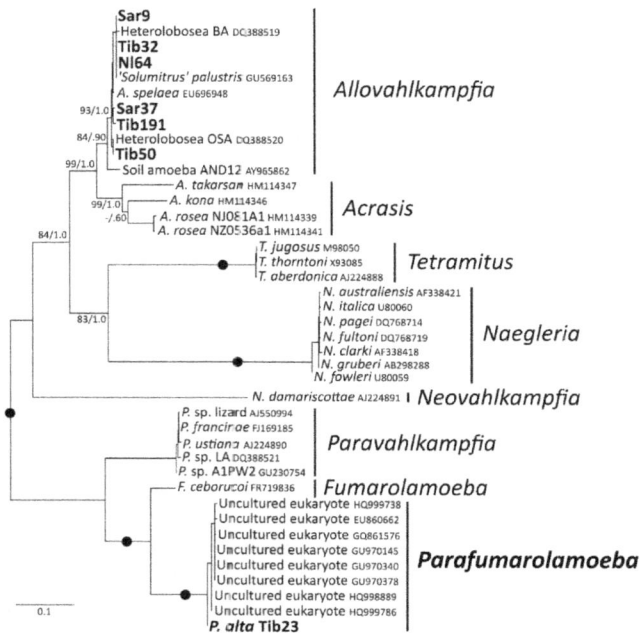

Fig. 42. Maximum likelihood tree based on SSU rDNA sequences. Support values at each node presented for RAxML / BI. RAxML ML bootstrap values and BI posterior probabilities equal to 100 % / 1.00, respectively, are represented by a black dot, whereas – is placed for unrecovered topologies and support values < 50 % (RAxML) / 0.50 (BI).
Note: Bootstrap support values within most genera omitted for convenience; relationships within *Allovahlkampfia* discussed in text. GenBank accession numbers of SSU sequences used in analyses listed next to taxon names. Scale bar represents evolutionary distance in changes per site.

Phylogenetic analyses based on the 5.8S rDNA resulted in a similar pattern (Fig. 43); the Allovahlkampfia group was most closely related to *Naegleria* and *Tetramitus*, while Tib23, *Fumarolamoeba* and *Paravahlkampfia* genera formed strongly supported branches with isolate Tib23. The six other new isolates formed a clade with *A. spelaea*, including '*S. palustris*'. The Allovahlkampfia clade was inversed; Sar 37 and Tib 191 were most diverging from the others placing those strains as most closely related to *Naegleria*. In contrast to the SSU rDNA phylogeny, where Tib50 occupied this position, the analysis on the 5.8 S

148

rDNA placed this strain deepest inside the *Allovahlkampfia* group. However, individual strains grouped similar to the SSU rDNA phylogeny, e.g. Sar9 as sister to both Nl64 and Tib32 and Sar37 closely related to Tib 191. An additional alignment of the entire ITS region of this clade was conducted to reveal relations within this clade which provided a higher resolution but in general supported the other analyses (Fig. 44).

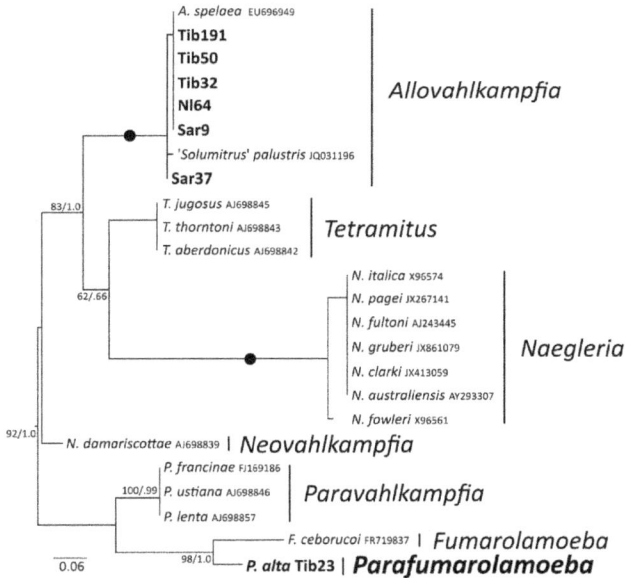

Fig. 43. Maximum likelihood tree based on 5.8S rDNA sequences. Support values at each node presented for RAxML / BI. RAxML ML bootstrap values and BI posterior probabilities equal to 100 % / 1.00, respectively, are represented by a black dot, whereas – is placed for unrecovered topologies and support values < 50 % (RAxML) / 0.50 (BI).
Note: *Acrasis* could not be included as the 5.8S are not available. Phylogenetic relations within most genera omitted for convenience; relationships within *Allovahlkampfia* discussed in text. GenBank accession numbers of the sequences used in analyses listed next to taxon names. Scale bar represents evolutionary distance in changes per site.

Sequence similarities for the SSU rDNA of our isolates most closely resembling *A. spelaea* in phylogenetic analyses ranged from 95.3 to 96.9 % (Table 8), indicating that they belong to the genus *Allovahlkampfia*, but that they should

149

be treated as different species. The high sequence similarity of the 5.8S rDNA, between 95.4 and 99.4 % (Table 8), confirmed that they belong to the same genus *Allovahlkampfia*. Heterolobosea BA and OSA had similar high sequence similarities of the SSU rDNA compared to *A. spelaea*, 95.9 and 97.1 %, respectively. On the other hand, '*S. palustris*' showed somehow lower sequence similarities for both molecules, 89.2 and 94.3 %respectively, but this could be due to the short sequence length of the SSU rDNA reported and the possibility of sequencing errors, as several mismatches were found in highly conserved regions in both genes. Strain AND12, with 90.3 % sequence similarity of the SSU rDNA compared to *A. spelaea*, and 89.2 to 91.2 % compared to our new isolates probably might represent a different genus other than *Allovahlkampfia* or a very basal species of *Allovahlkampfia*.

Strain Tib23 had sequence similarities of 85.9 % with *F. ceborucoi* and between 80.7 and 80.9 % with the *Paravahlkampfia* spp. for the SSU rDNA (Table 9) and even lower for the 5.8S rDNA, 80.7 and 70.4 %, respectively (Table 9). Therefore, strain Tib23 should be considered to represent a new genus.

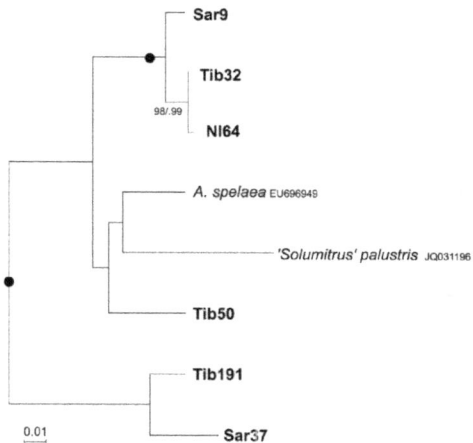

Fig. 44. Maximum likelihood tree based on the entire ITS region focusing on *Allovahlkampfia* strains. Support values at each node presented for RAxML / BI. RAxML ML bootstrap values and BI posterior probabilities equal to 100 % / 1.00, respectively, are represented by a black dot, whereas – is placed for unrecovered topologies and support values < 50 % (RAxML) / 0.50 (BI).

Table 9. Percentage identity matrix obtained with manually modified Clustal Omega alignments of the SSU rDNA (below the diagonal) and 5.8S rDNA (above the diagonal) in the *Fumerolamoeba-Paravahlkampfia* clade; NA = no comparison possible due to lack of a reference sequence

	Paraf. alta Tib23	F. ceborucoi	Parav. ustiana	Parav. francinae	Parav. sp. A1PW2	Parav. sp. LA	Parav. sp. li3	Parav. lenta
Paraf. alta Tib23	100	80.7	72.3	71.6	NA	NA	NA	70.4
F. ceborucoi	85.9	100	68.5	67.8	NA	NA	NA	68.1
Parav. ustiana	80.7	83.1	100	99.3	NA	NA	NA	97.9
Parav. francinae	80.9	83.2	99.7	100	NA	NA	NA	98.6
Parav. sp. A1PW2	80.7	82.9	98.3	98.7	100	NA	NA	NA
Parav. sp. LA	80.8	82.6	98.7	98.9	98.0	100	NA	NA
Parav. sp. li3	80.9	82.9	99.5	99.8	98.4	98.7	100	NA
Parav. lenta	NA	NA	NA	NA	NA	NA	NA	100

Generally, all strains showed distinct molecular sequence patterns in different parts of helix 23 of the SSU rDNA, i.e. the V4 region. Shared molecular signatures provided further evidence of the classification mentioned above; a clade separating the *Allovahlkampfia* clade including soil amoeba AND12 from *Acrasis* was supported by several molecular patterns, which are unique for the respective clades. For instance, the motif 5' - TACACTT - 3' in helix 8 of the SSU rDNA is specific for the *Allovahlkampfia* clade, whereas the motif in *Acrasis* is 5' - TACTCGT - '3. Another example is the addition of a C in helix 17 of the SSU rDNA specific in the *Allovahlkampfia* clade, where there is a gap in *Acrasis*.

Similar patterns were detected for all other clades denoted above, e.g. the separation of *Fumarolamoeba*, Tib23 and several uncultured environmental sequences from *Paravahlkampfia*, e.g. the profile 5' - AAGGTTTGG - 3' shared by *Fumarolamoeba*, Tib23 and the uncultured environmental sequences at the end of helix 17 is strikingly different from the pattern of all *Paravahlkampfia* spp., which share the sequence 5' – TCGGTTTGC - 3'. More sequence patterns distinguishing *Fumarolamoeba*, Tib23, uncultured sequences and *Paravahlkampfia* are shown in Table 10.

Table 10. SSU rDNA sequence motifs of *Parafumarolamoeba alta* Tib23, closely related uncultivated clones, *Fumarolamoeba ceborucoi,* and *Paravahlkampfia* spp.; NA : not available SSU region; Unc. euk. = Uncultured eukaryote

Helix #	Starting position in *F. ceborucoi*	*Paraf. alta* Tib23	Unc. euk. (HQ998889, HQ999738, HQ999786, EU860662, GU970145, GU970340, GU970378)	Unc. euk. GQ861576	*F. ceborucoi*	*Parav.* spp.
	44	GTCTT	GTCTT	NA	GCTAG	GCTCG
7	116	CTAGCTTCTTTT AT	CTAGCTTAAAG TTAA	NA	CTAGCTTCTTTT AT	CTAGTTTWTCT TAC
9-10	165	GAACCAAAGCT	GAACCAAAGCT	NA	GAATCAACGTC	GCTCAAAAGCC
11	268	GGA	GGA	NA	GAA	GAG
12	319	GGCCGTCA	GGCCGTCA	NA	GGCTATCA	GACACTTA
13	335	GAAAATTGGGG TTT	GAAAATTGGGG TTT	NA	GGGAATCAGT GTTT	GGGAATCAGT GTTC
15	381	T	T	NA	C	G
16	426	AT	AT	NA	AT	AATT
17	454	TCCTTC	TCCTTC	NA	TCCTTA	ACCTCA
18	505	CAAATT	CAAATT	NA	CAAACT	TAAATC
18-19	523	ACA	ACA	NA	ACA	TCG
32	2245	T	NA	T	T	C
35-36	1733	GCGGGA	NA	ACGGGA	GCGGGA	GCGGGG
36	1751	AT	NA	AT	AC	TT
36-38	2047	G	NA	G	G	T
37	1761	ATGAG	NA	ATGAG	GTTAA	GGAGA
37	1780	ATATG	NA	ATATG	ATTTG	GA
37	1799	TTTTGGAA	NA	TTTTGGAA	CTCTGGAA	CTTTGATT
43	1882	TTAAC	NA	TTAAC	TTAAC	CTAAT
43-44	1978	A	NA	A	A	G
44	2010	A	NA	A	A	G
45	2080	C	NA	C	T	T
45-46	2084	G	NA	G	C	A
47-48	2190	A	NA	A	G	G
48-49	2234	T	NA	T	T	C

Discussion

New isolates

We provide molecular evidence that the genus *Allovahlkampfia* shows high sequence heterogeneity as all SSU rDNA and ITS sequences obtained were dissimilar. They might represent different species but until a molecular definition is determined for *Allovahlkampfia* as was done in the genus *Naegleria* (De Jonckheere 2004), no new species names are proposed at this stage. Until now, *A. spelaea* was the only cultivated and described species in this genus (Walochnik and Mulec 2009). Profound differences to the original definitions of the genus *Allovahlkampfia* were found. Most strikingly were cyst sizes that in all of our strains reached only about a third of the dimensions previously described for *A. spelaea* and were therefore in a similar range of those reported for '*S. palustris*', but also the wide range of locomotive forms add to the typical "limax" form adopted by *A. spelaea* (Walochnik and Mulec 2009) and also '*S. palustris*' (Anderson et al. 2011). Also differences in temperature tolerances between *Allovahlkampfia* strains and a temperature tolerance of 37 °C, while that of *A. spelaea* was reported to be 30 °C (such as in strain NI64) add to the variability within *Allovahlkampfia*. Last, molecular sequence information obtained shows striking variability within the genus *Allovahlkampfia*. The abovementioned differences could indicate differences in ecological functioning or niche specialization between strains of *Allovahlkampfia*.

Morphology is not a reliable character to differentiate species and even genera in different clades of amoebae such as in the amoebozoan class Variosea (Berney et al. 2015) and also in the heterolobosean Vahlkampfiidae. In both clades, only sequence data has been shown to provide reliable information about identities of individual taxa. As such, morphologically similar species formerly united in the genus *Vahlkampfia* are now being placed in several distinct genera based on phylogenetic information (Brown and De Jonckheere 1999, De Jonckheere and Brown 2005). Sequence differences of SSU and even more ITS regions have also been described as reliable characters to distinguish species in other Vahlkampfiidae such as in *Tetramitus* (De Jonckheere and Brown 2005, Robinson et al. 2007). Examples of (nearly) identical SSU with profound differences in ecological functioning are reported for chrysophytes and cercozoan flagellates, and the authors therefore suggested the presence of cryptic species (Bass et al. 2007, Boenigk et al. 2007). Some morphologically different species of the Vannellidae located in the supergroup Amoebozoa vary by as little as 3 bp in the entire SSU (Smirnov et al. 2007) and different *Cochliopodium* spp. share identical sequences (Geisen et al. 2014c).

153

We have performed several independent PCRs and sequence reactions to build consensus sequences, always supporting sequence differences observed between the strains. Therefore, we are confident that the sequence differences observed are reliable positions to differentiate them, and not errors introduced by PCR or sequencing. Furthermore, the most closely related strains based on molecular information NI64 and Tib32 have differences in temperature tolerance, suggesting even functional differences between these two strains. These differences between strains could indicate that NI64 and Tib32 represent different species, which could be the case for the other isolates as these differ even more in their sequences as the former two.

In addition to the strains belonging to the genus *Allovahlkampfia*, we isolated a strain which cannot be included in any described genus. It forms a clade with several environmental sequences and the most closely related genera are *Fumarolamoeba* and *Paravahlkampfia*. We propose the genus name *Parafumarolamoeba*, in line with the nomenclature of other genera in Vahlkampfiidae (Brown and De Jonckheere 1999), as Tib23 most closely resembles *Fumarolamoeba* and the species name *alta*, indicating the high altitude from where the type strain originates. The large numbers of similar published environmental sequences indicate that species related to *Parafumarolamoeba alta* are common soil inhabitants, but strains or species within that genus seem to be negatively selected against in cultivation based efforts. Having no remaining culture *Parafumarolamoeba alta* undoubtedly is a disadvantage, but we obtained sequenced of the entire SSU and ITS region, have pictures of the morphology and have frozen DNA as a type material. Therefore, we conclude that a formal genus description of *Parafumarolamoeba alta* is useful in order to assign sequences obtained in the increasing number of environmental sequencing surveys (Lara et al. 2007a, Bates et al. 2013).

Published sequences

'*Solumitrus palustris*' has recently been erected as a new genus closely related to *Allovahlkampfia* (Anderson et al. 2011). Brown et al. (2012) put '*S. palustris*' between inverted commas and placed it into a single clade together with the genus *Allovahlkampfia*. They suggested that it could eventually be considered as a species of *Allovahlkampfia*, which was further supported by phylogenetic analyses of another study (Harding et al. 2013). Also our phylogenetic analyses of both 5.8S and SSU rDNA, which included several other related species, consistently placed '*S. palustris*' within the *Allovahlkampfia* clade (Fig. 42 and 43, respectively). Our new *Allovahlkampfia* isolates had 95.3 to 97.0 % sequence similarities with *A. spelaea*. '*S. palustris*' had lower sequence similarities of 89.2 % for the SSU rDNA (Table 8) and 94.3 % for the 5.8S rDNA (Table 8) compared to the type strain. As especially the beginning and end of the sequence of '*S. palustris*' showed several mismatches in highly conserved regions we suspect these to represent sequencing errors in both molecules. Only a re-sequencing of '*S. palustris*' will allow deciphering the true phylogenetic placement of this taxon among other strains inside *Allovahlkampfia* but we agree with other authors to that it should be included as a species of *Allovahlkampfia* (Brown et al. 2012, Harding et al. 2013).

In their trees Brown et al. (2012) and Harding et al. (2013) have also placed the Heterolobosea BA and OSA into *A. spelaea*. Our phylogenetic analyses confirm that they should be included within the genus *Allovahlkampfia*, as the SSU rDNA sequences of strains BA and OSA showed sequence similarities of 95.9 and 97.1 %, respectively to *A. spelaea* (Table 8). A sequence of the entire ITS region of these two strains is highly desirable in order to support phylogenetic relationship of other strains within the genus *Allovahlkampfia*.

Phylogenetic implications

Individual strains formed several distinct branches consistently in the SSU and 5.8S rDNA (and *Allovahlkampfia* focused ITS) phylogeny which might suggest the existence of several distinct species. As mentioned above, even the most closely related strains in our study based on phylogeny, i.e. Tib32 and NI64 are distinct not only by the presence of a long group I intron in NI64, but also in the differential growth ability at different temperatures. With the increasing use of high-throughput sequencing methods, the community structure of protists, such as heterolobosean allovahlkampfiids can be studied in more detail. However, not only the presence of introns, but also strongly diverging and generally long SSU rDNA shared by allovahlkampfiids negatively select against

155

these taxa (Geisen et al. 2015a)and these taxa will therefore remain undersampled. Only targeted approaches using (allo-)vahlkampfiid primers or primer-free metatranscriptomic approaches (Geisen et al. 2015b) will eventually allow studying the community structure of these undersampled protists in soils.

Soil amoeba AND 12 (Lara et al. 2007a) is sister to all *Allovahlkampfia* strains and could be a separate genus that bridges Allovahlkampfia and the genus Acrasis as proposed by De Jonckheere et al. (2011b). Soil amoeba AND 12 is of specific interest as it bridges the genus *Allovahlkampfia*, where only once a sorocarp formation has been reported in strain BA (Brown et al. 2012). No stage of soil amoeba AND12 that resembled an *Acrasis* slime mold has been described yet. Therefore it is possible that the ability to form sorocarps might have been present in the progenitor of all organisms within that family, but has been lost in several species inside the genus *Allovahlkampfia*. It is also possible that individual *Allovahlkampfia* species only rarely form sorocarps and under highly specific conditions which are not yet known. But as currently sorocarp formation in *Allovahlkampfia* seems rare, at least under laboratory conditions, and detailed information on species that bridge the genera *Acrasis* and *Allovahlkampfia* are lacking, further research is needed in order to provide further evidence for the current unification of *Acrasis* and *Allovahlkampfia* into the family Acrasidae (Brown et al. 2012). Interestingly, however, is that no *Allovahlkampfia* strain included in this study nor other previously cultivated ones consume fungi, while fungi are a common food source for *Acrasis* spp. (Brown et al. 2012), which could indicate functional differences between the genera *Acrasis* and *Allovahlkampfia*. As also most morphological and molecular characters differ strongly between both genera, the combination of both genera into the single family Acrasidae (Brown et al. 2012) might be worth investigating in more detail in the future studies, such as by obtaining ITS sequence information of *Acrasis* spp. to construct vahlkampfiid wide phylogenetic inferences.

Parafumarolamoeba alta undoubtedly represents a new genus based on phylogenetic information from both the 5.8S and SSU rDNA. Based on morphology *Parafumarolamoeba alta* shares several features with the closely related genera *Paravahlkampfia* and *Fumarolamoeba* such as high eruptive activity in the process of pseudopod formation and the presence of well-separated hyaline and granular regions (Fig. 41). The presence of pronounced uroidal filaments, a single nucleus and a limax shape during locomotion differentiates *Parafumarolamoeba alta* from *Fumarolamoeba* and makes it similar to *Paravahlkampfia*. Cyst sizes are, however, smaller than reported for any *Paravahlkampfia* species and in the lower range of cysts of *Fumarolamoeba*. Unfortunately, the strain went extinct before further morphological investigations and growths tests could be performed, so we are unable to compare specific features distinguishing it from other genera. However, morphological information rarely allows identifications in the Vahlkampfiidae, while the phylogenetic information from both the SSU rDNA and 5.8S genes prove that *Fumarolamoeba* is the sister genus of *Parafumarolamoeba*, and both are strongly different from the genus *Paravahlkampfia*.

Considerable sequence dissimilarity suggests that *Parafumarolamoeba alta* and several uncultured sequences (Valster et al. 2009, Valster et al. 2010, Valster et al. 2011, Farhat et al. 2012) should be considered as a separate genus, most likely containing several independent species.

Ecological implications

Strains Tib32, Tib50 and Tib191, and the new genus *Parafumarolamoeba alta* were isolated from extreme altitudes in Tibet. Interestingly, De Jonckheere (2006) detected distinct species in the genus *Naegleria* that were identical in the arctic and sub-Antarctic, but have not been detected anywhere else. We suspect that these particular *Naegleria* spp. might also be present at high altitude with similar low temperatures, but we were unable to isolate any *Naegleria* strain from the high altitude in Tibet. Maybe this is due to the isolation method used in our investigation, which is not the common procedure for isolating *Naegleria* strains. Also Robinson et al. (2007) described two closely related *Tetramitus* species both from cold and alpine sites.

Species of the class Heterolobosea have often been isolated from extreme environments such as low pH (Amaral-Zettler et al. 2002, Baumgartner et al. 2009), high salinity (Park and Simpson 2011) and extreme heat (De Jonckheere et al. 2011a) or cold (De Jonckheere 2006). While *Allovahlkampfia* strains Tib32, Tib50 and Tib191 were isolated at high altitudes, closely related strains Sar9 and Sar37 originated from hot and dry soil on Sardinia. Also nearly sequence identical Nl64 and Tib32 were isolated at very distant locations from soils and the apparently closely related 'S. *palustris*' was isolated from temperate freshwater.

Also taxa in the genus *Tetramitus* were described from extremely diverse environments such as T. *thermoacidophilus* (Baumgartner et al. 2009), which is an extreme thermophile (54 °C) and acidophile (pH 1 to 5), while most of the closely related *Tetramitus* spp. grow under 'temperate' conditions. Therefore it seems that closely related heterolobosean strains have a remarkably wide tolerance to diverse and often extreme conditions.

Twin-ribozyme introns in the genus *Allovahlkampfia*

A twin-ribozyme with a His-Cys box, similar to that found in the genus *Naegleria*, has been described in *Allovahlkampfia* strain BA (Tang et al. 2014). Alignment of the intron found in strain NL64 with that of strain BA (Fig. 45) showed that it is a similar twin-ribozyme with a length of 256 AA, compared to 252 AA in strain BA, and differing by 98 AA to that in strain BA. The similarity of 62 % is much lower than the similarities within the genus *Naegleria* where it varies from 81 % to 100 %. Since the twin-ribozymes seem to be an ancestral acquisition which evolves with the species differentiation (De Jonckheere 1994), NL64 might in fact be considered a different species from BA, as is likely to be the case for the other *Allovahlkampfia* isolates.

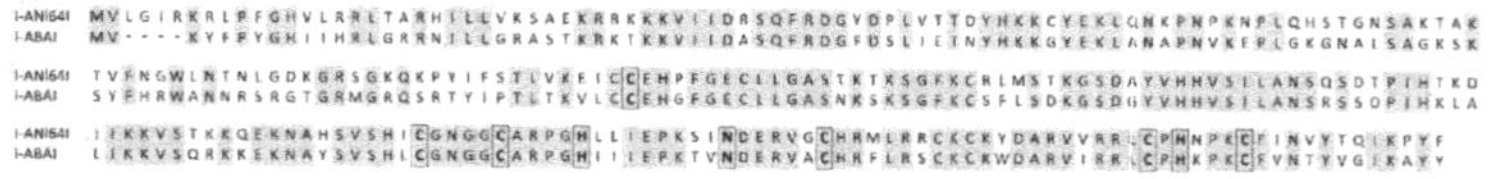

Fig. 45. Alignment of the His-Cys homing endonucleases encoded by strain BA and NL64 of the genus *Allovahlkampfia*. Identical residues are boxed, in grey and deletions are indicated by dashes. Functionally important C, H and N residues involved in zinc binding and catalysis (His-Cys box) are boxed.

Diagnosis

Genus *Allovahlkampfia*, emended

Uninucleated amoebae, without a known flagellate stage, sharing eruptive pseudopodia formed during locomotion. Locomotive form limax or flabellate with eruptive pseudopodia. Variable uroidal filaments of different lengths often present during locomotion. Size and shape of amoebae strongly differing even within strains. Also the spherical cysts without pores produced in culture vary strongly in size. Usually no distinct multicellular stage formed.

No growth observed at 37 °C.

Food: bacterivorous

Habitat: Soil, freshwater or tree bark

Molecular patterns: sequences of both SSU and 5.8S rDNA with > 95 % similarity between strains, with diverse molecular sequence patterns such as motif 5' - TACACTT - 3' in helix 8 only shared by members in this clade. Closest phylogenetic relationship based on the SSU rDNA with soil amoeba AND12 (~ 10 % bp difference) and the genus *Acrasis* (> 15 % bp differences). All species show sequence differences in the V4 region of the SSU rDNA.

Type species *A. spelaea* Walochnik and Mulec 2009

Genus *Parafumarolamoeba*, new genus

Uninucleated amoebae, no known flagellate stage, exhibiting high eruptive activity while forming pseudopodia. Locomotive form limax. Uroidal filaments, sometimes branching, formed during locomotion, often longer than actual cell body. Cysts of almost equal diameter but of different shape varying from round to wrinkled forms.

Food: bacterivorous

Habitat: Soil

Type species *Parafumarolamoeba alta*, new species

159

Trophozoites 14.8 - 31.1 µm (average 20.9 µm) in lengths and 3.9 - 13.4 µm (average 7.9 µm) in widths. Locomotion with highly eruptive pseudopodia in limax forms. Pronounced uroidal filaments formed. Cysts differing little in size (diameter 5.3 - 6.2 µm, average 5.7 µm), but differing in shape (wrinkled to round). Very distinct molecular sequencing profiles both in the entire ITS region, including the 5.8S rDNA, and SSU rDNA (EBI accession N°: KF547913 and KF547920, respectively).

Type locality: type strain Tib23 isolated from a high altitude soil in Tibet (N29°52', E92°34) at 4149 m altitude.

Reference material: the culture went extinct, but frozen DNA as a type material is available upon request.

Acknowledgements

Supported with research grants involved in the EU-project 'EcoFINDERS' No. 264465.

Part 2

Diversity analyses characterizing soil protist communities with four different techniques

Part 2 – Chapter 5

Soil water availability strongly alters the community composition of soil protists

Geisen Stefan[1,5], Bandow Cornelia[2,3,4], Römbke Jörg[2,3], *Bonkowski Michael[1]*

[1] Department of Terrestrial Ecology, Institute for Zoology, University of Cologne, Zülpicher Str. 47b, 50674 Cologne, Germany

[2] LOEWE Biodiversity and Climate Research Centre BiK-F, Frankfurt/Main, Germany

[3] ECT Oekotoxikologie GmbH, Flörsheim, Germany

[4] Goethe University Frankfurt, Department Aquatic Ecotoxicology, Frankfurt/Main, Germany

[5] present address: Department of Terrestrial Ecology, Netherlands Institute for Ecology (NIOO-KNAW), 6708 PB Wageningen, The Netherlands

Abstract

Drought and heavy rainfall are contrasting conditions expected to result from increasingly extreme weather during climate change; and both scenarios will strongly affect the functioning of soil systems. However, little is known about the specific responses of soil microorganisms, whose functioning is intimately tied to the magnitude of the water-filled pore space in soil. Soil heterotrophic protists, being important aquatic soil organisms are considered as key-regulators of microbial nutrient turnover. We investigated the responses of distinct protist taxa to changes in soil water availability (SWA) using a modified enumeration technique that enabled quantification of protist taxa up to genus level. Our study revealed a non-linear shift of protist abundance with decreasing SWA and this became apparent at a maximum water-filled pore size of $\leq 40\,\mu m$. Generally, taxa containing large specimen were more severely affected by drought, but responses to either drought or rewetting of soils were not uniform among taxa. Changes in water availability may thus affect the functioning of key taxa and soil ecosystems long before aboveground "drought" effects become apparent

Introduction

Global temperature has increased and is expected to further increase in the coming century, with annual daily maximum temperatures rising by about 3°C by mid-21st century and by about 5°C by the late 21st century, resulting in more frequent and extreme drought events in many parts of the world (IPCC 2012, Sherwood et al. 2013). The term 'drought' is generally associated with the damage of plants due to the lack of soil water (Kramer 1983), but limited soil water availability (SWA) can impair the function of soil ecosystems long before symptoms become visible aboveground. This is especially true for processes performed by microbial soil organisms, whose functions are intimately tied to the magnitude and connectivity of water films around soil particles. We are, however, still largely ignorant of specific responses of soil organisms to these global change phenomena (de Vries et al. 2012a, de Vries et al. 2012b, Bradford 2013). However, this topic has gained more interest in the last decade and soil microbial communities with altered SWA were investigated, mostly showing changes in bacterial and fungal communities (Fierer et al. 2003, Evans and Wallenstein 2012, Evans and Wallenstein 2014, Fuchslueger et al. 2014).

In contrast to bacteria and fungi, their microbial predators, i.e. soil protists are largely ignored in these studies. Due to their high biomass and with estimated annual production rates of more than 100 kg ha^{-1} (Bouwman and Zwart 1994), protists are assumed to play a key role in carbon (C) and nutrient cycling in soils (Schröter et al. 2003, Christensen et al. 2007, Crotty et al. 2012b). Protists are the most basal microbial consumers, being a fundamental source for C transfer to higher trophic levels in the soil food web (Crotty et al. 2012a). Direct effects of protists result from their high grazing impact on microbial communities, but more important appear to be indirect effects of protists that lead to a stimulation of microbial turnover and respiration (Clarholm 1985, Bonkowski 2004, Anderson 2008) and plant performance (Koller et al. 2013). For example laboratory experiments with planted soil have shown that consumption of microbial biomass by protists led to a 20 - 40 % increased microbial activity and CO_2-C release (Alphei et al. 1996, Rosenberg et al. 2009) and microbial liberation of CO_2 from decomposing plant litter increased up to 100 % in presence of protists (Bonkowski et al. 2000a).

Despite living in soil, protists are aquatic organisms, and their function ultimately depends on the availability of water in the three-dimensional pore space (Anderson 2000, Griffiths et al. 2001). Decreasing SWA has been shown to reduce protist replication rates due to limited mobility of protist grazers in the microvolumes of soil water and hence reduced accessibility to bacterial prey (Darbyshire 1976). This can result in significantly negative effects on soil nutrient cycling and plant growth (Kuikman et al. 1991). Protists, are extremely diverse and individual taxa differ fundamentally based on phylogenetic relatedness, morphology and behaviour (e.g. Cavalier-Smith 1998, Adl et al. 2012, Saleem et al. 2013). The size of individual taxa can differ by as much as three orders of magnitude in soil (Foissner 1998, Finlay 2002, Glücksman et al. 2010). Consequently, taxon-specific dependencies on SWA are likely, simply because large, free-swimming taxa will be more vulnerable to desiccation than small, surface-associated forms. However, it is largely unknown how complex, natural protist communities respond to altered soil moisture regimes.

Taxonomic studies on natural populations of soil protists have mainly been restricted to groups with larger specimens that dominate the upper humus layers and share fixed, readily determinable morphological characters, such as testate amoebae and ciliates (e.g. Bamforth 1971, Foissner 1987a, 1999, Bamforth 2007, Krashevska et al. 2007). In comparison, knowledge on the taxonomic composition of communities of flagellates and naked amoebae is extremely limited, despite these groups containing a huge diversity of soil species, and vastly outnumbering other protist groups in the mineral soil horizons (e.g. Elliott and Coleman 1977, Finlay et al. 2000, Scharroba et al. 2012, Domonell et al. 2013)   . Unlike suspension and filter feeders such as many ciliates, and free-swimming flagellates, naked amoebae and amoeboid flagellates directly graze on bacterial colonies and biofilms attached to substrates (Darbyshire et al. 1989, Parry et al. 2004, Weitere et al. 2005, Böhme et al. 2009), and their flexible bodies seem particularly suited to survive in the tiny water films around mineral particles. It has been shown that they still can access prey in water-filled soil pores of only 2 μm in diameter with help of their elongate pseudopodia (Elliott et al. 1980a, Darbyshire 2005), but their overall activity is expected to decline concomitantly as the connectivity between soil aggregates decreases at reduced SWA (Ritz and Young 2011). Profoundly rapid changes in abundance of flagellates and naked amoebae were reported with increasing SWA using cultivation-based enumeration studies (Clarholm 1981, Anderson 2000, Bischoff 2002), but none of these studies aimed at resolving the taxonomic composition of the protist communities. Recent high-throughput soil surveys confirmed strong impacts of moisture on the community composition of protists in various soils investigated (Baldwin et al. 2013, Bates et al. 2013), but these sequence based approaches generally fail to provide quantitative information on protist abundance (Medinger et al. 2010, Pawlowski et al. 2011, Weber and Pawlowski 2013, Stoeck et al. 2014). Further, it is unclear if protist abundance will decline in a linear manner. Since the individual size of different protist species in soil commonly spans over three orders of magnitude, a linear decline of total protist abundance with pore space might be assumed. Therefore, detailed quantitative studies distinguishing the responses of specific protist taxa to changes in SWA regimes are needed.

The purpose of our study was to obtain quantitative estimates on the abundance of flagellate and amoeboid soil protists with high taxonomic resolution, and to relate the expected changes in community composition to specific moisture conditions. The study was performed as part of a controlled semi-field experiment in Terrestrial Model Ecosystems (TME) (Knacker et al. 2004). We expected higher abundances of protists at higher SWA as habitat space and connectivity between particles increases with increasing water films. Further, we hypothesized stronger changes in community composition and shifts towards smaller species at decreasing SWA.

Materials and Methods

Study site and experimental setup

A controlled semi-field experiment in terrestrial model ecosystems (TME) consisting of undisturbed soil cores was set up close to Flörsheim, Germany (N50°04'; E8°40') in order to manipulate SWA and to evaluate effects on soil meso- and microfauna (www.bik-f.de). The soil cores (30 cm diameter, 40 cm depth) were excavated from a meadow on alluvial clay. Soil texture was a silty clay with 9.9 % sand, 41.9 % clay and 28.2 % silt, that contained 2.93 % organic matter (pH ($CaCl_2$) = 6.9) and had a water holding capacity of 58.7 %.

After the TME cores had been excavated in the field they were incubated on temperature-controlled carts under laboratory conditions (see Knacker et al. 2004 for details of the TME approach). The temperature of the soil cores was kept at 18 - 24 °C and light intensity was 300 ± 50 µE m^{-2} s^{-1}. The TME cylinders could be drained at the bottom and the volumetric SWA of each TME was monitored in the upper 6 cm using hydra probes (accuracy ± 3 % v/v, ecotech Umwelt-Messsysteme GmbH, Bonn, Germany). This monitoring was used to adjust the daily irrigation volumes individually for each TME replicate. Depending on the watering volume either a pump disperser or "rain heads" (i.e. acrylic glass vessels with micropipettes at their bottom) were used for irrigation (Knacker et al. 2004). Moisture manipulation started at the beginning of the experiment (start: March 29th, 2013) in all TMEs, aiming for SWA of 30 %, 50 % and 70 % of the water holding capacity (WHC). The desired WHC levels were reached after 31, 9, and 2 days after starting the study, respectively. One week before harvest (harvest: July 19th, 2013), the irrigation scheme was changed in half of the TMEs in order to simulate heavy rain events. All individual moisture treatments were replicated three times (Fig. 46).

To estimate the diameter of water filled pores at a given moisture level, actual WHC values were converted into water potentials using the SPAW graphical interface (Saxton and Rawls 2006). The pore neck diameter of the largest water-filled pores (µm) was then calculated using the formula $D = 300 * P^{-1}$, where D is the pore neck diameter of the largest water-filled pores (µm), and P

is the water potential (kPa) according to Carson et al. (2010). By calculating maximum water filled pore sizes at any given soil availability level, it might be possible to infer the habitable pore space and to generalize results on protist performance for soils with different texture.

After destructive sampling the TMEs soil samples were stored at 5 °C for up to one week until analyses. SWA determined after oven drying of a small subsample for 2 days at 80 °C. The final mean SWA levels are depicted in Fig. 46.

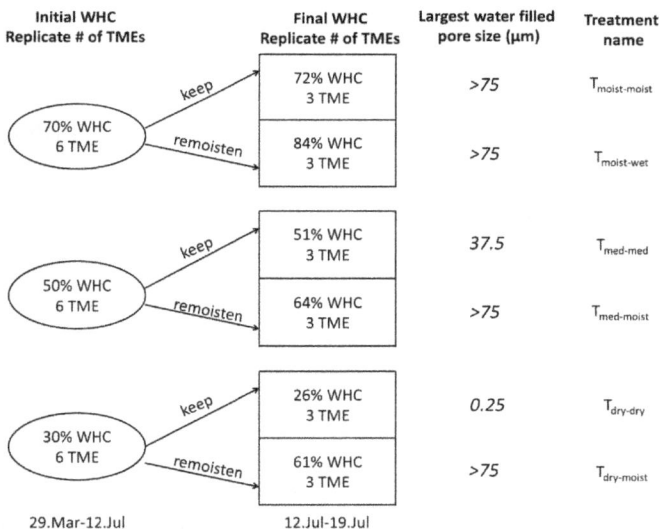

Fig. 46. Left: Schematic diagram of the experimental design (TME: Terrestrial Model Ecosystem; WHC: Water Holding Capacity); largest water filled pore sizes (µm) estimated according to SPAW graphical interface; Treatment names as they are used in the present manuscript; they reflect the aspired soil moisture during the first 15 weeks and during the last week of the experiment.

Enumeration of protists

Protist numbers were determined by a liquid aliquot method (LAM) according to Butler and Rogerson (1995) with the following modifications to simplify the procedure: Briefly, 20 g fresh weight soil of each replicate was homogenized by sieving for 10 s. A subsample of 1 g dry wt was suspended in 200 ml Neff's Modified Amoeba Saline (NMAS; Page 1976)   by vigorous shaking on an orbital shaker (Köttermann, Germany) at 100 rpm for 10 min followed by inverting the suspension and shaking vigorously for 10 s to detach protists from soil particles. The homogenized samples were left to settle for 5 min, before 5 µl aliquots from the centre of the suspension were inoculated, each to one of 144 wells of two 96 well microtitre plates (flat bottom, Sarstedt, Germany), whose wells were filled with 195 µl 0.15 % wheat grass medium (WGM) (Weizengras, Sanatur, Singen, Germany). Plates were sealed with Parafilm and incubated at 15 °C in the dark. Each well was checked twice after 14 and 28 days for presence of protists with an inverted microscope (Nikon Eclipse TS100) at 100 x and 200 x magnification.

Protists were determined to morphogroup level according to Lee et al. (2000), Smirnov and Brown (2004), Smirnov et al. (2011b) and Jeuck and Arndt (2013). Naked amoebae were identified to genus level whenever possible (Fig. 47). The most recent cumulative phylogeny of amoebae (largely from the supergroup Amoebozoa) was then used to combine individual genera together into higher taxonomic levels (Smirnov et al. 2011b). Fig. 47 gives an overview of all amoeba genera identified with their respective phylogenetic classification. Small amoebae (< 7 µm) could not reliably be identified by light microscopy and were grouped together as "Nanoamoebae", which consequently most likely represents an artificial taxonomic assembly of amoebae.

Finally, total numbers of flagellates and amoebae were calculated from the cumulative abundances in the 96 well-plates, the relative proportions of individual groups were determined and corrected using a Poisson distribution.

Supergroup	Subphylum	Class	Subclass	Order	Family	Genus	Morpho-group
Amoebozoa	Lobosa	Tubulinea		Euamoebida	Hartmannellidae		Monopodial
				Echinamoebida	Vermamoebidae	*Vermamoeba*	Monopodial
		Discosea	Flabellinia	Vannellida	Vannellidae		Fan-shaped
				Himatismenida	Cochliopodiidae	*Cochliopodium*	Lens-like
			Longamoebia	Thecamoebida	Thecamoebidae	*Stenamoeba*	Lingulate
				Centramoebida	Acanthamoebidae	*Acanthamoeba*	Acanthopodial
	Conosa	Variosea			Filamoebidae	*Filamoeba*	Acanthopodial
						Flamella	Flamellian
					Acramoebidae	Diverse	Branched
Excavata		Heterolobosea					Eruptive
ND – „Nanoamoebae"						Diverse	Diverse

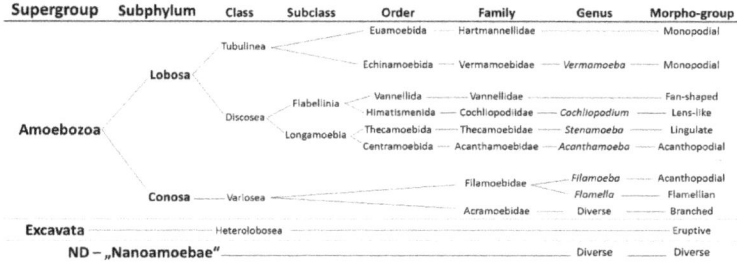

Fig. 47. Phylogenetic classification of the identified amoebae according to Smirnov et al. (2011b) and morphogroups according to Smirnov and Brown (2004).

Statistical analyses

Because relative ratios of protist taxa were not independent of each other, the data were analysed by multivariate analysis of variance (MANOVA, Roy's Greatest Root). In case of significant MANOVA followed a one-way analyses of variance (ANOVA), i.e. 'protected ANOVA' (Scheiner and Gurevitch 2001).

Relative contributions of individual protist taxa were arcsinus transformed to approximate homogeneity of variance. All ANOVAs were performed in R version 3.0.2 (R foundation for statistical computing; available at http://www.R-project.org). Spearman rank correlations between protist taxa were performed to investigate if taxa responded similar to changes in moisture levels. Correlations were tested for significance using GraphPad Prism software, version 5 (GraphPad Software, San Diego, CA, USA). The Shannon–Weaver index was applied to compare protist diversity between treatments (Shannon and Weaver 1949),

$$H = -\sum_i p_i * \ln p_i$$

where p_i is the proportion of the ith morphogroup.

Evenness of protist communities was based on the Shannon-Weaver-index (H)

$$E_H = {}^H/_{Hmax} = {}^H/_{\ln(S)}$$

169

where S represents the total number of species in the community (richness) and H the Shannon's diversity index.

Results

Abundance

As expected, total protist numbers were significantly affected by SWA ($F_{[5, 12]} = 3.39$, $p < 0.05$). Total protist abundance differed 8-fold between the two most extreme moisture treatments, i.e. $T_{dry\text{-}dry}$ and $T_{moist\text{-}wet}$ (Fig. 48). Total flagellate numbers tended to increase with increasing SWA ($F_{[5, 12]} = 2.64$, $p = 0.08$), but the higher number of total protists was mainly caused by an increased abundance of amoebae ($F_{[5, 12]} = 3.59$, $p < 0.05$), which reached 7.2-fold higher numbers in $T_{moist\text{-}wet}$ compared to $T_{dry\text{-}dry}$ (Fig. 48).

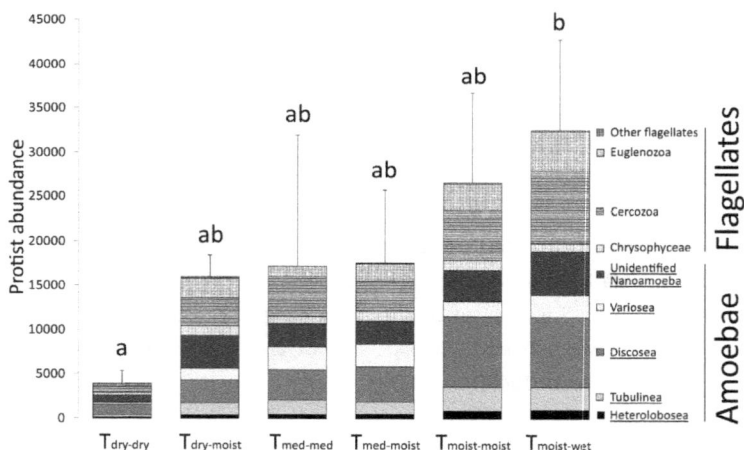

Fig. 48. Total abundance of protists (individuals g-1 soil dry weight) showing the abundance of major clades of flagellates and amoebae with increasing soil moisture levels. For an explanation of moisture treatments see Fig. 46. Underlined: Classes of amoebae; Means ± SD. Bars with different letter are statistically significant (Tukey's HSD test; $p \leq 0.05$).

Generally, protist numbers responded positively to increasing SWA. Overall both flagellates and amoebae increased, but numbers of amoebae increased relatively more strongly than flagellates as shown by the lower than 1:1 slope in Fig. 49.

Although a general uniform positive response of protist abundance to increasing SWA was observed, correlations revealed two distinct clusters of protists that responded more uniformly to each other. The Spearman's rho revealed a statistically positive relationship between the amoeba classes Discosea with both Tubulinea (ρ = 0.79, R^2 = 0.70, p < 0.001) and Variosea (ρ = 0.69, R^2 = 0.46, p < 0.001). Similarly Heterolobosea abundance changed in parallel with the flagellate phyla Euglenozoa (ρ = 0.7, R^2 = 0.45, p < 0.001) and Cercozoa (ρ = 0.6, R^2 = 0.38, p < 0.01), and the flagellate class Chrysophyceae (ρ = 0.5, R^2 = 0.26, p < 0.01).

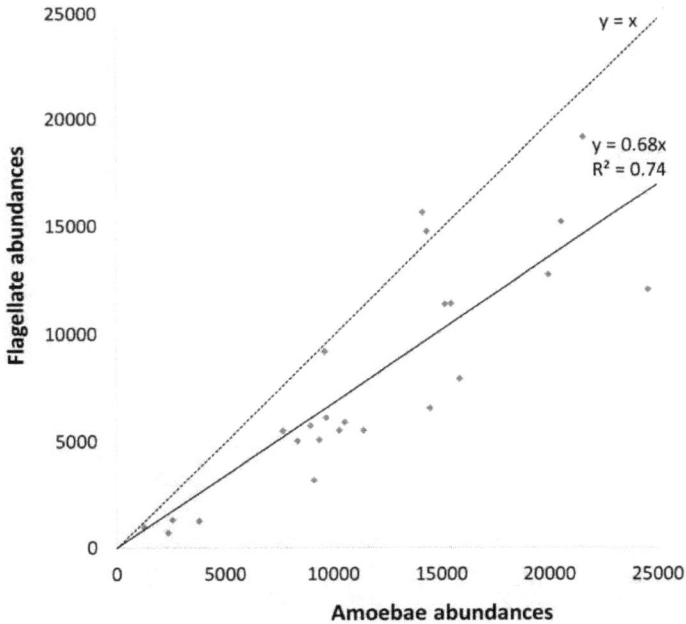

Fig. 49. Correlation of the abundances of flagellates to amoebae; dotted line: y = x.

171

Almost all protist taxa could still be detected even in the most extreme moisture treatment, but significant shifts in the relative abundance of specific clades were detected in amoebae. For instance, amoebae of the class Discosea were most abundant in moist soils ($T_{moist-wet}$ and $T_{moist-moist}$) averaging 7,900 ind g^{-1} dry wt soil. Numbers decreased to 33 % and 15 % in $T_{dry-moist}$ and $T_{dry-dry}$, respectively ($F_{[5, 12]} = 6.49$, $p < 0.01$). Within Discosea, numbers of Longamoebia decreased most strongly with decreasing SWA ($F_{[5, 12]} = 5.02$, $p < 0.05$), and among Longamoebia it was the genus *Stenamoeba* that responded most sensitive to decreasing SWA ($F_{[5, 12]} = 13.92$, $p < 0.001$; Fig. 50), while the other dominant genus *Acanthamoeba*, tended to decrease only in the treatment of lowest SWA ($F_{[5, 12]} = 2.39$, $p = 0.09$; Fig. 50). Heterolobosea increased in $T_{moist-wet}$ and decreased in $T_{dry-dry}$ compared with $T_{med-med}$, albeit with marginal significance ($F_{[5, 12]} = 2.74$, $p < 0.07$).

Among flagellates, both Cercozoa and Euglenozoa tended to increase in $T_{moist-wet}$ compared to $T_{med-med}$ with 1.8- and 3.6-fold numbers, respectively, while decreasing to 15 % and 22 % of $T_{med-med}$ in $T_{dry-dry}$, respectively (Cercozoa: $F_{[5, 12]} = 2.63$, $p = 0.08$; Euglenozoa: $F_{[5, 12]} = 2.55$, $p = 0.09$).

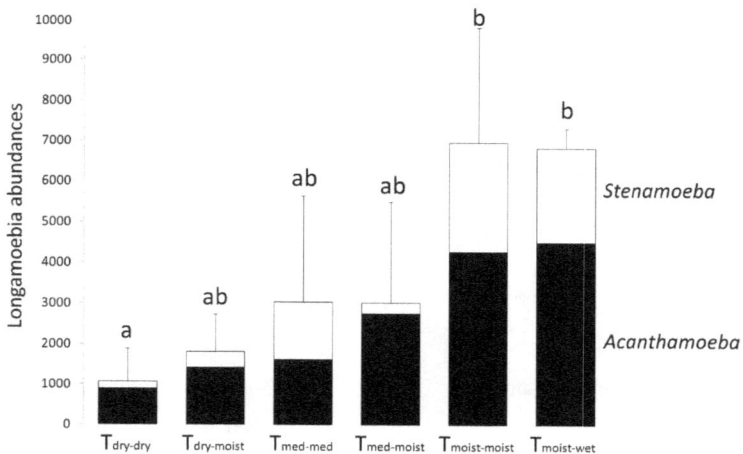

Fig. 50. Abundance of the genera *Stenamoeba* and *Acanthamoeba* within the order Longamoebia; Means ± SD. Bars with different letter are statistically significant (Tukey's HSD test; p ≤ 0.05).

Community structure

Amoebae always slightly dominated the protist community comprising between 59 and 69 % of total protist abundance. Within amoebae, the class Discosea represented always the numerically dominant group (28 - 49 % of all amoebae [OA], 16 - 31 % of all protists [OP]) with the exception of unidentified Nanoamoebae, which were dominant in $T_{dry-moist}$ (44 % OA, 26 % OP) and $T_{med-moist}$ (36 % OA, 24 % OP). Nanoamoebae represented the second most abundant clade comprising 21 - 44 % OA (15 - 26 % OP). The classes Variosea (6 - 29 % OA, 4 - 19 % OP), Tubulinea (10 - 21 % OA, 6 - 9 % OP,) and Heterolobosea (3 - 5 % OA, 2 - 4 % OP) were less abundant. At the genus level, *Acanthamoeba* represented always the most dominant genus of all identified amoebae comprising between 11 to 40 % of all amoebae, while the genera *Cochliopodium* and *Flamella* were always rare ($\leq$ 3 %).

Flagellates comprised 33 - 43 % OP. The dominant flagellate taxa were Cercozoa (39 - 79 % of all flagellates [OF], 19 - 24 % OP), followed in descending order by Euglenozoa (12 - 33 % OF, 5 - 13 % OP), Chrysophycea (7 - 21 % OF, 3 - 7 % OP) and others (0 - 12 % OF, 0 - 4 % OP). On a finer taxonomic resolution, cercomonads were most abundant (32 - 77 % OF, 16 - 24 % OP), followed by bodonids (11 - 30 %, 4 - 12 % OP) and chrysomonads (7 - 21 % OF, 3 - 7 % OP).

Diversity

Neither overall protist diversity nor evenness were altered by moisture treatments in the range of 26-84 % WHC (Fig. 51). Only the diversity of flagellates was significantly reduced in $T_{dry-dry}$ compared to $T_{med-med}$ from 0.92 to 0.61 ($p < 0.05$).

Similarly, the overall ratio of amoebae to flagellate numbers was not affected by SWA. Some individual groups, however, responded with a proportional shift in abundance to changes in SWA. The contribution of the class Variosea to the protist community peaked in $T_{med-med}$ and reached its minimum in $T_{dry-dry}$ ($F_{[5, 12]} = 3.11$, $p < 0.05$), which was caused by a reduction of larger, branched amoebae ($F_{[5, 12]} = 4.47$, $p < 0.05$). The genus *Stenamoeba* was strongly reduced to 1 % OP (2 % OA) in $T_{med-moist}$ and a to 3 % OP (4 % OA) in $T_{dry-dry}$, while *Stenamoeba*

173

contributed most to the abundance of protists (12 %) and amoebae (18 %) in $T_{moist-moist}$ ($F_{[5, 12]}$ = 4.26, p < 0.05). None of the flagellate groups were affected by SWA.

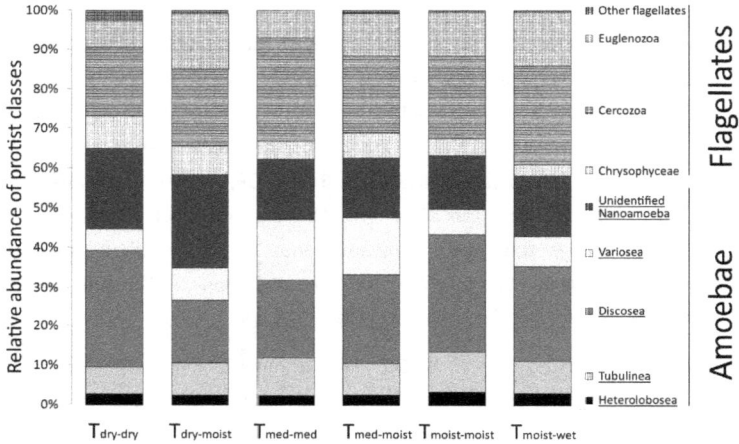

Fig. 51. Relative abundance of major classes of flagellates and amoebae. Underlined: Classes of amoebae.

Habitable pore space and protist performance

As expected, reduced water availability reduced protist numbers, but surprisingly, strong reductions in protist abundance occurred only in the most extreme drying treatment when the maximum size of water filled pores (Pmax) had dropped to ~ 0.25 μm. Protist community composition was very resilient, and did almost not change even in the $T_{dry-dry}$ treatment. Soil protists can readily form desiccation-resistant cysts and are therefore well adapted to changing conditions of SWA (Darbyshire 1994), even though the activity of protists was severely reduced in the $T_{dry-dry}$ treatment. Moistening the dry soil for one week to Pmax > 75 μm, however, increased total protist numbers fourfold to exactly the same level as in $T_{med-med}$ where Pmax was ~ 38 μm. Moistening $T_{med-med}$ to Pmax > 75 μm for one week, however, did not affect

174

protist abundance at all. Significantly higher protist numbers were only observed at an initial Pmax $\geq$ 75 µm, as in $T_{moist\text{-}moist}$ and $T_{moist\text{-}wet}$ (Fig. 47).

Highest protist abundance was reached when all, even the largest soil pores, were water-filled. When the habitable pore space became smaller ($\sim$ 75 µm), the abundance of the protist community slightly decreased, but dropped almost by half, when the size of water-filled pores became lower than $\sim$ 50 µm. Extrapolating these data to a Pmax of $\sim$ 1 µm indicated that the protist community will already sharply decline at a Pmax of $\sim$ 10 µm (Fig. 52). However, individual taxa reacted dissimilarly to differences in water availability. Some taxa reached highest numbers only in fully water-saturated soils and readily decreased when the habitable pore space became smaller, e.g. *Stenamoeba* (Fig. 52-2). Other taxa appeared to be more resistant to decreasing water and only decreased at a later stage of drought (Pmax < 60 µm), e.g. the abundant genus *Acanthamoeba* (Fig. 52-1). Finally, some clades such as Variosea became relatively more abundant when Pmax was still $\sim$ 30 - 40 µm when relative numbers of *Stenamoeba* and *Acanthamoeba* were already in decline (Fig. 52-3). Overall, as hypothesized, the largest protist species decreased with increasing soil dryness, but in particular Nanoamoebae dominated in the dry soil and not flagellates as we initially expected.

Interesting, individual responses of protist taxa to increasing SWA strongly differed from their responses to decreasing SWA. After rewetting soils for one week, Variosea showed maximum numbers at a small Pmax $\sim$ 10 - 40 µm, but did not further increase with increasing SWA, while *Acanthamoeba* showed a constant increase in abundance with increasing water availability. Numbers of *Stenamoeba* on the other hand only started to increase when Pmax was larger than $\sim$ 50 - 80 µm.

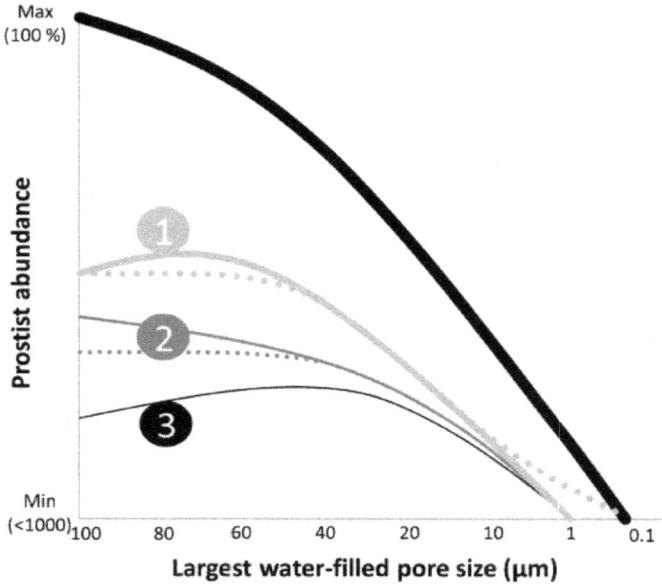

Fig. 52. Conceptual graph of drying and re-wetting on relative abundance of three different taxa of amoeboid protists: *Acanthamoeba* (1), *Stenamoeba* (2) and Variosea (3) at different soil moisture levels, expressed as largest water-filled pore size (LWPS). *Acanthamoeba* peaked below 70 LWPS, Variosea at ~ 30 μm LWPS. *Stenamoeba* had highest relative numbers at highest LWPS, and gradually decreased with decreasing LWPS. All Amoebae strongly declined below ~ 25 μm LWPS. Active trophozoits could still be detected at 0.1 μm LWPS.

Discussion

Total protist densities

Protist densities in soil are mainly regulated by the availability of water and food. As expected, protist abundance decreased strongly from 32,500 to 4,000 ind. g^{-1} soil dry wt from wet to dry soil, respectively. These numbers are well within the range reported for protists in natural soils (Darbyshire 1994, Finlay et al. 2000, Domonell et al. 2013). However, measures of protist community composition, such as diversity and evenness did barely change within the range of 84 % to 26 % WHC of our soils. Amoebae benefited stronger from increasing

SWA (Fig. 49) and were always slightly more numerous than flagellates in all treatments.

As protists essentially depend on the water layer connecting soil pores to move, feed and multiply, the habitat size of these organisms will increase or shrink with changing SWA (Ritz and Young 2011). Consequently, maximum protist numbers are often found during moist seasons and after rainfall (Clarholm 1981, Anderson 2000, Bass and Bischoff 2001, Rodriguez-Zaragoza et al. 2005). Interestingly, protist abundance did not increase in our study, when soils were kept at 50 % WHC or brought from 50 % to 70 % WHC ($T_{med-med}$ and $T_{med-moist}$). Foissner (1987) explained this effect with the accumulation of microbial antibiotics in moist soils, and meanwhile a strong chemical warfare between bacteria and protist grazers has been well confirmed (Matz and Kjelleberg 2005, Bonkowski and Clarholm 2012, Jousset 2012). Drought inactivates this inhibitory effect (Foissner 1987) and it is well known that rewetting of air-dried soils will lead to strongly enhanced abundance of protists (Foissner 1987) preying on the quickly recovering microbial community (Fierer and Schimel 2003, Borken and Matzner 2009, Meisner et al. 2013, Fuchslueger et al. 2014). Consequently, Anderson (2000) found precipitation, but not SWA availability, to be the most important environmental factor influencing the abundance of amoebae in a long-term study. Similarly, precipitation after a prolonged drought led to a 20-fold enhanced abundance of naked amoebae four days after rainfall (Clarholm 1981), and 2 - 6 days after rewetting was repeatedly reported to yield in highest protist numbers (Finlay et al. 2000, Saetre and Stark 2005). These findings support our results where protist numbers essentially recovered within one week in the $T_{dry-moist}$ treatment, but the increase was not uniform. Individual taxa responded with different proportions than during decreased water availability and also the increase of protist abundance was not uniform with increasing SWA in the different treatments. Thus protist responses to drying are different from those after rewetting soils, and appear to be non-linear.

Individual groups

This is the first soil ecological study classifying individual amoebae according to recent phylogenetic classifications (Smirnov et al. 2011b, Adl et al. 2012). The limiting pore size for protist activity in soil has estimated to be 3 - 6 µm, which may be reached at a water potential of -0.15 MPa, corresponding to a pF-value of 3.2 (Alabouvette et al. 1981), but more recent investigations have set this threshold of Pmax even to 2 µm (Darbyshire 2005). Our results indicate that significant shifts in protist community composition at decreasing SWA may occur well before these extreme values are reached. Since even closely related protist taxa exert different feeding strategies (Page 1977, Boenigk and Arndt 2002, Glücksman et al. 2010), drought-induced shifts in the relative abundance of specific protist taxa can be expected to influence the composition and functioning of microbial communities (Saleem et al. 2012).

For example *Acanthamoeba* spp. have been found to represent the most abundant group in deserts (Rodriguez-Zaragoza et al. 2005) and grassland soils (Elliott and Coleman 1977, Brown and Smirnov 2004). In line with the classic note of Page (1988) that *Acanthamoeba* is "the most frequently isolated and probably the most common genus of gymnamoebae, possibly even the most common free-living protozoon", *Acanthamoeba* was also the most abundant genus in our study. *Acanthamoeba* can tolerate long periods of dehydration (Rodriguez-Zaragoza et al. 2005) and correspondingly, we found that acanthamoebae made up the highest proportions among protists in the driest soils, indicating their superior resistance to environmental stress, which could be due to an enormous diversity of *Acanthamoeba* strains (Risler et al. 2013, Geisen et al. 2014b) and an enormous biosynthetic capacity even within strains (Anderson et al. 2005, Clarke et al. 2013b). Surprisingly, the activity of Acanthamoebae in our study rapidly decreased below a Pmax of 40 µm. This is important since *Acanthamoeba* can strongly control bacterial abundance, shape the bacterial community composition and positively influence plant growth (Bonkowski 2004, Kreuzer et al. 2006, Rosenberg et al. 2009, Bonkowski and Clarholm 2012, Koller et al. 2013). In addition, several *Acanthamoeba* strains can cause human infections and host potentially pathogenic bacteria

178

(Schuster and Visvesvara 2004, Khan 2006, Visvesvara et al. 2007, Lagkouvardos et al. 2014), revealing that targeted analyses of this genus are essentially needed to evaluate in more detail their functional roles and niches of survival in soils.

The group most strongly affected by decreasing SWA was the genus *Stenamoeba*. Its abundance peaked in treatments with the highest average SWA, indicating that these amoebae fundamentally depend on high water contents. For a long time only a single species was known and grouped within the genus *Platyamoeba* (now *Vannella*), and only recent studies revealed that it is an independent genus within a completely different subclass in Amoebozoa (Smirnov et al. 2007). Four new species of *Stenamoeba* have since been described (Dyková et al. 2010b, Geisen et al. 2014d) and it can be expected that the species richness of *Stenamoeba* is way higher. Characteristics of *Stenamoeba* in cultures are rapid growth and low attachment to the substratum. The latter might partly explain its decline at decreasing SWA.

Since both *Stenamoeba* and *Acanthamoeba* responded strongly to SWA, the whole subclass and class combining both genera, i.e. the Longamoebia and Discosea were affected by SWA. However, the above findings clearly show that the level of taxonomic resolution can fundamentally influence the outcome of a study, and only detailed taxonomic information allowed us to separate the different responses of *Stenamoeba* and *Acanthamoeba*.

It was impossible to differentiate the small Nanoamoebae by light microscopy. They probably represent different clades such as heteroboseans, *Echinamoeba*, *Nolandella* and most likely unknown and overlooked taxa, such as the recently described *Micriamoeba* (Atlan et al. 2012). Nanoamoebae comprised the majority of amoebae which is in line with other studies that reported highest numbers of small amoebae in soil (Elliott and Coleman 1977, Anderson 2000). However, due to their slow motion compared to flagellates and extremely small size, Nanoamoebae are easily overlooked using cultivation-based enumeration techniques. Their share was disproportionately high in the rewetted treatments, suggesting that these small amoebae have faster recovery rates than flagellates and larger amoebae. Their proportion increased

179

particularly in the rewetted dry soil, where their fast reproduction rates might have provided them with a decisive head start compared to larger protists such as *Acanthamoeba* (~ 3 to 5-fold individual biomass) and Variosea (~ 5 to 100-fold individual biomass). The latter, containing the largest protist species found this study, were most abundant in soils with medium and high initial moisture and therefore appeared to be more desiccation resistant.

Habitable pore space and protist performance

Based on measurements of protist abundances of each soil desiccation treatment, we proposed a model of protist community responses (Fig. 52) showing that individual protest taxa respond differently to SWA. Therefore we strongly argue that ecological studies investigating soil protist communities should aim at identifying protists according to real phylogenetic relatedness and to the deepest level possible, avoiding common morphotype determinations, because ecological effects on the protist community are otherwise likely to be masked.

In conclusion, amoebae and amoeboflagellates dominated the protist community. As expected, protist abundance depended on SWA, but the decrease in numbers with decreasing habitable pore space was not linear. While overall protist diversity was quite resilient over the whole investigated SWA gradient, specific taxa appeared to be highly susceptible when the SWA decreased, and community composition of protists might already strongly change at a Pmax of < 40 µm, which corresponds to the most common size of protists in soil. Also, the dynamics of community changes to desiccation differed from those of rewetting soils. Especially Nanoamoebae and not flagellates as expected became dominant members of the protist community soon after rewetting of soils. As protists represent important components within soil ecosystems playing fundamental roles in nutrient cycling in soil food webs, changes in protist community composition may affect the functioning of soil ecosystems well before effects of drought might be noticed aboveground. More detailed studies are now needed to help uncover the links between soil function and shifts in complex protist communities.

Acknowledgements

The present study was funded by the research funding programme "LOEWE – Landes-Offensive zur Entwicklung Wissenschaftlich-ökonomischerExzellenz" of Hesse's Ministry of Higher Education, Research, and the Arts. This project (HA project no. 155/08-17) is financially supported in the framework of Hesse ModellProjekte, financed with funds of LOEWE - Landes-Offensive zur Entwicklung Wissenschaftlich-ökonomischer Exzellenz, Förderlinie 3: KMU-Verbundvorhaben (State Offensive for the Development of Scientific and Economic Excellence). SG was supported by the European Comission through the project 'EcoFINDERS' (FP7-264465).

Part 2 – Chapter 6

Acanthamoeba everywhere: high diversity of *Acanthamoeba* in soils

Geisen Stefan[a,c*], Fiore-Donno Anna Maria[a], Walochnik Julia[b] and Bonkowski Michael[a]

[a] Department of Terrestrial Ecology, Institute of Zoology, University of Cologne, Zülpicher Str. 47b, 50674 Köln, Germany
[b] Molecular Parasitology, Institute of Specific Prophylaxis and Tropical Medicine, Centre for Pathophysiology, Infectiology and Immunology, Medical University of Vienna, Kinderspitalgasse 15, A-1090 Vienna, Austria
[c] Department of Terrestrial Ecology, Netherlands institute for Ecology (NIOO-KNAW, 6708 PB Wageningen, Netherlands

Abstract

Acanthamoeba is a very abundant genus of soil protists with fundamental importance in nutrient cycling, but several strains can also act as human pathogens. The systematics of the genus is still unclear: currently 18 small-subunit (SSU or 18S) ribosomal RNA sequence types (T1-T18) are recognized, which sometimes contain several different morphotypes; on the other hand, some morphological identical strains belong to different sequence types, sometimes appearing in paraphyletic positions. Here we cultivated 65 *Acanthamoeba* clones from soil samples collected under grassland at three separate locations in the Netherlands, in Sardinia and at high altitude mountains in Tibet. We obtained 24 distinct partial sequences, which predominantly grouped within sequence type T4 followed by T2, T13, T16 and "OX-1" (in the T2/T6 clade). Our sequences were 98-99 % similar, but none was identical to already known *Acanthamoeba* sequences. The community composition of *Acanthamoeba* strains differed between locations, T4 being the dominant sequence type in Sardinia and Tibet, but represented only half of the clones from soils in the Netherlands. The other half of clones from the Dutch soils was made up by T2, T16 and "OX-1", while T13 was only found in Sardinia and Tibet. None of the sequences was identical between localities. Several T4 clones from all three localities and all T13 clones grew at 37 °C while one T4 clone was highly cytopathogenic.

Introduction

The amoebozoan genus *Acanthamoeba* shows an ubiquitous worldwide distribution ranging from aquatic to terrestrial environments, and it might even be one of the most dominant protists in soil (Page 1988, Rodríguez-Zaragoza 1994). In these habitats, acanthamoebae are important grazers of the bacterial biomass, thereby not only controlling the abundance and turnover, but also the diversity of bacterial communities in soil and plant rhizsopheres (Griffiths et al. 1999, Rønn et al. 2002b, Kreuzer et al. 2006, Rosenberg et al. 2009). In the soil microbial loop acanthamoebae liberate nutrients bound in the microbial biomass, ultimately benefiting plant growth (Bonkowski and Clarholm 2012).

In the past decades special attention was also given to *Acanthamoeba* from a medical point of view as several species can cause human diseases such as *Acanthamoeba* keratitis and granulomatous amoebic encephalitis (Rodríguez-Zaragoza 1994, Schuster 2002, Schuster and Visvesvara 2004).

Acanthamoeba belongs to the phylum Amoebozoa (Cavalier-Smith 1998) and the family Acanthamoebidae (Sawyer and Griffin 1975, Pussard and Pons 1977). Molecular phylogenies based on the small-subunit (SSU) ribosomal RNA gene provided further evidence for the monophyly of the genus *Acanthamoeba* with *Balamuthia* as sister genus (Amaral-Zettler et al. 2000, Smirnov et al. 2011b).Traditionally, species of *Acanthamoeba* were classified into three groups based on cyst morphology (Pussard and Pons 1977). Investigations based on molecular data confirmed the monophyly of group I (with the apomorphy of stellate endocysts) but not of the remaining two groups (group II: various endocysts, ectocyst wrinkled; group III: round endocyst, smooth ectocyst) (Pussard and Pons 1977). Although group II and III are related, sufficient divergence within these groups and of both groups to group I led to the progressive recognition of 18 SSU sequence types (T1 – T18), based on ≥5 % sequence dissimilarity between sequence types (Gast et al. 1996, Stothard et al. 1998, Horn et al. 1999, Hewett et al. 2003, Corsaro and Venditti 2010, Nuprasert et al. 2010, Qvarnstrom et al. 2013). Current systematics within *Acanthamoeba* is obscured by several polyphyletic species, such as *Acanthamoeba castellanii* and *A. polyphaga* and an unavailability of type strains for the described species (Gast et al. 1996, Qvarnstrom et al. 2013). Furthermore, T16 has independently been erected twice (Łanocha et al. 2009, Corsaro and Venditti 2010), with both isolates described as T16 being in fact unrelated (Corsaro and Venditti 2011), but none has been abandoned yet.

Lastly, neither morphological groups nor sequence types are indicators of pathogenicity, as pathogenic *Acanthamoeba* have been detected in all morphological groups and most sequence types (Qvarnstrom et al. 2013, Risler et al. 2013). The majority of human infections have been caused by amoebae belonging to the most prevalent sequence type T4 that is however intermingled with non-pathogenic ones (Gast et al. 1996, Stothard et al. 1998, Booton et al. 2002, Maciver et al. 2013, Risler et al. 2013).

The main objective of this study was to assess the occurrence and diversity of the dominant cultivable *Acanthamoeba* spp. in soil samples from three spatially distinct locations, i.e. the Netherlands, Sardinia and Tibet. In total, we cultivated 65 strains of *Acanthamoeba* and obtained 24 distinct partial SSU rDNA sequences. We discuss the phylogenetic results in the light of ecology, biogeography and potential medical relevance.

Materials and Methods

Establishing cultures

The top 20 cm of mineral soil from were collected from grassland sites in the Netherlands (Hedlund et al. 2003), in the Berchidda-Monti long term observatory on Sardinia, Italy (Bagella et al. 2013), and in mountain meadows at high altitudes of >4100m from Mila mountain in Tibet (Table 11). After 2 mm sieving the soil samples were transferred in thermo-isolated containers to the laboratory. A soil suspension was prepared by mixing 50 g of dry wt soil with 50 ml of sterile distilled water. After gently shaking the soil suspension for 20 min, soil particles were allowed to settle for 15 min. From each site, 20 enrichment cultures were established by transferring 100 µl of the soil suspension each, into 10 standard Petri dishes (9 cm) filled with Prescott-James (PJ) medium (Page 1991), enriched with 0.15 % wheat grass (WG) (Weizengras, Sanatur GmbH, Germany) to a final concentration of 0.15 %, and into 10 Petri dishes filled with PJ medium containing 1.5 % agar, respectively. Each dish was carefully examined twice (at days 10-14 and 24-28) with an inverted Nikon Diaphot phase contrast microscope at 100x and 400x magnifications. To establish clonal cultures, single amoebae were transferred with a glass pipette to new 6 cm Petri dishes filled with PJ medium enriched with WG (WG medium). Clonal cultures of amoebae were morphologically identified according to Pussard and Pons (1977), Page (1988), Smirnov and Brown (2004) and Smirnov et al. (2011b).

Table 11. Description of soil samples used in this study

Code used in this study	Locality	Soil origin and land use	Geographic coordinates	Altitude (m)
NlMY	The Netherlands, Veluwe	Ex arable field; 2 years abandoned	N52°21', E5°82'	16
NlMM	The Netherlands, Veluwe	Ex arable field; 9 years abandoned	N52°01', E5°99'	47
NlMO	The Netherlands, Veluwe	Ex arable field; 22 years abandoned	N52°03', E5°80'	18
NlLui	The Netherlands, Veluwe	Ex arable field; 34 years abandoned	N52°06', E6°00'	57
Sar	Italy, Sardinia, Berchidda-Monti	Grassland	N40°46', E9°10'	181
TibE	Tibet, Mila mountain east slope	Meadow	N29°52' E92°33'	4149
TibT	Tibet, Mila mountaintop	Meadow	N29°49' E92°20'	5033
TibW	Tibet, Mila mountain west slope	Meadow	N29°42' E92°10'	4149

Testing for pathogenicity-related characters

Temperature tolerance was tested by growing *Acanthamoeba* strains at temperatures of 34 °C, 37 °C and 42 °C. 50 amoebae were inoculated into each well of a 24-well plate containing 1 ml of WG medium. Growth of trophozoites was estimated every 24 hours for one week. Further, we tested temperature tolerance of all *Acanthamoeba* clones that multiplied at 37 °C by incubating them at 42 °C for one day followed by an incubation at 37 °C for one week using the same settings as described above.

All strains that showed growth ≥ 34 °C (temperature of the human cornea; Efron, 1989) were subjected to cytopathogenicity testing as described earlier (Walochnik et al. 2000). In brief, cysts were harvested from plate cultures, washed in sterile saline solution, and incubated in 3 % HCl overnight to eliminate co-existing bacteria. After centrifugation (500 g, 10 min) the pellet was washed once again in sterile saline solution and transferred into sterile-filtrated PYG medium (proteose peptone-yeast extract-glucose medium in a 4:2:1 mixture) (axenised suspension). In parallel, HEp-2 cells were cultured in a 1:1 mixture of PC-1 and CO_2-independent medium (Life Technologies, Ltd., Paisley, Scotland) supplemented with L-glutamine (2 mM) in 75-cm^2 tissue culture flasks (Corning/Costar, Bodenheim, Germany) at 37 °C under sterile conditions. Subsequently, 1 ml of a 10^5-cell/ml axenised suspension of each isolate was inoculated into a culture flask containing a HEp-2 cell monolayer. Co-cultures were incubated at 34 °C and monitored for 72 hours. All experiments were carried out in duplicates and repeated in an independent set-up.

DNA extraction, amplification and sequencing

Genomic DNA was isolated from fresh cell cultures using the guanidine isothiocyanate protocol (Maniatis et al. 1982a). Briefly, WG medium was discarded, Petri dishes washed twice with sterile WG medium, which was subsequently replaced by 100 µl guanidine isothiocyanate. Amoebae were scraped off using a sterile metal cell scraper, and transferred to 2 ml centrifuge tubes. Subsequent steps were performed according to the cited protocol.

The complete SSU rDNA was amplified from all strains using the universal eukaryotic primers RibA (all primers sequences written in the order 5' - 3': ACC TGG TTG ATC CTG CCA GT) and RibB (TGA TCC ATC TGC AGG TTC ACC TAC) (Cavalier-Smith and Chao 1995, Pawlowski 2000). PCR settings consisted of an initial denaturation at 95 °C for 5 min, followed by 35 cycles of denaturation at 95 °C for 30 s, annealing at 50 °C for 45 s and elongation at 72 °C for 1.5 min with a final 5 min elongation at 72 °C. PCR products were enzymatically purified by adding 0.15 µl Endonuclease I (20 U/µl, Fermentas GmbH, St. Leon-Rot, Germany), 0.9 µl Shrimp Alkaline Phosphatase (1 U/µl, Fermentas, Germany) and 1.95 µl H$_2$O. The mixture was incubated for 30 min at 37 °C, followed by 20 min at 85 °to stop the reaction. Sequencing was carried out at GATC (Konstanz, Germany) using the following sequencing primers: RibB, 12r (AAC GGC CAT GCA CCA CC) and JDP2 (CTC ACA AGC TGC TAG GGG AGT CA) (Dyková et al. 1999).

Alignments and phylogenetic analyses

We obtained 65 sequences of which 24 were unique (the remaining 41 showed 100 % sequence identity; Table 12). These were deposited in GenBank under the accession numbers KF928933 to KF928956. Unique sequences were manually aligned in Seaview 4 (Gouy et al. 2010) together with representatives of the 18 currently described *Acanthamoeba* sequence types (Gast et al. 1996, Stothard et al. 1998, Horn et al. 1999, Hewett et al. 2003, Corsaro and Venditti 2010, Nuprasert et al. 2010, Qvarnstrom et al. 2013). We also used our new sequences to perform a BLASTn search against the NCBI nucleotide database online (http://blast.ncbi.nlm.nih.gov/; last accessed October 24th 2013).): five best hits for each sequence were added to our alignment. Five sequences of *Balamuthia* spp. and two of *Protacanthamoeba* spp. were added as outgroups resulting in a total of 145 sequences.

For phylogenetic analyses, 1,771 unambiguously aligned positions were retained, excluding ambiguous positions and several positions in variable regions especially in V2 (helices 9-11), V4 (helix E23), V5 (helix 29), V7 (helix 43), V8 (helices E45-1 and 46) and V9 (helix 49). Maximum likelihood phylogenetic analyses were run using RAxML v. 7.2.6 (Stamatakis 2006) with the GTR+γ+I model of evolution, as proposed by jModeltest v. 2.1.3 under the

Akaike Information Criterion (Darriba et al. 2012), the γ approximated by 25 categories. 1000 non-parametric bootstrap pseudoreplicates were run. Bayesian phylogenetic analyses were run using Mr Bayes v. 3.2.1 with GTR+γ+I model of evolution and 8 categories (Huelsenbeck and Ronquist 2001). Two runs of four simultaneous Markov chains were performed for 4,000,000 generations (with the default heating parameters) and sampled every 100 generations; convergence of the two runs (average deviation of split frequencies < 0.01) was not reached. Analyses of the two independent consensus trees revealed that both trees diverged in tree regions that were also unresolved in the maximum likelihood analysis. Therefore we decided to discard the first 30,000 trees and built a consensus tree from the remaining 10,000 trees, where the average deviation of split frequencies between runs was < 0.02. In order to improve the resolution between closely related sequences, we performed another phylogenetic analysis using the same settings as above, focusing on the *Acanthamoeba* clones obtained in this study. In total, 116 sequences with 1,947 sequences unambiguously aligned positions were included in maximum likelihood and Bayesian analyses as described above. Convergence of the two Bayesian runs was reached after 2,610,000 generations. The remaining 1,390,000 trees were used to build a consensus tree.

Results

Cultures

In total, 65 *Acanthamoeba* clones were retrieved from the enrichment cultures of all eight soils (32 clones from the Netherlands, 18 clones from Tibet and 15 clones from Sardinia; Table 12). Most cultures of *Acanthamoeba* (80 %) were obtained from agar medium, the others from liquid culture. Our clones belonged to morphogroups II and III, and none to morphogroup I.

Table 12. Name, origin, SSU type, primers used, length of partial sequence obtained, temperature tolerance and cytopathogenicity of the Acanthamoeba strains obtained in this study; ND = not determined; Seq. len. (bp) = Sequence length obtained (bp); Cytopat. = Cythopathogenic

Strain	Additional clones (Strain)	Soil Origin	SSU type	Sequencing primers	Seq. len. (bp)	Growth (37 °C)	Growth (37 °C) after 24h (42 °C)	Cytopat. on HEp-2 cells
NI2	1 (NI76)	NIMM, NIMO	"OX1"	RibB, 12r, JDP2	1869	No	No	No
NI4	1 (NI8)	NIO	T4	RibB, 12r	1709	Yes	Yes	No
NI5	0	NIY	T4	RibB	912	ND	ND	ND
NI9	2 (NI41, NI72)	NIMY	T2	RibB, 12r	1559	ND	ND	ND
NI14	0	NIM	T4	RibB, 12r	1473	No	No	No
NI21	2 (NI35, NI49)	NIY,NIMM	T4	RibB, 12r	1371	ND	ND	ND
NI24	1 (NI70)	NIMY	T16 (C&V)	RibB, 12r	1269	No	No	No
NI123	1 (NI150)	NILui	T4	RibB, 12r	1591	Yes	Yes	Yes
NI130	5 (NI141, NI142, NI145, NI151, NI156)	NILui	T16 (C&V)	RibB, 12r	1441	No	No	No
NI134	0	NILui	T4	RibB, 12r	1515	Yes	Yes	No
NI135	7 (NI144, NI149, NI153, NI154, NI166, NI172, NI180)	NILui	T4	RibB, 12r	1759	Yes	Yes	No
NI152	0	NILui	T16 (C&V)	12r	699	ND	ND	ND
Sar43	5 (Sar46, Sar60, Sar65, Sar73, Sar91)	Sar	T4	RibB, 12r, JDP2	1939	No	No	No
Sar44	3 (Sar47, Sar55, Sar84)	Sar	T4	RibB, 12r	1338	Yes	No	No
Sar45	0	Sar	T4	RibB, 12r, JDP2	1941	No	No	No
Sar48	0	Sar	T13	RibB, 12r	1501	Yes	No	No
Sar63	2 (Sar76, SarSC2)	Sar	T4	RibB, 12r	1290	Yes	Yes	No
Tib1	2 (Tib75, Tib170)	TibE, TibW	T13	RibB, 12r	2195	Yes	No	No
Tib22	4 (Tib29, Tib 116, Tib125, Tib186)	TibT,Tib E,TibW	T4	RibB, 12r	1331	No	No	No
Tib79	1 (Tib157)	TibW	T4	12r	656	ND	ND	ND
Tib121	1 (Tib185)	TibE	T4	RibB, 12r	1264	No	No	No
Tib122	2 (Tib127, Tib171)	TibE,Tib W	T4	12r	676	Yes	Yes	No
Tib128	0	TibE	T13	12r, JDP2	1041	No	No	No
Tib142	1 (Tib160)	TibE	T4	12r	632	ND	ND	ND

The tests on pathogenicity-related characters of 18 of the 24 *Acanthamoeba* clones that differed in their SSU rDNA sequences showed that all isolates grew well at 34 °C, half of the isolates at 37 °C, but none at 42 °C (Table 12). More specifically, several clones from each location grew at 37 °C, i.e. four from the Netherlands (clones Nl4, Nl123, Nl134 and Nl135), three from Sardinia (Sar44, Sar48 and Sar63) and two from Tibet (clones Tib1 and Tib122). Six clones (Nl4, Nl123, Nl134, Nl135, Sar63 and Tib122) tolerated 42 °C for 1 day and subsequently grew at 37 °C (Table 12). Only clone Nl123 rapidly multiplied on HEp-2 cell monolayers that were entirely destroyed within 48 hours. No other clone tested showed cytoplasmic effects on cell layers. However, morphological features of Nl123 did not reveal any morphological characters that distinguished this clone from the others.

Sequences and phylogenetic analyses

We obtained 24 different SSU rDNA sequences from the 65 clones. None of our sequences was identical to already published sequences, the similarities with each of the best hit using BLASTn ranged from 98 % to 99 %. Phylogenetic analyses of an alignment of 145 sequences and 1,771 positions resulted in several strongly supported clusters with a clade containing T7-T8-T9-T17-T18 in the most basal position within the genus *Acanthamoeba* (Supplementary Fig. 3). This clade comprised species assigned to the morphogroup I.T5 containing *A. lenticulata* formed a strongly supported clade, while the other clades were not highly supported using a Centramoebida-wide (*Acanthamoeba*, *Protacanthamoeba* and *Balamuthia*) phylogenetic analysis (Supplementary Fig. 4).

The second analysis aimed to assign the phylogenetic positions of our new sequences more precisely. Therefore outgroups and the two most basal clades were omitted, and the remaining group was resolved into distinct clades with a deep dichotomy (Fig. 53, Supplementary Fig. 3). The first clade, T1-T2-T10-T12-T13-T14-T15-T16, contained two described T16 sequence types (Łanocha et al. 2009, Corsaro and Venditti 2010), accordingly we added the initials of the respective authors for clarity, i.e. T16 (C&V) for the type described by Corsaro and Venditti (2010) and "T16" (Ł) for that described by Łanocha et al. (2009).

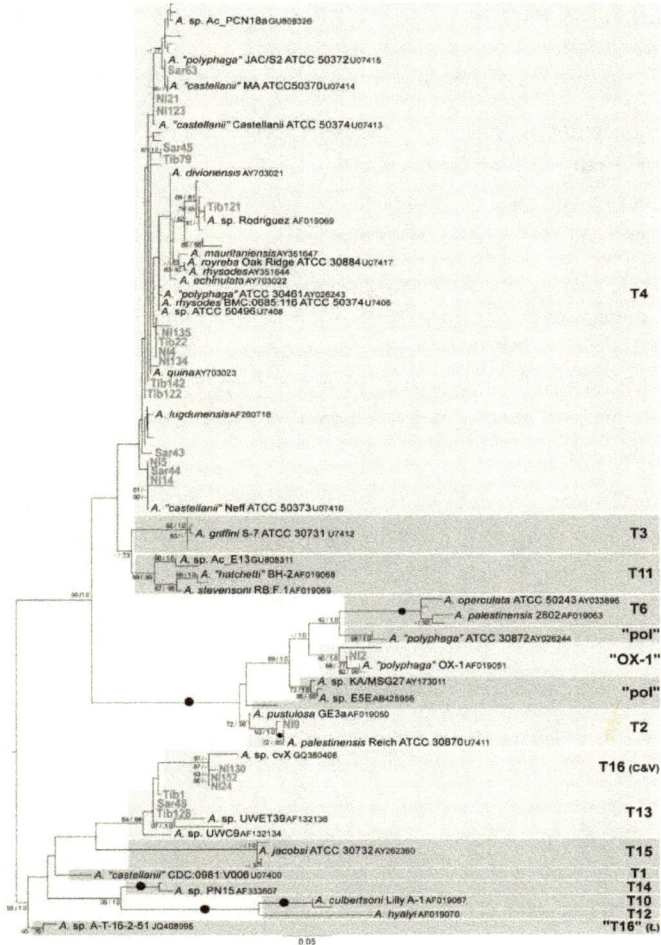

Fig. 53. Phylogenetic analysis focusing on sequence types and sequences associated with the new sequences obtained in this study. In total, 116 sequences with 1,947 unambiguously aligned positions were used, with only single representative sequences shown for individual sequence types and strains; the tree is unrooted; values higher than 60 for maximum likelihood analyses (left) and 0.60 for Bayesian analyses (right) are shown. Black circles represent full support; red: strains obtained in this study; boxes: associated sequence types with bright boxes containing clones obtained in this study; the entire figure showing all strains included in the analysis is shown in Supplementary Fig. 4.

191

However, we place the latter in quotation marks to indicate that this sequence type has been described based on a partial sequence only, leading Corsaro and Venditti to abandon "T16" (Ł) as a sequence type (Corsaro and Venditti 2011).

The other clade showed a dichotomy between T3-T4-T11 and T2-T6. The latter further split into different groups, named "OX-1", "pol", T2 (containing *A. palestinensis* "Reich") and T6, as previously suggested (Corsaro and Venditti 2010, Corsaro and Venditti 2011, Risler et al. 2013).

All of our sequences branched into groups composed of known sequence types. The majority of our clones grouped at different positions within the largest sequence type T4. Three consensus sequences retrieved from 9 clones grouped in T16 (C&V), 3 consensus sequences from 5 clones in T13, one consensus sequence grouped in T2 (3 clones) and one consensus sequence grouped with "OX-1".

Growth performance varied between and within clades. Most *Acanthamoeba* clones from T4 (75 %) and all T13 grew at 37 °C, while T16 (C&V) and "OX-1" did not grow at high temperatures. T13 clones did not tolerate 42 °C while all except one of the T4 clones growing at 37 °C tolerated a 24 h incubation at this high temperature (Table 12).

Comparison of *Acanthamoeba* community composition in soils

Despite the high number of isolated clones from each site, we never found a similar clone from distinct locations. Instead, a diverse community of at least five different sequences occurred at each location, i.e. the Netherlands, Sardinia and Tibet (Table 12). Almost all isolated clones from Sardinian and Tibetan belonged to T4 (93 % and 78 %, respectively), the remaining to T13 (7 % and 22 %, respectively) (Table 12).

The dominant community of acanthamoebae from soils in the Netherlands was richer in distinct isolates and differed in composition from the other locations with only approximately half of the clones belonging to T4 (56 %), followed by T16 (C&V) (28 %), T2 (9 %) and "OX-1" (6 %), while T13 was not recovered (Table 12).

192

Discussion

We retrieved a wide variety of different *Acanthamoeba* clones from all samples, confirming the potential ubiquity, but also a significant diversity of these common amoebae in soils (Page 1988, Rodríguez-Zaragoza 1994). The phylogenetic analyses using the Centramoebida-wide alignment revealed well-supported branches especially in basal *Acanthamoeba* sequence types (Supplementary Fig. 3), while the topology of the more derived SSU types was only recovered in more focused phylogenetic analyses (Fig. 53, Supplementary Fig. 4). Combined information from the two phylogenetic analyses show stable clades such as T10-T12-T14, T13-T16 (C&V) and T2-T6 (Corsaro and Venditti 2010, Corsaro and Venditti 2011, Risler et al. 2013). As previously described, T2-T6 was split into independent lineages (Corsaro and Venditti 2010, Corsaro and Venditti 2011, Risler et al. 2013). The two T16 SSU types T16 (C&V) and T16 (Ł) represent distinct sequence types as suggested by Corsaro and Venditti (2011) and T13-T16 (C&V) formed a separate clade, but without bootstrap support. Perhaps the clade was destabilized by the inclusion of the shorter sequences Tib1, Sar48 and Tib128, which branched intermediate between T13 and T16 (C&V); an analysis including more sites aiming at placing Tib1, Sar48 and Tib128 revealed their unambiguous placement in T16 (C&V) (data not shown). Full sequences would likely have increased the phylogenetic resolution within T13-T16 (C&V) when more divergent sequence types are included in phylogenetic analyses, as shown by Corsaro and Venditti (2011).

Our results confirm the predominance of the T4 sequence type, both in clinical and environmental samples (Gast et al. 1996, Stothard et al. 1998, Booton et al. 2002, Maciver et al. 2013, Risler et al. 2013). Several formally described species based on morphological features have been located in T4 such as *A. royreba*, *A. hatchetti*, *A. divionensis*, *A. echinulata*, *A. polyphaga*, *A.rhysodes*, *A. lugdunensis* and *A. castellanii,* making a subdivision of this genotype desirable as suggested by Booton et al. (2002), Maciver et al. (2013) and Risler et al. (2013). Several of our T4 clones showed clear differences in pathogenicity-related characters, indicating diverse environmental adaptations within this clade (Maciver et al. 2013). Similarly, the T2-T6 clade contains several named species, i.e. *A. operculata*, *A. hatchetti*, *A. polyphaga*, *A. pustulosa* and *A. palestinensis*. This clade has also been divided based on sequence information (Corsaro and Venditti 2010, Corsaro and Venditti 2011, Risler et al. 2013). This subdivision is confirmed by our isolates NI9 resembling T2 and NI2 resembling "OX-1", both not matching any sequences in GenBank. These two examples suggest that a more precise definition of *Acanthamoeba* species and subspecies is needed based on a combination of morphological features, multi-gene phylogenies and information on ecological functions or potential pathogenicity.

Several *Acanthamoeba* clones of the T4 and T13 sequence types were isolated from soils at high altitudes in Tibet. These environments are characterized by harsh, often extremely cold conditions, which are in strong contrast to temperatures in the human body. Nevertheless, strain Tib1 proliferated at 37 °C revealing a broad ecological tolerance. As several other strains of *Acanthamoeba* from all locations grew at high temperatures and several exhibited cytopathogenicity on human cell monolayers, soils may act as a major source for potentially pathogenic *Acanthamoeba*. Generally, *Acanthamoeba* is known to have a broad ecological niche and has been found to dominate protist communities even in extreme environments such as deserts (Rodriguez-Zaragoza et al. 2005) or polluted soils (Lara et al. 2007a) and our findings of different *Acanthamoeba* isolates from high altitude soils add to this spectrum. The generally high numbers of *Acanthamoeba* in soils, the robustness of its cysts and their ease of spread with dust particles, and its growth on various substrates explains why potentially pathogenic *Acanthamoeba* strains are so widely distributed in the environment.

The Dutch soils appeared to contain a different community of *Acanthamoeba* sequence types compared with soils from Sardinia and Tibet, including the only cytopathogenic clone NI123. As none of the individual sequences were identical between locations the real *Acanthamoeba* species diversity and their functional adaptations are likely underestimated. Although we cannot deliver an ultimate proof for the population structure of *Acanthamoeba* due to a too shallow sequencing depth to allow reliable statistical analyses, our study indicated the existence of quite distinct communities of *Acanthamoeba* at each of the three sampling sites. Future studies are needed to investigate environmental niches, community structures and determine potential (a)biotic factors that shape *Acanthamoeba* populations in soils, such as recently conducted for other groups of soil protists (Heger et al. 2013, Vannini et al. 2013).

Taken together, our study indicates that communities of *Acanthamoeba* spp. in soils are highly diverse and that potentially pathogenic strains might be present in many soils, even on remote mountains at extreme altitudes.

Acknowledgements

This work was supported by a research grant from the EU-project 'EcoFINDERS' No. 264465.

Supplementary Files

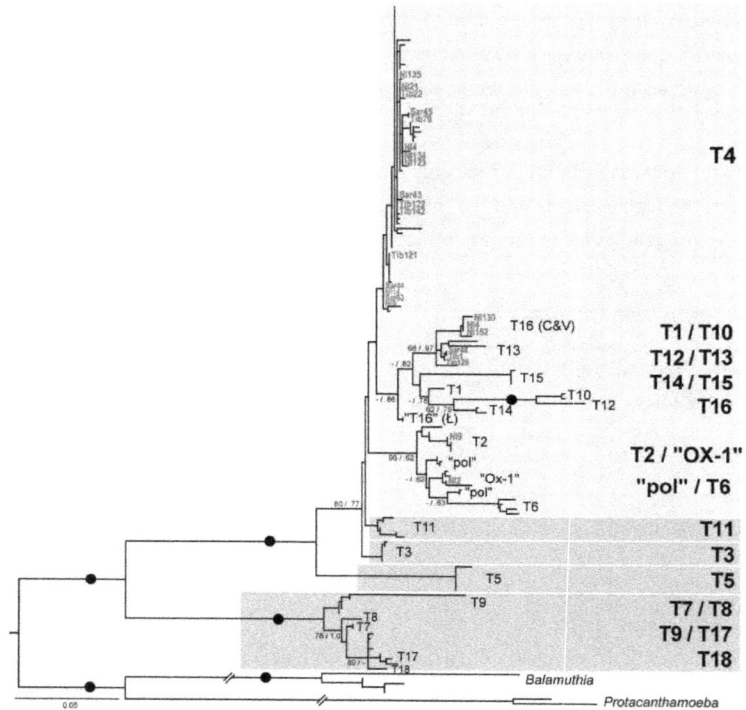

Supplementary Fig. 3: Phylogenetic analysis of Centramoebida based on 145 SSU sequences (all 18 described SSU types of *Acanthamoeba* represented) and 1,771 positions, with *Balamuthia* and *Protacanthamoeba* as outgroups. Major clades and T4 are boxed. Only sequence types are named, while individual sequences were omitted for clarity, except for new sequences obtained in this study (in red). Support values >50 % (maximum likelihood, left) or > 0.5 (posterior probability, right) are shown for branches leading to major clades. Filled circles represent full support in both analyses; boxes: clades of sequence types with bright boxes containing clones obtained in this study.

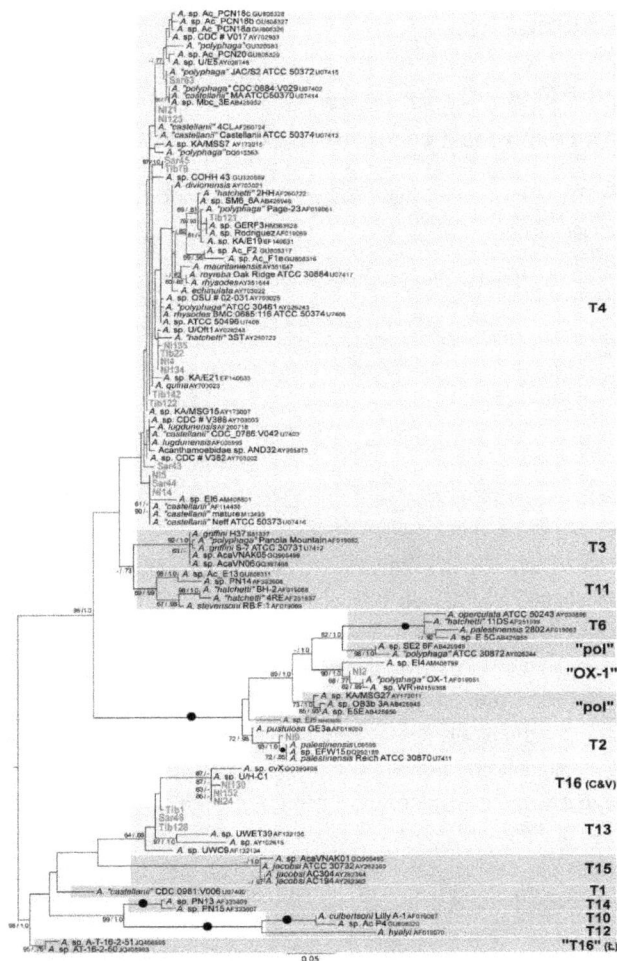

Supplementary Fig. 4: Phylogenetic analysis focusing on sequence types and sequences associated with the new sequences obtained in this study. In total, 116 sequences with 1,947 unambiguously aligned positions were used; the tree is unrooted; values higher than 60 for maximum likelihood analyses (left) and 0.60 for Bayesian analyses (right) are shown. Black circles represent full support; red: strains obtained in this study; boxes: associated sequence types with bright boxes containing clones obtained in this study;

Part 2 – Chapter 7

Profound effects of geographic location and use intensity on cercozoan communities in European soils

Geisen Stefan[1], Tisserant Emily[2], Buée Marc[2], Martin Francis[2], Bass David[3], Bonkowski Michael[1]

[1] Department of Terrestrial Ecology, Institute for Zoology, University of Cologne, Germany
[2] Institut National de la Recherche Agronomique (INRA), Nancy, France
[3] Department of Life Sciences, The Natural History Museum, London, UK

Abstract

Cercozoa are among the most abundant and diverse groups of soil protists. However, their diversity, community composition and the influence of environmental shaping that shape those remain largely unknown. We applied a high-throughput sequencing (HTS) approach using primers optimized to select for Cercozoa to evaluate potential differences in soil cercozoan communities and decipher factors that shape those. Cercozoan communities in a total of 122 samples from at least fourfold replicates of two land use intensity (LUI) treatments in five geographically distant sites across Europe were analysed and compared. All geographically distant sites differed profoundly in cercozoan community composition and LUI also influenced cercozoan diversity in some cases. However, the differences observed depended on the taxonomic level of analysis, with slight differences shown at high taxonomic resolution (class) and most profound differences at the lowest taxonomic level (OTU). Distinct OTUs showed strong association to treatment or site that were driving overall differences in the community composition of Cercozoa. The strong biogeographic patterns of soil cercozoan communities underline urgently needed follow-up studies to determine environmental factors that shape biogeographic patterns of soil cercozoan communities and those of protists in general.

Introduction

Soil single-celled heterotrophic protists are key components in terrestrial soil food webs due to their high abundance, turnover, diversity and their functional roles as dominant bacterivores (Clarholm 1985, Ekelund and Rønn 1994, Adl et al. 2012). Cercozoa are among the most abundant soil protists are Cercozoa, confirmed both by traditional (Ekelund et al. 2001, Bass and Cavalier-Smith 2004, Bass et al. 2009b, Howe et al. 2009, Scharroba et al. 2012, Domonell et al. 2013) and recent high-throughput sequencing (HTS) studies (Urich et al. 2008, Baldwin et al. 2013, Bates et al. 2013). The phylum Cercozoa was only erected in 1998 being the first phylum described based solely on molecular phylogenetic information (Cavalier-Smith 1998, Bass and Cavalier-Smith 2004). Several studies have now proven the tremendous diversity of taxa in the phylum Cercozoa, which include naked (e.g. vampyrellids) and testate amoebae (e.g. euglyphids), amoeboflagellates (e.g. cercomonads), flagellates (e.g. glissomonads), and plant / stramenopile parasitic protists (e.g. plasmodiophorids), with total estimated species numbers considerably exceeding a thousand (Bass and Cavalier-Smith 2004, Bass et al. 2007, Bass et al. 2009b, Howe et al. 2009, Howe et al. 2011a, Adl et al. 2012, Berney et al. 2013, Neuhauser et al. 2014). Little, however, is known about the overall composition of soil Cercozoa and if communities differ between soils.

Similarly, knowledge on biogeographic patterns of protists in general is scarce and has been debated intensely, especially in the last decade (Baas-Becking 1934, Finlay 2002, Foissner 2006, Martiny et al. 2006). Recent evidence supports a wide distribution of dominant morphospecies of protists, while some species display a clearly more limited distribution (Foissner 2006, Bass et al. 2007, Heger et al. 2013). Undebated is, however, the high abundance and enormous diversity of protists. Especially soil protists show nearly a continuum of morphologically defined species, with molecular information often introducing a finer continuum within morphologically indistinguishable taxa (Bass et al. 2007, Epstein and López-García 2008, Adl et al. 2012, Bachy et al. 2013). The functional importance of this vast diversity of soil protists remains, however, largely unknown; the high diversity of bacteria, for instance, is suggested to act as an insurance for soil functioning to long term changes (Nielsen et al. 2011, Reich et al. 2012), and increasing resistance to pathogen invasion (van Elsas et al. 2012). Likewise, protist diversity might be important for the resilience of soil functions, e.g. due to differential feeding on bacterial species (Glücksman et al. 2010, Saleem et al. 2012, Saleem et al. 2013), which in turn might alter plant growth and other ecosystem services (Clarholm 1985, Bonkowski 2004, Bonkowski and Clarholm 2012).

Molecular tools, such as HTS techniques, enable the analysis of high sample numbers at the same time. Since many protists cannot be cultivated, molecular methods targeting DNA has resulted in the discovery of many novel protist lineages and avoids laborious cultivation (Moon-van der Staay et al. 2001, Dawson and Hagen 2009, Lejzerowicz et al. 2010, Bates et al. 2013).

Until today, the few studies targeting soil protists with HTS techniques have focused on the broad taxonomic patterns of soil protist community composition, but detailed studies of specific taxonomic groups of free-living soil protists are virtually lacking with the exception of plant parasitic *Phytophthora* spp. (Vannini et al. 2013).

In this study we developed new specific primers targeting protists of the phylum Cercozoa in a HTS approach. We hypothesised that the community composition of Cercozoa would differ between geographically distinct sites and that change in LUI further influence the cercozoan community structure.

Materials and Methods

Soil locations, sampling and DNA extraction

Five well-characterized long term observatories (LTOs) were appointed within the EU project EcoFINDERS, which were located at distant sites across Europe and represented a variety of typical land management types of those regions (Table 13). Each LTO contained two contrasting types of LUI to discriminate local effects of land management form regional differences in species diversity. The following treatments were differentiated: Grasslands were sampled in

England; one treatment was improved by application of fertilizers (Ei), the other unimproved (Eu), both represented by 15-fold replicated samples. Permanent grasslands (Fg) and grassland in a permanent agricultural rotation (Fc) were sampled in France, each fourfold replicated. Three treatments were differentiated in Italy, with intensively managed grassland (Ii), samples from grass patches (Ig) and under trees (It) taken from wooded land, all three replicated ninefold. Recently abandoned fields, two-five years after agricultural use (Nl) were differentiated from long-term abandoned (20 - 25 years) fields in the Netherlands (Nh), replicated 9- and 12-fold, respectively. The last treatments were taken from fertilized (Sf) and unfertilized coniferous forests (Su) in Sweden, both being replicated 18-fold. A summary of these site and treatments characteristics are shown in Table 13.

Table 13. Site and treatment characteristics; $\bar{X}$ = mean value

Site location	Sampling time	H_2O (g/kg)	C_{org} (g/kg)	Total N (g/kg)	C/N	Organic matter (g/kg)	pH	P (P_2O_5) OLSEN) (g/kg)	Land use (Label; sample # before /after quality filtering])
Lusignan, France	March 2011	NA	9 – 12; $\bar{X}$ = 10.7	1.0 – 1.3; $\bar{X}$ = 1.2	8.9 – 9.6; $\bar{X}$ = 9.3	16 – 21; $\bar{X}$ = 18.5	NA	NA	Permanent agriculture (Fc; 4 / 4) Permanent grassland (Fg; 4 / 4)
Lancester, England	June 2011	21 – 46; $\bar{X}$ = 33.1	43 – 95; $\bar{X}$ = 59.3	2.8 – 8.7; $\bar{X}$ = 5.3	9.8 – 15.1; $\bar{X}$ = 11.3	74 – 165; $\bar{X}$ = 102.7	4.8 – 6.6; $\bar{X}$ = 5.4	0.009 – 0.170; $\bar{X}$ = 0.040	improved grassland (Ei; 14 / 15) unimproved grassland (Eu; 13 / 15)
Sardinia, Italy	May 2011	9 – 23; $\bar{X}$ = 15.1	16 – 35; $\bar{X}$ = 24.0	0.9 – 2.1; $\bar{X}$ = 1.5	12.5 – 21.5; $\bar{X}$ = 15.9	28 – 60; $\bar{X}$ = 41.4	5.2 – 6.4; $\bar{X}$ = 5.9	0.005 – 0.045; $\bar{X}$ = 0.015	intensive grassland (Ii; 8 / 9) wooded land grass patch (Ig; 9 / 9) wooded land under tree (It; 8 / 9)
Veluwe, Netherlands	July 2011	7 – 15; $\bar{X}$ = 11.2	17 – 48; $\bar{X}$ = 29.3	1.0 – 2.3; $\bar{X}$ = 1.4	16.0 – 25.5; $\bar{X}$ = 21.5	30 – 83; $\bar{X}$ = 50.7	4.1 – 6.0; $\bar{X}$ = 5.4	0.090 – 0.360; $\bar{X}$ = 0.227	recently abandoned fields (Nl; 9 / 9) long abandoned fields (Nh; 8 / 12)
Lamborn, Sweden	July 2011	5 – 42; $\bar{X}$ = 22.3	21 – 178; $\bar{X}$ = 68.6	0.6 – 4.7; $\bar{X}$ = 1.9	31.3 – 49.8; $\bar{X}$ = 36.8	36 – 308; $\bar{X}$ = 118.7	4.0 – 4.9; $\bar{X}$ = 4.6	0.005 – 0.042; $\bar{X}$ = 0.017	fertilized forest (Sf; 11 / 18) unfertilized forest (Su; 12 / 18)

Sampling was conducted between March and June 2011. The upper 10 cm of the organic soil horizon were sieved, roots and stones removed and DNA was extracted according to a standardized ISO 11063 protocol (Plassart et al. 2012).

Primer design and amplicon preparation of 18S rRNA gene

Amplicons of ~1,200 bp were generated from each sample using the cercozoan-specific primer combination 25F (5' - CAT ATG CTT GTC TCA AAG ATT AAG CCA - 3') and 1256R (5' - GCA CCA CCA CCC AYA GAA TCA AGA AAG AWC TTC - 3'; Bass and Cavalier-Smith 2004) in a first round (94 °C for 1 min, 35 cycles of 94 °C for 30 s, 70 °C for 60 s and 72 °C for 2 min with a final extension for 5 min at 72 °C). PCR reactions were carried out in 31 µl volume consisting of 0.6 µl of each primer (10 µM), 0.6 µl nucleotides (10 mM), 1.0 µl template DNA, 24.5 µl H$_2$O, 3 µl GreenTaq Buffer and 0.15 µl GreenTaq polymerase (5 U * µl^{-1}) (Fermentas, St. Leon-Rot, Germany). 1.0 µl aliquots of the resulting PCR products were used as template for a hemi-nested PCR step using the same reverse primer with the new forward primer "PreV4" (5' - GYT GCA GTT AAA AAG CTC GTA GTT G - 3'; this study) at the 5' end of the SSU V4 region, judged to be the most informative SSU barcoding region, because of its high variable nature and sufficient length for phylogenetic analyses (Pawlowski et al. 2012), while producing an amplicon of ~500 bp, appropriate for 454 sequencing. PreV4 was designed *in silico* and tested for specificity by sequencing environmental DNA from 14 clones extracted from Dutch soil. All sequences obtained were of cercozoan taxa (data not shown).

PCR conditions of the second step were the same, except an annealing temperature of 66 °C and elongation for 90 s were used. A five bp long MID-identifier was incorporated onto the 5' ends of both PreV4 and 1256R before the nested PCR step. Both PCRs were replicated twice for each sample to increase product yield and to reduce PCR errors. Duplicates from each sample were pooled, purified by gel extraction using the Agarose GelExtract Mini Kit (5PRIME, Hilden, Germany) and quantified by NanoDrop spectrophotometry (NanoDrop Technologies, Wilmington, USA).

All 122 samples were divided into eight batches by equimolar pooling PCR products of 15 - 16 random samples. Pooled libraries were sent for pyrosequencing using the standard protocol (titanium chemistry) on a Genome Sequencer FLX system (Beckman Coulter, Fullerton, USA).

Bioinformatics analyses

Sequences obtained were first demultiplexed according to their multiplex identifier (MID) using the sffinfo command of Mothur v.1.22.2 (Schloss et al. 2009), allowing one mismatch per MID. Fasta and quality files were converted into fastq file using the faqual2fastq.py script of Usearch v7.0.1001 (Edgar 2013). All sequences were then labelled with a sample name and pooled.

The raw sequences were filtered and trimmed using the fastq_filter command of Usearch with the option fastq_truncqual 10, such as sequences are truncated at the first position having a quality score ≤ 10. Sequences were then truncated to the length of 300 bp, with shorter sequences discarded using the option fastq_trunclen. Sequences from forward and reverse primers were sorted according to their primer sequences using the trim.seqs and split.groups commands of Mothur, allowing two mismatches. Sequences from reverse primers were removed from the analysis. Trimmed sequences were dereplicated to remove duplicated sequences using derep_fulllength command of Usearch. Dereplicated sequences were sorted by decreasing abundance and singletons were discarded using sortbysize command of Usearch. Operational taxonomic units (OTUs) were generated from abundance-sorted sequences using the cluster_otus command of Usearch for 99 %, 98 %, 97 %, 96 %, 95 % and 90 % similarity thresholds. For each similarity threshold considered, trimmed sequences (including singletons) were mapped against the OTU representative sequences using usearch_global of Usearch. Based on these mapping results, matrices containing the sequence abundances of different OTUs in each soil sample were generated using uc2otutab.py script of Usearch. To make comparable samples with different number of sequences, matrices were subsampled with an identical sequences number (n = 500) for each soil sample using the sub.sample command of Mothur. This sequence number was determined to conserve a minimum of four independent replicates for each treatment considered in the study.

Taxonomic assignation was determined for each OTU representative sequences using the Basic Local Alignment Search Tool (BLAST) algorithm v 2.2.23 (Altschul et al. 1990) against the Protist Ribosomal Reference Database PR2 (Guillou et al. 2013). All assignations were determined using an e-value cut-off of $1e^{-5}$, an identity cut-off of 90 % and a coverage cut-off of 80 % of the query sequence covered in the alignments.

Only sequences assigned to sequences specific for the phylum Cercozoa were subject to downstream comparisons and statistical analyses. The cercozoan OTUs were assigned to the different taxonomic levels class, order, family, genus, species and OTU level.

Statistics

Statistical analyses were performed as suggested by Anderson and Willis (2003) including unconstrained ordination using principal coordinate analyses with Bray-Curtis distance matrices (Gower 2005), constrained analyses using canonical analysis of principal coordinates (Anderson and Willis 2003), statistical tests using analysis of similarities (ANOSIM) (Clarke 1993) and permutational multivariate analyses of variance (PERMANOVA) (Anderson 2001) and characterization of OTUs that shape the multivariate analyses using detrended correspondence analysis (DCA) and linear discriminant analyses (LDA). Further, two clustering methods (paired group (UPGMA, using Bray-Curtis similarity index) and Ward's method (Euclidian similarity index) were applied. Global beta richness was assessed by combining all samples from each of the five distinct geographic origins. All analyses were performed in PAST (Hammer et al. 2001). Differences were considered as significant when a threshold of $p < 0.05$ was reached.

After the initial taxonomic binning of 966 cercozoan OTUs several verification / refinement steps were carried out. A custom database was built incorporating as many taxonomically assigned cercozoan sequences as possible: the PR2 database (Guillou et al. 2013), unpublished cercomonad and glissomonad sequences (David Bass laboratory), and sequences from recent publications not yet incorporated into PR2. The 957 OTUs were locally BLASTn-searched against the custom database. All query sequences returning a BLASTn sequence identity result > 95 % were accepted as belonging to the genus indicated. The remaining 536 were aligned with a representative pared-down cercozoan alignment to identify sequence clusters in the OTUs and their broad taxonomic affiliations. Representative OTUs from the clusters were re-aligned with a comprehensive sequence dataset of the group to which they belonged. The genus-level affiliation of all the OTUs in the cluster was inferred from this. Where a genus-level affiliation was not possible due to the divergence of the sequence, order level was used instead, or a clear definition of an environmental clade. Our convention for reporting the taxonomic affiliations is to associate a genus with the OTU number that blasts to it (e.g. Cercomonas_[OTU]12), as robust affiliation to the species level is difficult to prove (possibly apart from the rare cases where the blast sequence identity match is 100 %); we have devised an approach that can be consistently applied across the dataset as a whole.

Results

Amplicon analyses

A total of 224,179 sequences were obtained from 122 sites after quality filtering. 28 samples with sequence numbers lower than 500 were discarded leaving 94 samples with an average of 2,343 sequences (546 - 7,994 sequences). OTU clustering at different levels revealed an exponential increase of assigned OTUs up to a similarity threshold of 98 % (r^2 = 0.975). Clustering to 99 % strongly outreached exponential growth and reducing the r^2 to 0.87 (Fig. 54). Similarly, reads were remapped confidently up to an OTU clustering of 98 % assuming an exponential decrease of assignable OTUs (r^2 = 0.98), while when including clustering at 99 % strongly deflated the successful re-assignment of reads (r^2 = 0.83, Fig. 54). Therefore, subsequent analyses focused on an OTU clustering level of 98 %.

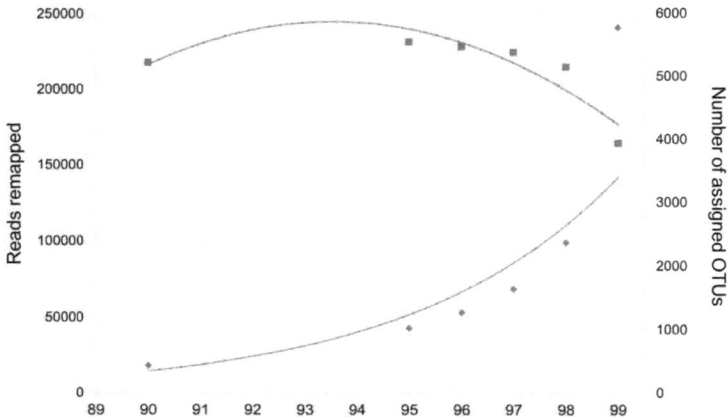

Fig. 54. Number of OTUs (diamonds and line in blue; right y-axis) and reads remapped (squares and polynomial line in red, left y-axis) at OTU assignment of 90 % and 95 - 99 %. Exponential fitting best explained increasing numbers of OTU (blue, r^2 = 0.87) and decreasing numbers of reads remapped (red, r^2 = 0.83) with increasing OTU clustering levels.

The primers proved to be highly cercozoan specific with 81.9 % (1,919 sequences per sample) assigned as Cercozoa. 11.7 % of the sequences remained unassigned, 4.2 % were fungi and 1.7 % streptophytes. Other assigned phyla were negligible (> 0.4 %). OTU clustering at 98 % sequence similarity resulted in a total of 1,636 OTUs of which 957 were cercozoan-specific, while 559 could not be assigned against the protist PR2 database. OTU numbers levelled off, but did not reach saturation either at sites (Fig. 55) and only rarely in individual samples (Supplementary Fig. 5).

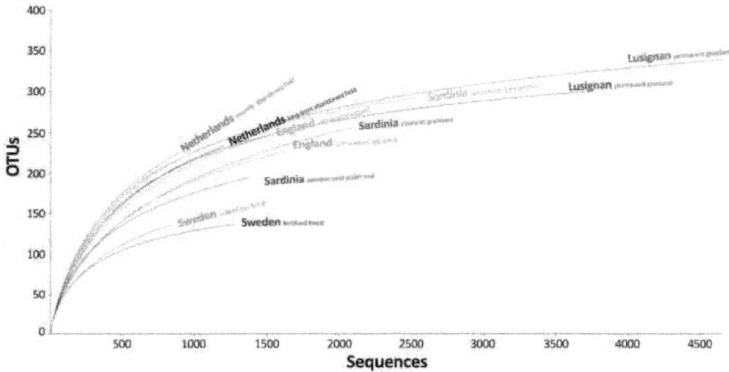

Fig. 55. Rarefaction curves of amplicon data sets from all combined OTUs combining all samples of respective treatments.

The 957 cercozoan sequences were assigned to 9 classes, 24 orders, 42 families, 65 genera and 110 species. Re-blasting using our own database revealed that most sequences could not reliably be assigned on the species level, often not even on genus or family level (Supplementary Table 7). The majority of OTUs showed identities < 99 % on the species level, strongly suggesting that these sequences were not necessarily equivalent to the assigned species. A subset of 47 OTUs is shown as an example for focused phylogenetic analyses of cercomonads and glissomonads in Supplementary Figs. 6 and 7.

We therefore focus our respective analyses on the lowest (i.e. OTU) taxonomic level and compare those to highest (i.e. class) levels.

Cercozoan community structure and site comparisons

We found strongly differing cercozoan community compositions (CCCs) between sites and treatments. The CCCs in coniferous forest sites in Sweden differed most strongly from grassland and arable field sites, suggesting that coniferous forests contain distinct CCCs. The level of taxonomic classification had a profound impact on the results. When CCCs were analysed on the deepest level (OTU level), CCCs differed between all sites revealing similarities

207

between 0.24 (Sweden and France) to 0.51 (England and France; Fig. 56 and 57). The differences in CCCs were less pronounced when analysing the cercozoan communities at higher taxonomic levels, especially class level. Here, similarities ranged from 0.74 (England and Sweden) to 0.97 (Netherlands and Sweden; Supplementary Fig. 8 and 9). However, multivariate analyses demonstrated that the CCCs clearly differed between locations irrespective of phylogenetic resolution (class: R = 0.43, p < 0.001 (ANOSIM) and F = 14.9, p < 0.001 (PERMANOVA); OTU: R = 0.63, p < 0.001 (ANOSIM) and F = 8.7, p < 0.001 (PERMANOVA); Table 13).

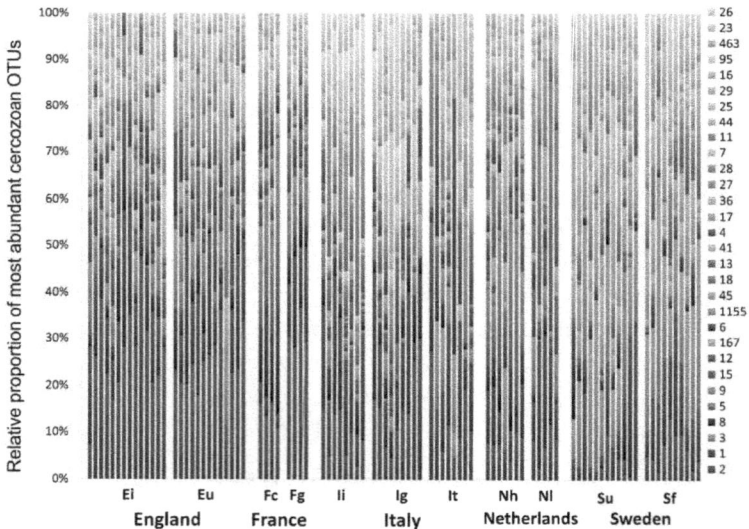

Fig. 56. Rarified relative abundances of the 30 most abundant cercozoan OTUs in all samples at respective sites and treatments (x-axis); (see Table 13 for details on soil characteristics and treatment abbreviations).

In general, LUI of specific land management types had a lower influence on CCCs than geographic location, indicating the existence of specific CCCs in different European soils. In line, the majority of treatments from respective sites clustered together, especially at lower taxonomic resolution (Fig. 57; Supplementary Fig. 8). Most similar were both English grassland treatments (class level: 0.97; OTU level: 0.71), while Swedish fertilized forest samples were least similar to those of French culture at OTU level (similarity = 0.14; Fig. 57), while the lower resolution on CCC differences on class level revealed largest differences from intensive grassland in Italy with culture in France (similarity of 0.63; Supplementary Fig. 8). Despite being grasslands, the CCCs in the respective treatments in England, France, the Netherlands and Italy clearly differed. Interestingly, the CCCs of recently abandoned agricultural fields in the Netherlands clustered with agricultural fields in France, suggesting that agricultural management provokes distinct and long-lasting changes in protist communities. Both multivariate statistical analyses (ANOSIM and PERMANOVA) proved that cercozoan communities differed between treatments (class: R = 0.49, p < 0.001 (ANOSIM) and F = 10.9, p < 0.001 (PERMANOVA); OTU: R = 0.67, p < 0.001 (ANOSIM) and F = 5.2, p < 0.001 (PERMANOVA); Table 15).

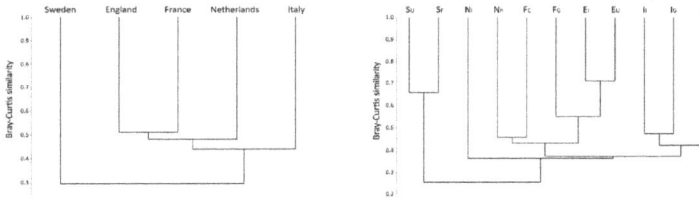

Fig. 57. Similarity of cercozoan communities between sites (left) and treatments (right) based on OTU level using the paired group (UPGMA) analyses with Bray Curtis similarity; see Table 13 for details on soil characteristics and treatment abbreviations.

Table 14. Differences of Cercozoan communities on the OTU level between geographic sites based on ANOSIM (ASIM) and PERMANOVA (POA)

	England		France		Italy		Netherlands		Sweden	
	ASIM	POA	ASIM	POA	ASIM	POA	ASIM	POA	ASIM	POA
England			0.0012	0.0001	0.0001	0.0001	0.0001	0.0001	0.0001	0.0001
France	0.0012	0.0001			0.1505	0.0001	0.0016	0.0002	0.0001	0.0001
Italy	0.0001	0.0001	0.1505	0.0001			0.0036	0.0001	0.0001	0.0001
Netherlands	0.0001	0.0001	0.0016	0.0002	0.0036	0.0001			0.0001	0.0001
Sweden	0.0001	0.0001	0.0001	0.0001	0.0001	0.0001	0.0001	0.0001		

Table 15. Differences between treatments based on OTU level using the multivariate statistical analyses ANOSIM (ASIM) and PERMANOVA (POA); see Table 12 for details on soil characteristics and treatment abbreviations

Treatment	Method	Ei	Eu	Fc	Fg	Ii	Ig	It	Nh	Nl	Su	Sf
Ei	ASIM		0,242	0,000	0,104	0,000	0,000	0,000	0,000	0,001	0,000	0,000
Ei	POA		0,433	0,001	0,003	0,000	0,000	0,000	0,000	0,000	0,000	0,000
Eu	ASIM	0,242		0,001	0,211	0,000	0,000	0,000	0,000	0,001	0,000	0,000
Eu	POA	0,433		0,001	0,001	0,000	0,000	0,000	0,000	0,000	0,000	0,000
Fc	ASIM	0,000	0,001		0,030	0,002	0,004	0,154	0,049	0,014	0,000	0,001
Fc	POA	0,001	0,001		0,029	0,003	0,002	0,002	0,006	0,007	0,000	0,001
Fc	ASIM	0,104	0,211	0,030		0,002	0,053	0,488	0,003	0,016	0,001	0,001
Fc	POA	0,003	0,001	0,029		0,002	0,001	0,002	0,004	0,007	0,001	0,001
Ii	ASIM	0,000	0,000	0,002	0,002		0,000	0,000	0,000	0,001	0,000	0,000
Ii	POA	0,000	0,000	0,003	0,002		0,000	0,000	0,001	0,001	0,000	0,000
Ig	ASIM	0,000	0,000	0,004	0,053	0,000		0,000	0,000	0,001	0,000	0,000
Ig	POA	0,000	0,000	0,002	0,001	0,000		0,000	0,000	0,001	0,000	0,000
It	ASIM	0,000	0,000	0,154	0,488	0,000	0,000		0,002	0,013	0,000	0,000
It	POA	0,000	0,000	0,002	0,002	0,000	0,000		0,000	0,001	0,000	0,000
Nh	ASIM	0,000	0,000	0,049	0,003	0,000	0,000	0,002		0,010	0,001	0,000
Nh	POA	0,000	0,000	0,006	0,004	0,001	0,000	0,000		0,009	0,002	0,000
Nl	ASIM	0,001	0,001	0,014	0,016	0,001	0,001	0,013	0,010		0,001	0,001
Nl	POA	0,000	0,000	0,007	0,007	0,001	0,001	0,001	0,009		0,000	0,000
Su	ASIM	0,000	0,000	0,000	0,001	0,000	0,000	0,000	0,001	0,001		0,031
Su	POA	0,000	0,000	0,000	0,001	0,000	0,000	0,000	0,002	0,000		0,001
Sf	ASIM	0,000	0,000	0,001	0,001	0,000	0,000	0,000	0,000	0,001	0,031	
Sf	POA	0,000	0,000	0,001	0,001	0,000	0,000	0,000	0,000	0,000	0,001	

For convenience restrain our analysis to the OTU level in the analysis of individual soil samples. Cercozoan communities followed the general trends observed on combined site and treatment analyses, with CCCs largely being separated between sites and treatments (Fig. 58). All Swedish samples clustered distantly from other samples (brown symbols), English grassland sites grouped together (red symbols) and French agriculture (black circles) grouped with recently abandoned agricultural fields in the Netherlands (blue triangles; Fig. 58). The most divergent communities according to LUI were found in the abandoned agricultural fields in the Netherlands (Fig. 58).

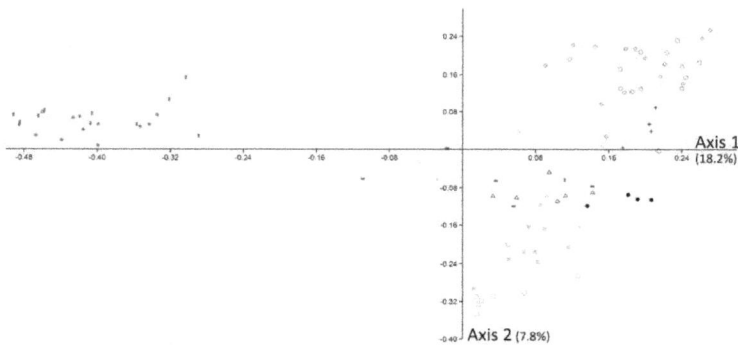

Fig. 58. PCoA showing sample specific differences in the cercozoan communities on the OTU level; geographically distant sites are illustrated with different colours, treatments with different symbols; black dots: French culture (Fc); black plus: French grassland (Fg); green empty squares: Italian intensive grassland (Ii); green filled squares: Italian wooden land, grass patch (Ig); green X: Italian wooden land under tree (It); red circle: English improved grassland (Ei); red diamond: English unimproved grassland (Eu); blue triangle: Dutch recently abandoned fields (Nh); blue dash: Dutch long-term abandoned fields (Nl); brown bars: Swedish unfertilized forest; brown stars: Swedish fertilized forest.

Taxon-specific patterns

On the class level, differences in cercozoan communities were largely caused by changes in Thecofilosea. This class comprised the highest relative abundance of 40.7 % in English grasslands and lowest in Swedish forests (14.3 %), but a general dependence on land management, such as an association with grasslands could not be detected (Supplementary Fig. 9; Supplementary Table 6). The dominant class Sarcomonadea was with 64.3 % of all cercozoan sequences significantly higher represented in France compared to Italy (45.8 %) and England (43.6 %; Supplementary Fig. 9; Supplementary Table 6), while Imbricatea was higher in Italy (27.0 %) and Sweden (21.7 %) compared to England (10.2 %) and France (8.8 %; Supplementary Fig. 9; Supplementary Table 6).

Especially on the OTU level the Swedish forests were clearly different to the other sites (Fig. 56; Supplementary Table 8). Striking examples were *Corythion* OTU3 and Glissomonad_Z OTU 36 being highly dominant in Swedish sites unlike at other sites (Fig. 56; Supplementary Table 8). *Rhogostoma* OTU2 represented by far the most abundant species in England, while it was less dominant in other sites, especially in Sweden. *Euglypha* OTU11 and *Trinema* OTU17 were more dominant in Italian compared with English soils, while Glissomonad_U OTU15 showed an inverse association with these two sites. *Neoheteromita* OTU5 was most dominant in French and Dutch soils especially in comparison to Italian soils (Fig. 56; Supplementary Table 8). Other site difference patterns of the most abundant species are shown in Supplementary Table 8.

Discussion

This study reveals clearly distinct CCCs in different geographic locations, with Swedish coniferous forests most strongly differing, but even grasslands soils harbouring different cercozoan communities. Grassland sites were replicated across Europe with treatments in England, France, Italy and the Netherlands, but CCCs differed between them. Similarly, no common OTU was found that was indicative for grasslands, suggesting that a complex combination of geographic distance, land management and other abiotic factors shape cercozoan- and protist communities. In line, climatic factors such as moisture have been shown to strongly influence protist communities (Bischoff 2002, Bates et al. 2013, Heger et al. 2013).

LUI had much less effect than geographic distance, but slight changes between low and high LUI were observed. These were most pronounced in France, the site with highest differences in LUI between treatments. CCCs in the intensively managed permanent agricultural fields in France more closely resembled those from recently abandoned fields in the Netherlands, suggesting a dominant and long lasting legacy effect of agriculture (Foissner 1997). We could, however, not detect a reduction in diversity or absence of major OTUs, indicating that synchronous shifts in nearly all OTUs might be underlying reasons rather than changes in only few groups. Testate amoebae have been suggested to be strongly reduced by agriculture, while species richness of ciliates sometimes even being higher in agricultural soils (Foissner 1997). Therefore, the entire CCCs might more closely resemble those of ciliates. We have not detected significant changes in testate amoebae, but as the resolution to species level was not possible, we might have missed species-specific changes and losses in some larger, k-strategist testate amoebae might have been compensated for by closely-related fast-growing r-strategists (Wanner et al. 2008).

We found that differences in CCCs become much more pronounced with increasing taxonomic resolution, resulting in largest differences on the OTU level. This likely explains why cultivation-based studies that are often limited to low taxonomic identification levels fail to detect substantial differences in community compositions of Cercozoa and protist in general (Finlay et al. 2000, Domonell et al. 2013). Interesting, however, is that differences in CCCs were even observed at the highest taxonomic levels, i.e. class level. Therefore, cercozoan and protist communities in general seem to be much more divergent in soils as suggested before (Finlay et al. 2000, Esteban et al. 2006).

The total of 957 distinct cercozoan specific OTUs placing at distant positions in phylogenetic analyses (Supplementary Figs. 6 and 7), provide strong support for the enormous diversity of soil cercozoans (Bass and Cavalier-Smith 2004, Bass et al. 2009b, Howe et al. 2009, Howe et al. 2011a). Such an analysis depth within a single study can only be accomplished by HTS methods and confirms that HTS will eventually become the gold-standard in studying (soil) protist diversity (Bates et al. 2013, Stoeck et al. 2014).

The results of different OTU clustering levels with an appropriate similarity threshold of 98 % is in line with a recent 454 study that targeted the V4 region in ciliates (Stoeck et al. 2014). Therefore it seems that this OTU clustering level when targeting the hypervariable region V4, as the suggested barcode region for protists (Pawlowski et al. 2012), is appropriate when analysing high-throughput data obtained with 454 pyrosequencing platforms. The need of OTU clustering to counteract sequencing errors and the non-existence of truly general, unbiased primers (Epstein and López-García 2008, Adl et al. 2014) currently lead to an underestimation of taxon richness in HTS. The use of recently developed PR2 database strongly reduced false OTU assignments that commonly introduce biases in HTS (Medinger et al. 2010, De Jonckheere et al. 2012, Stoeck et al. 2014). However, significant OTU numbers could not be assigned as protist sequence information is still highly underreported in databases (Pawlowski et al. 2012, Stoeck et al. 2014). Intensive efforts using cultivation-based approaches are needed to fill the immense gaps in public databases with reliable sequence information.

Little is known about soil Cercozoa, their biogeography and environmental factors shaping cercozoan soil communities, especially on the species level. Therefore, classical indicator or flagship species (Foissner 2006, 2009) could not reliably be assigned. However, we detected individual clades, especially at the OTU level, preferentially being associated with certain sites and LUI. Cercozoan spp. associated with moss were dominant in Swedish coniferous forest soils, confirming strikingly different communities of other protist groups present in coniferous forest soils (Bamforth 1980, Wanner 1991, Foissner 1998, Bobrov 2005). The testate amoebae *Corythion* spp. were among the dominant OTUs in Swedish forest, which have been shown to be dominant in moss-covered, grassland and forest soils (Heal 1965, Wanner 1991, Bamforth 2010, Carlson et al. 2010). The specious and widespread testate amoeba genus *Euglypha* is commonly found in mosses and litter layers (Wanner 1991, Bamforth 2010). However, *Euglypha laevis* was reported in high abundances in a range of orchards (Wanner 1991), which is in line with the high fraction of *Euglypha* among Cercozoa in wooden sites in Italy. In contrast, *Neoheteromita* OTU5 comprised only a small fraction among Italian and Swedish cercozoans. The genus *Neoheteromita* formerly belonged within the abundant and geographically widespread species complex *"Heteromita" globosa* (Howe et al. 2009), geographic location seems not to be the driving factor impacting this genus. Glissomonad_Z OTU 36 was another OTU dominating in Swedish forest soil. The sequences of as yet uncultivated is known from forest soils in the USA (Lesaulnier et al. 2008) and might be associated with forest soils. *Rhogostoma* is suggested to be abundant in freshwater and soil habitats (Howe et al. 2011b), rendering explanations of the dominance in English grassland sites difficult.

Taken together, we reveal that the community structure of Cercozoa shows biogeographic patterns with soil management also affecting the CCC, by the broadest study targeting soil protists to date. The enormous diversity sequences assigned to non-described taxa further suggests that a plethora of cercozoan species still is undiscovered, encouraging future studies to combine cultivation-based and molecular tools to broaden the understanding of taxon-specific functions within communities of Cercozoa and protist in general.

Supplementary Files

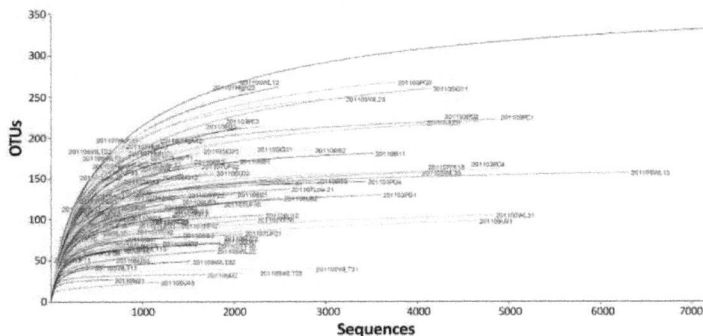

Supplementary Fig. 5. Rarefaction curves of amplicon data sets from all individual soil samples.

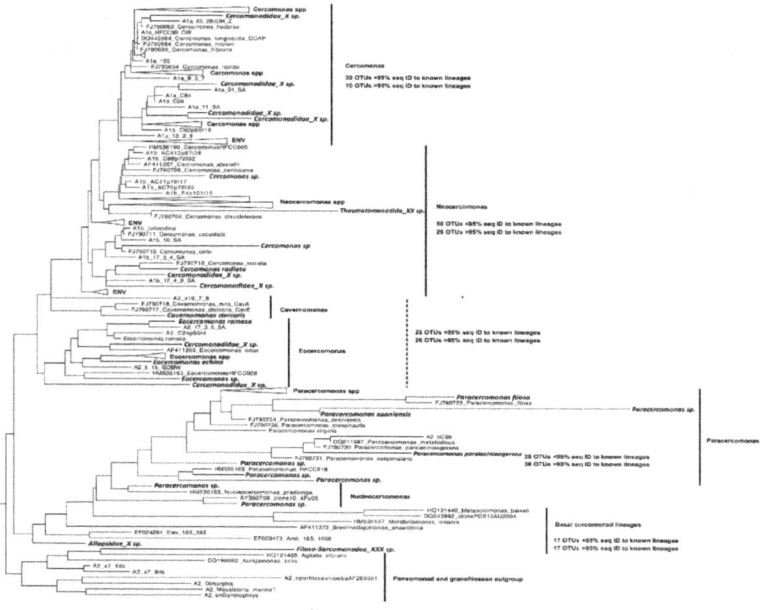

Supplementary Fig. 6. Maximum likelihood phylogenetic analyses placing 28 cercomonad OTUs (top). Bold italic: OTUs obtained in this study.

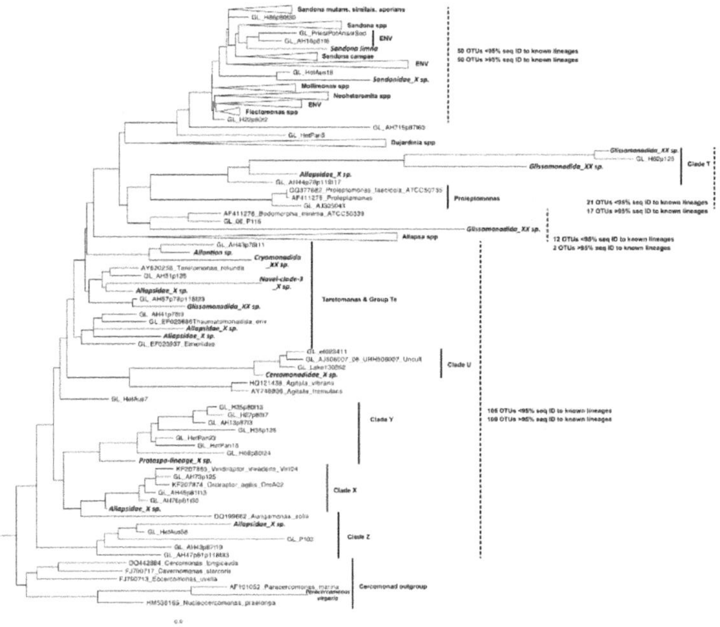

Supplementary Fig. 7. Maximum likelihood phylogenetic analyses placing 19 glissomonad OTUs. Bold italic: OTUs obtained in this study.

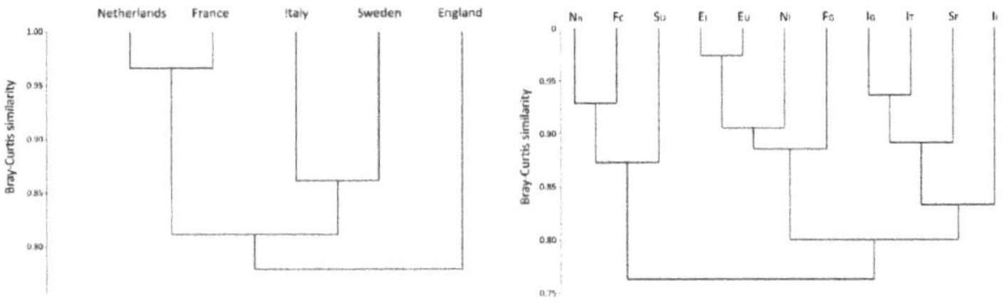

Supplementary Fig. 8. Similarity of cercozoan communities between sites (left) and treatments (right) based on class level using the paired group (UPGMA) analyses with Bray Curtis similarity; see Table 13 for details on soil characteristics and treatment abbreviations.

217

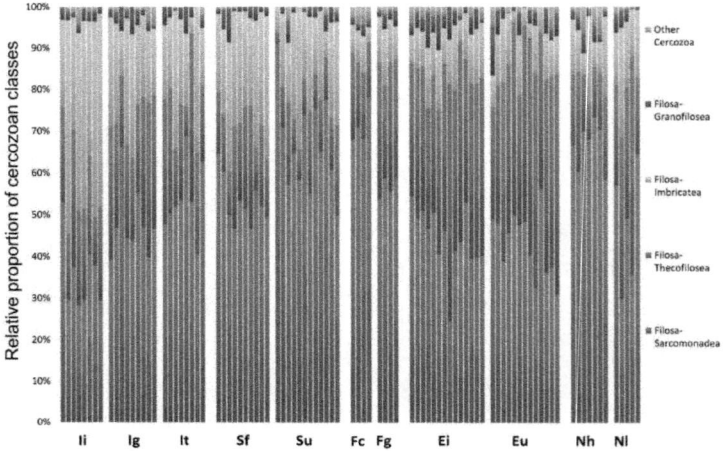

Supplementary Fig. 9. Rarified relative abundances of cercozoan classes in all samples at respective sites and treatments (x-axis); (see Table 12 for details on soil characteristics and treatment abbreviations).

Supplementary Table 6. Between-site comparisons of differences on cercozoan class level; E = England, F = France, I = Italy, N = Netherlands, S = Sweden; Stars indicate significant differences (* = q < 0.05, ** = q < 0.01)

Class	E-F	E-I	E-N	E-S	F-I	F-S	I-S
Imbricatea		**		**	*	*	
Sarcomonadea	*				*		
Thecofilosea	*	**	*	**			**

Supplementary Table 7. Between-site comparisons of differences on cercozoan OTU level; E = England, F = France, I = Italy, N = Netherlands, S = Sweden; Stars indicate significant differences (* = q < 0.05, ** = q < 0.01)

OTU	E-F	E-I	E-N	E-S	F-I	F-N	F-S	I-N	I-S	N-S
Glissomonad_Z OTU 36				**			**		**	*
Cercomonas OTU216				**			*		**	
Cercomonas OTU12				**					**	
Glissomonad_Y OTU103				**						
Corythion OTU3				**			**		**	*
Teretomonas_Te OTU45				**						
Eocercomonas OTU7									**	
Euglypha OTU11		**							**	
Euglypha OTU14		*							**	
Euglyphida OTU21		**						*	**	
Nudifila-relative OTU 293				**			**		**	
Cercomonad_undet OTU29				**					**	
Glissomonad_U OTU15 sp.		**		**						
Limnofila OTU93		*								
Neoheteromita OTU5					*		**	*		
Neoheteromita OTU34								**		
Paracercomonas OTU6				**				**		
Peregrinia OTU63						*	*			
Rhogostoma OTU2	*	**	*	**			*		**	
Glissomonad_NF-D OTU167		**								
Thaumatomonas OTU32						*	**		**	
Trinema OTU17				**			**			
Trinema OTU13	*	**			*	*	**		**	

Supplementary Table 8. Overview of 30 most abundant cercozoan OTUs obtained in this study; blast from original; T. = Thecofilosea; I. = Imbricatea; S. = Sarcomonadea

OTU	Class	Genus	Blast	Re-blast including own database	%IDENT	Genus
2	Filosa-T.	Rhogostoma	HM628668	HQ121430	97,9	Rhogostoma
1	Filosa-T.	Rhogostoma	AY620264	HQ121430	91,0	Rhogostoma
3	Filosa-I.	Corythion	EF456751	EF456751	96,2	Corythion
8	Filosa-S.	Flectomonas	AY965866	SCCAP_H251	100	Flectomonas
5	Filosa-S.	Sandona	U42447	ATCC50780p145at4____4173_bp	100	Sandona
9	Filosa-S.	Eocercomonas	AF372741	A2_244_25 2309_bpDA_2	100	Eocercomonas
15	Filosa-S.	Glissomonad_U	EF024447	AJ506007_06_URH506007_Ucultur	99,7	Glissomonad_U
12	Filosa-S.	Neocercomonas	FJ790711	Wyth3_8p68p138at6LS_2	100	Neocercomonas
167	Filosa-S.	Glissomonad_NF-D	JN207873	H63p80t19____2871_bp_____DA	97,9	Glissomonad_NF-D
6	Filosa-S.	Paracercomonas	AY884342	WA42p142at4	100	Paracercomonas
115 5	Filosa-T.	Rhogostoma	AB534325	HQ121430	97,5	Rhogostoma
45	Filosa-T.	Teretomonas_Te	EF024805	DQ303924	87,1	Teretomonas_Te
18	Filosa-T.	Teretomonas_Te	EF024805	AH51p125____3002_bp_____DA	86,2	Teretomonas_Te
13	Filosa-I.	Trinema	EF023588	EF456752	96,1	Trinema
41	Filosa-T.	Rhogostoma	EU798722	HQ121430	92,3	Rhogostoma
4	Filosa-T.	Rhogostoma	EF023854	HQ121430	92,6	Rhogostoma
17	Filosa-I.	Trinema	AJ418792	AJ418792	97,5	Trinema
36	Filosa-S.	Glissomonad_Z	EU709266	HetAus17____2820_bp_____DA	97,2	Glissomonad_Z
27	Filosa-T.	Rhogostoma	AY620303	HQ121430	94,4	Rhogostoma
28	Filosa-S.	Sandona	EU709147	AH15p125	100	Sandona
7	Filosa-S.	Eocercomonas	EF023536	C24p65t4	99,7	Eocercomonas
11	Filosa-I.	Euglypha	GQ330594	EF456753	95,4	k
44	Filosa-S.	Glissomonad_NF-E	EU709178	AH11p125	99,3	Glissomonad_NF-E
25	Filosa-T.	Rhogostoma	EF023854	DQ303924	93,0	Rhogostoma
29	Filosa-S.	Cercomonad_undet	AB695519	x7_4its	97,3	Cercomonad_undet
16	Filosa-S.	Paracercomonas	FJ790723	B1a_198_25f____2211_bp_____DA	100	Paracercomonas
95	Filosa-I.	Trinema	EF024634	AJ418792	97,2	Trinema
463	Filosa-S.	Glissomonad_Z	EU709266	HetAus17____2820_bp_____DA	98,9	Glissomonad_Z
23	Filosa-S.	Eocercomonas	AB534320	Z70.14_AH	88,1	Eocercomonas
26	Filosa-T.	Rhogostoma	EU798722	HQ121430	92,3	Rhogostoma

Part 2 – Chapter 8

Metatranscriptomic census of active protists in soils

Stefan Geisen[1,2], Alexander T. Tveit[3], Ian M. Clark[4], Andreas Richter[5], Mette M. Svenning[3], Michael Bonkowski[1*], Tim Urich[6]

[1] Department of Terrestrial Ecology, Institute of Zoology, University of Cologne, Germany
[2] Department of Terrestrial Ecology, Netherlands Institute for Ecology, (NIOO-KNAW), Wageningen, The Netherlands
[3] Department of Arctic and Marine Biology, UiT The Arctic University of Norway, Tromsø, Norway
[4] Department of AgroEcology, Rothamsted Research, Harpenden, Herts, UK
[5] Department of Microbiology and Ecosystem Sciences, University of Vienna, Austria
[6] Department of Ecogenomics and Systems Biology, University of Vienna, Austria

Abstract

The high numbers and diversity of protists in soil systems have long been presumed, but their true diversity and community composition have remained largely concealed. Traditional cultivation based methods miss a majority of taxa, while molecular barcoding approaches employing PCR introduce significant biases in reported community composition of soil protists. Here we applied a metatranscriptomic approach to assess the protist community in twelve mineral and organic soil samples from different vegetation types and climatic zones using small subunit ribosomal RNA transcripts as marker. We detected a broad diversity of soil protists spanning across all known eukaryotic supergroups and revealed a strikingly different community composition than shown before. Protist communities differed strongly between sites, with Rhizaria and Amoebozoa dominating in forest and grassland soils, while Alveolata were most abundant in peat soils. The Amoebozoa were comprised of Tubulinea, followed with decreasing abundance by Discosea, Variosea and

Mycetozoa. Transcripts of Oomycetes, Apicomplexa and Ichthyosporea suggest soil as reservoir of parasitic protist taxa. Further, Foraminifera and Choanoflagellida were ubiquitously detected, showing that these typically marine and freshwater protists are autochthonous members of the soil microbiota. To the best of our knowledge, this metatranscriptomic study provides the most comprehensive picture of active protist communities in soils to date, which is essential to target the ecological roles of protists in the complex soil system.

Introduction

Soils harbour a spectacular microbial diversity. Among the least studied soil microorganisms are single-celled protists. They display a high diversity of fundamentally different taxa, based on both morphological features and phylogenetic relatedness (Adl et al. 2012, Pawlowski 2013). Despite being microscopic, protist biomass in soils has been estimated to exceed that of most soil animal taxa (Schaefer and Schauermann 1990, Zwart et al. 1994, Schröter et al. 2003). Protists play important ecological roles in controlling bacterial turnover and community composition, recycling of nutrients, and plant growth promotion (Clarholm 1985, de Ruiter et al. 1993, Bonkowski 2004). However, we still lack basic knowledge of the protist communities in soils, and therefore a comprehensive understanding of their distribution and ecological functions in different soil systems has not yet been achieved.

The pervasive lack of knowledge on soil protist communities is mainly caused by the need to establish enrichment cultures, as the majority of protist taxa are difficult to extract and cultivate from soils, and the opaqueness of soil particles that prevents direct microscopic observation of the majority of taxa (Foissner 1987, Clarholm et al. 2007). Expert knowledge is needed for the time-consuming microscopic identification (Foissner 1987, Smirnov et al. 2008, Fenchel 2010, De Jonckheere et al. 2012). Therefore these traditional attempts to describe the full diversity of protist taxa in natural soils are rare and identification is usually only possible to a rather shallow taxonomic level (Finlay et al. 2000, Bamforth 2007, Geisen et al. 2014a). Cultivation-based approaches also introduce bias, as different growth media select for different species, and only a subset of taxa likely is cultivable (Ekelund and Rønn 1994, Foissner 1999b, Smirnov and Brown 2004).

The advent of molecular techniques along with a revised species concept that usually includes molecular information such as the universal protist barcode, the small subunit ribosomal RNA (SSU rRNA) gene (Pawlowski et al. 2012), has fundamentally altered the view on the "protist world". Consequently, the classification of and relatedness between protist taxa are constantly being revised (Adl et al. 2012, Pawlowski 2013). Further, environmental sequencing studies based on the SSU rRNA gene have revealed a huge diversity of previously unknown protists (Bass and Cavalier-Smith 2004, Berney et al. 2004, Lara et al. 2007a, Lejzerowicz et al. 2010, Bates et al. 2013). Despite molecular tools having diminished some of the problems associated to deciphering the community structure of soil protists, they introduce new biases that still obscure the true protist diversity in soils. Fundamental problems include (i) the lack of SSU rRNA reference sequences for a substantial number of protist species and genera, (ii) numerous mislabelled sequences in public databases, and (iii) a large seed bank of dormant protist cysts that may survive for decades (Goodey 1915a, Moon-van der Staay et al. 2006, Epstein and López-García 2008, Smirnov et al. 2008, Adl et al. 2012, De Jonckheere et al. 2012). Further biases are introduced by the PCR step of SSU rRNA gene studies (e.g. Bachy et al 2013). The often applied "general" eukaryotic primers to decipher the community structure of protists are in fact far from being truly universal (Adl et al. 2014). Thus, a strongly biased view of the protist community in soils is being depicted as only a subset of its diversity can be recovered (Jeon et al. 2008, Hong et al. 2009), while on the other hand some taxa within this subset will be overrepresented due to preferential PCR amplification (Berney et al. 2004, Medinger et al. 2010, Stoeck et al. 2014). Taxa of the common supergroup Amoebozoa as one of the dominant soil protists for instance, are notoriously underrepresented in molecular surveys due to long SSU rRNA sequences, frequent mismatches in primer regions and presence of introns (Berney et al.

2004, Fiore-Donno et al. 2010, Pawlowski et al. 2012). Ciliates on the contrary are highly overrepresented due to their shorter SSU rRNA sequences that ease amplification, and the presence of extremely high SSU rRNA gene copy numbers (Gong et al. 2013).

Most of these obstacles are avoided when directly targeting SSU rRNA transcripts instead of genes, especially by random hexamer-primed reverse transcription as in metatranscriptomic approaches. These rRNA transcripts are indicative of ribosomes and thus are likely derived from metabolically active cells and can be considered markers for living biomass (Urich and Schleper 2011). The generated cDNA fragments originate from different regions of the SSU rRNA molecule unlike PCR primed specific sites, and are therefore insensitive to the presence of introns or primer mismatches (Urich et al. 2008). These cDNA fragments can further be assembled into longer fragments, or even full-length SSU rRNA molecules for phylogenetic analyses (Urich et al. 2014). However, also metatranscriptomics has biases; the reported community composition is influenced e.g. by the different accessibility of SSU rRNA regions to primers and reverse transcriptase and also the physiological status of cells resulting in varying ribosomal content. Using this approach, Urich et al. (2008) generated cDNA from the total extracted RNA of soil communities. The cDNA was subjected directly to high throughput sequencing (HTS) without any SSU rRNA gene PCR steps. Although the fraction of SSU rRNA originating from protists is comparably small in metatranscriptomes, recent studies showed that the sequencing depth even with 454 pyrosequencing yielded sizable datasets of protist SSU rRNAs (Urich et al. 2008, Turner et al. 2013, Tveit et al. 2013).

We used this PCR-free metatranscriptomic approach to reveal the diversity of the active soil protist communities within five different natural soil systems in Europe, including forest, grassland and peat soils as well as beech litter. We annotated all protist SSU rRNA sequences to a reference database consisting of manually curated, published protist SSU rRNA sequences and revealed high protist diversity, with communities strongly differing between sites. We show that Rhizaria and Amoebozoa are the most abundant protists groups and detect abundant plant and animal pathogens. Further, we perform for the first

224

time an in-depth molecular analysis of the amoebozoan community structure. Finally, we report the widespread presence of protist clades that are usually associated with marine and freshwater environments, but not considered typical soil inhabitants.

Materials and methods

Soil sampling and processing

Arctic peat soils were sampled as described in Tveit et al. (2013). The grassland site (Park Grass, untreated control plot 3d, Rothamsted) was sampled by coring, with intact cores being brought to the laboratory, topsoil (5-10 cm) sieved (5mm mesh size) and subsequently flash-frozen in liquid nitrogen. Beech (*Fagus sylvatica* L) forest soils were sampled as described in Kaiser et al. (2010), the top soil (5-10 cm) sieved (5mm mesh size) and subsequently flash-frozen in liquid nitrogen. Beech litter from the same site was homogenized with a sterilised coffee grinder for 10 seconds and flash-frozen in liquid nitrogen.

Nucleic acid extraction, cDNA synthesis and sequencing

Nucleic acids were extracted from five gram of mineral soil and 1.2 gram of litter and peat and processed as previously described (Urich et al. 2008). cDNA synthesis was performed as described before (Radax et al. 2012, Tveit et al. 2013). 454-pyrosequencing was done either with FLX (forest soil and litter) or FLX Titanium (grassland, peat soil) chemistry. Sequencing was carried out at the CEES at the University of Oslo (Norway).

Sequence processing and analysis

Raw reads were processed as described in Tveit et al. (2013). Sequences were first filtered using LUCY (Chou and Holmes 2001), removing short (<150 bp) and low-quality sequences (>0.2% error probability). Small subunit (SSU) ribosomal RNA sequences of eukaryotes were identified by MEGAN analysis of BLASTn files against a 3-domain SSU rRNA reference database (Lanzén et al 2011; parameters: min. bit score 150, min. support 1, top percent 10; 50 best blast hits). All identified eukaryotic SSU rRNAs were reanalysed with CREST (Lanzén et al. 2012) using the Silvamod database and MEGAN with LCA parameters min bit score 250, top percent 2 (50 best blast hits) for classification of protist sequences. Correct taxonomic assignment of rRNA reads was verified by manual BLASTn searches against the NCBI GenBank nt database. For the high-resolution taxonomic annotation of Amoebozoa rRNA sequences, a custom-made database was constructed consisting of 1164 sequences from Silva (www.arb-silva.de) and sequences of in-house cultivated and newly described species (Geisen et al. 2014b, Geisen et al. 2014c, Geisen et al. 2014d). The taxonomy was set according to the most recent taxonomy of Amoebozoa (Smirnov et al. 2011b, Adl et al. 2012, Lahr et al. 2013) to enable high-resolution taxonomic placement of sequences. Reference database and taxonomy were then generated with the CREST scripts (Lanzén et al. 2012), and Amoebozoa sequences were classified using the same parameters in MEGAN as described above. The database can be obtained from TU or SG upon request.

SSU rRNA sequences of Choanoflagellida and Foraminifera were assembled into ribo-contigs using CAP3 (Huang and Madan 1999), performing two subsequent rounds of assembly with (1) a minimum overlap of 150bp with a minimum similarity threshold of 99% and mismatch and gap scores of -130 and 150, and (2) minimum overlap of 150bp and minimum 97% similarity threshold, respectively (Radax et al. 2012).

Phylogenetic analysis

Assembled choanoflagellate contig sequences were subjected to BLASTn searches against the NCBI nt database and manually aligned with their respective five best hits in Seaview 4 (Gouy et al. 2010). Additionally, SSU rRNA sequences of described choanoflagellates were added to this alignment. For phylogenetic analyses, 1,438 unambiguously aligned positions of a total of 71 unique sequences were retained, excluding ambiguous positions and several positions in variable regions especially in V4 (helix E23). Maximum likelihood phylogenetic analyses were run using RAxML v. 7.2.6 (Stamatakis 2006) using the GTR+γ+I model of evolution, as proposed by jModeltest v. 2.1.3 under the Akaike Information Criterion (Darriba et al. 2012), γ approximated by 25 categories. 1000 non-parametric bootstrap pseudoreplicates were performed.

Subsequently, Bayesian phylogenetic analyses were run in Mr Bayes v. 3.2.1 with GTR+γ+I model of evolution and 8 categories (Huelsenbeck and Ronquist 2001). Two runs of four simultaneous Markov chains were performed for 2,000,000 generations (with the default heating parameters) and sampled every 100 generations; convergence of the two runs (average deviation of split frequencies < 0.01) was reached after 590,000 generations. Therefore we discarded the first 5,900 trees and built a consensus tree from the remaining 14,100 trees.

Unweighted Pair Group Method (UPGMA) as a cluster analysis was applied to evaluate differences between the protist community composition in all samples (Sokal 1961).

Data deposition

The sequence data generated in this study was deposited in the Sequence Read Archive of NCBI under accession number SRP014474.

Results

Community composition of protist supergroups

In total, 32,808 SSU rRNA transcripts of protists were obtained from twelve soil metatranscriptomes (Table 16). In all cases, the biological replicates of each site yielded a very similar community composition (Fig. 59) and grouped together in a cluster analysis (Fig. 60). All five protist supergroups according to Adl et al. (2012) were found at each site (Table 16; Fig. 59). The SAR group, consisting of the formerly independent supergroups Stramenopiles, Alveolata and Rhizaria, dominated the active communities at all sites with sequences of Rhizaria being most numerous. We observed a clear dichotomy in protist community composition (Fig. 60) with dominance of Alveolata in peat soils with their high moisture and high organic matter content, versus dominance of Rhizaria and Amoebozoa in grassland and forest soils, including forest litter (Fig. 59).

227

Table 16. Soil sample description, protist SSU rRNA sequence numbers (±SD) and relative abundances obtained from each site; NA = Data not available; in italics: eukaryotic supergroups.

Sample name	Grassland Soil (Gs)	Forest Soil (Fs)	Forest Litter (Fl)	Peatland soil "Knutsen" (PsK)	Peatland soil "Solvatn" (PsS)
# of replicates	2	4	2	2	2
pH	4.9	4.5-5.1	NA	7.3	7.6
Moisture (% soil dry weight)	33	43-64	18	1010	900
Substrate type / Horizon	Mineral soil / A horizon	Mineral soil / A horizon	Litter /organic (L/F) horizon	Organic peat / Top layer	Organic peat / Top layer
Vegetation	Grassland	Beech forest	Beech forest	Fen wet land, moss dominated	Fen wet land, moss dominated
Climatic zone	Temperate	Temperate	Temperate	Arctic	Arctic
Location	Rothamsted, UK	Vienna woods, Austria	Vienna woods, Austria	Svalbard, Norway	Svalbard, Norway
# of Protist SSU rRNAs (%)	253±21 100%	961±470 100%	4862±2579 100%	2722±773 100%	6631±210 100%
SAR	163±15 64%	662±287 69%	3494±1735 72%	2058±844 76%	5265±187 79%
·Stramenopiles	24±8 9.5%	77±26 8.0%	234±238 4.8%	294±71 10.8%	731±25 11.0%
·Alveolata	24±1 9.5%	96±34 10.0%	870±542 17.9%	1295±756 47.6%	3797±144 57.3%
·Rhizaria	115±7 46%	489±236 50.9%	2391±954 49.2%	470±160 17.3%	737±18 11.1%
Amoebozoa	65±4 26%	250±87 26%	1009±633 20.8%	322±6 11.8%	586±29 8.8%
Excavata	7±1 2.8%	25±11 2.6%	233±141 4.8%	148±20 5.4%	420±23 6.3%
Archaeplastida	12±6 4.7%	11±7 1.1%	47±19 1.0%	144±93 5.3%	222±10 3.3%
Opisthokonta	6±4 2.4%	19±10 2.0%	82±52 1.7%	50±7 1.8%	139±6 2.1%

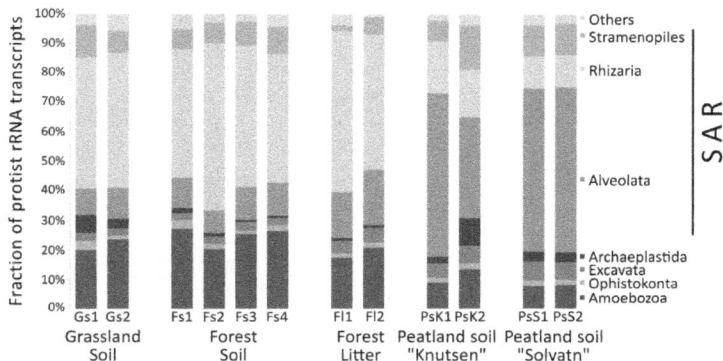

Fig. 59. Community composition of protist supergroups in the investigated soils. For detailed information on soil parameters see Table 16.

The Rhizaria were almost exclusively comprised of Cercozoa (Fig. 61A) with dominance of rRNAs of the small flagellates belonging to the class Filosa-Sarcomonadea (formerly combined into Cercomonadida) and both, flagellated and amoeboid Cercozoa in the class Filosa-Imbricatea (formerly Silicofilosea). Foraminifera SSU rRNAs occurred in relatively constant, albeit low abundances in all samples (Fig. 61A) except for the peatland site "Knutsen", where they were more abundant and comprised 22 % of all Cercozoa. Alveolata SSU rRNAs were quite variable among samples and dominated in the peat soils (Fig. 59) with the phylum Ciliophora and its main orders Spirotrichea, Colpodea and Oligohymenophorea being most abundant, while the exclusively parasitic Apicomplexa still represented up to 11% of all Alveolata (Fig. 61B). The third group in SAR, Stramenopiles, was less abundant, with Oomycetes representing the dominant stramenopiles in forest soils (Fig. 61C), while abundant transcripts of the photosynthetic Bacillariophyta were characteristic in the waterlogged peatland soils. Chrysophyceae SSU rRNAs were abundantly found in grassland soils and litter, as were transcripts of the Bicosoecida. The supergroup Amoebozoa represented up to 30% of SSU rRNAs, with highest abundances in the grassland and forest soil samples (Fig. 59 and Table 16). The other protist supergroups, i.e. Excavata, Archaeplastida and Opisthokonta

229

(excluding fungi and animals) were generally less abundant (Fig. 59 and Table 16).

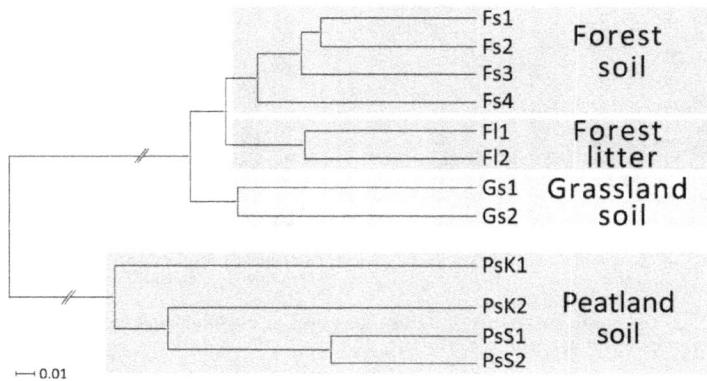

Fig. 60. UPGMA clustering analysis to evaluate differences between soil protist communities. For detailed information on soil characteristics and abbreviations see Table 16.

SSU rRNA transcripts of protist clades highly represented in cultivation based approaches were analysed to evaluate if these clades were also recovered in our metatranscriptomes. Among the clades targeted were flagellates of the supergroups SAR (Glissonomadida and Cercomonadida (Sarcomonadea; Rhizaria), Chrysophyceae and Bicosoecida (Stramenopiles) and Excavata (Bodonidae and Euglenida), as well as amoebae in the supergroup Excavata (Heterolobosea). All clades were found at each location, with cercomonads being highly abundant in grassland and forest habitats. Glissomonads were abundant especially in grasslands (5 % of all protists). Bodonids were also common (1.9 % of all protists) while euglenids, chrysophytes, bicosoecids and heteroloboseans comprised only about 1 % of all protist transcripts. Exhaustive analyses of the most abundant clades of amoebae are presented in the next section.

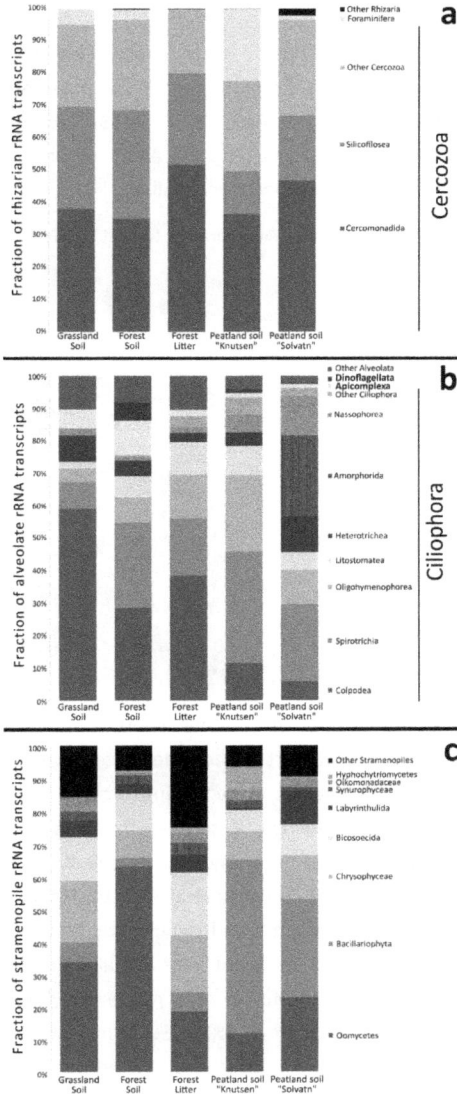

Fig. 61. Community composition within the SAR supergroup independently showing the community compositions within the individual clades of SAR, i.e. Rhizaria (a), Alveolata (b) and Stramenopiles (c). Shown are means of biological replicates. Labels and sites as described in Table 16 and Figs. 59 and 2.

Community composition of Amoebozoa

The supergroup Amoebozoa was of particular interest, since no comprehensive molecular taxonomic analysis of this protist supergroup in soil exists to date due to PCR-primer biases (Baldwin et al. 2013, Bates et al. 2013). For this purpose we constructed a high-resolution reference database and taxonomy of Amoebozoa (see Table 17 and Materials and Methods for details).

Table 17 Overview of taxa and taxonomic classification of Amoebozoa in the CREST reference database and taxonomy

Supergroup	Subphylum	Class	Subclass	Order	Suborder	Family	Genus
Amoebozoa	Lobosa	Tubulinea		Euamoebida		Unresolved	
				Arcellinida	Arcellina	Unresolved	
					Difflugina	Unresolved	
					Phryganellina	Unresolved	
				Nolandida		Nolandellidae	*Nolandella*
				Echinamoebida		Vermamoebidae	*Vermamoeba*
						Echinamoebidae	*Echinamoeba*
				Leptomyxida		Leptomyxidae	*Leptomyxa* *Rhizamoeba*
		Discosea	Flabellinia	Vannellida		Vannellidae	*Vannella*
				Himatismenida		Cochliopodiidae	*Cochliopodium*
				Dactylopodida		Vexilliferidae	*Vexillifera*
						Paramoebidae	*Korotnevella*
				Thecamoebida		Thecamoebidae	*Stenamoeba* *Thecamoeba*
			Longamoebia	Dermamoebida		Mayorellidae	*Mayorella*
						Dermamoebidae	*Paradermamoeba*
				Centramoebida		Acanthamoebidae	*Acanthamoeba*
						Balamuthiidae	*Balamuthia*
	Conosa	Variosea		Varipodida		Filamoebidae	*Filamoeba* *Flamella*
						Acramoebidae	Unresolved
				Schizoplasmodiida		Unresolved	Unresolved
		Mycetozoa		Unresolved			

Using this taxonomic assignment approach, the rather short SSU rRNA sequence reads could reliably be classified at least to the order, often even to the genus level. Amoebozoa were highly diverse at each sampling site, with rRNAs assigned to four classes, i.e. Tubulinea, Discosea, Variosea and Mycetozoa with major orders Euamoebida, Leptomyxida, Arcellinida and Centramoebida (in the subclass Longamoebia) (Smirnov et al. 2011b) (Fig. 62). The community composition within Amoebozoa differed significantly between sites. For instance, sequences assigned to the dominant class Tubulinea made up between 27.4 and 69.2% in Solvatn peat and forest soils, respectively, and were inversely related to Discosea with 41.6 to 12.7% in Solvatn peat and forest soils. Similar to the patterns observed at the class level, the proportional distribution within classes differed between sites. Among Tubulinea, the dominant order Euamoebida reached highest relative abundances in grasslands and forest mineral soils, while testate amoebae of the order Arcellinida became more abundant in organic rich substrates of litter and peat soils, and Leptomyxida were characteristic of forest habitats. The remaining tubulinean orders Echinamoebida and Nolandida were generally rare. Among the class Discosea, SSU rRNAs assigned to the subclass Longamoebia were almost entirely (96.1%) composed of the order Centramoebida. Sequences of the discosean subclass Flabellinia were generally less abundant (7.0 % oas) and mostly derived from the order Vannellida (50.1% of Flabellinia SSU rRNAs). Sequences assigned to the subphylum Conosa could only reliably be assigned to the class level, as taxonomy and the phylogenetic affiliations, especially of protists in the class Variosea, are still largely unresolved (Adl et al. 2012). Variosea was the dominant conosan class in grassland, forest and the Solvatn peat soil, while Mycetozoa were more abundant in forest litter and Knutsen peat soil (Fig. 62).

Widespread were amoebozoan SSU rRNA sequences with high sequence identity to potential parasites. At all sites occurred diverse sequences related to facultative human pathogens of the genus *Acanthamoeba* (≥ 97% sequence identity; 1.1 – 4.4% oas) while sequences most closely resembling *Balamuthia* were discovered in low relative abundance (0.1 – 0.5% oas) in all except the grassland soils. Transcripts related to groups of non-Amoebozoan parasitic taxa were also found, such as sequences most closely resembling the human pathogen *Naegleria fowleri* (Heterolobosea in the supergroup Excavata; ≥ 97% sequence identity) in all four arctic peat samples. Further, opisthokont Ichthyosporea (mainly animal parasites) were ubiquitously found (0.4 – 3.3% oas) as well as predominantly plant-parasitic *Plasmodiophora* (up to 0.8 % oas).

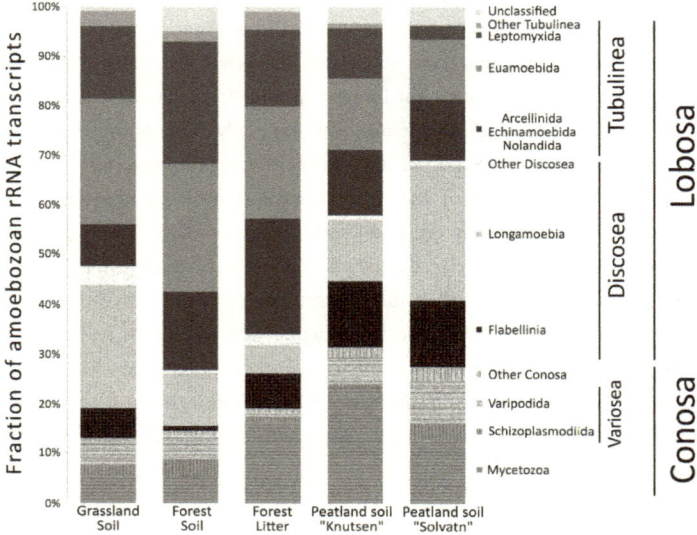

Fig. 62. Community composition within the supergroup Amoebozoa in the investigated soils. Shown are means of biological replicates. Labels and sites as described in Table 16 and Figs. 59 and 60.

Widespread presence of Foraminifera and Choanoflagellida in soils

SSU rRNA transcripts of the typically marine groups Foraminifera and Choanoflagellida revealed their general presence and activity in all samples (Fig. 63) and for the first time allowed to estimate the relative abundance of these taxa in soils. They comprised between 0.1 and 3.5% of all protist SSU rRNAs.

234

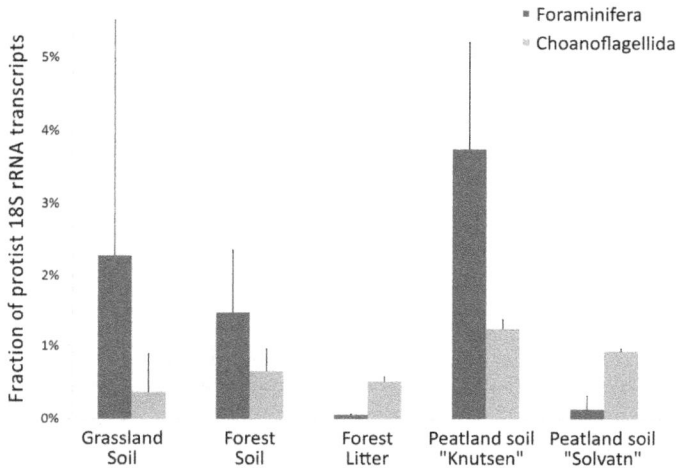

Fig. 63. SSU rRNA transcript abundance ± SD of Foraminifera and Choanoflagellida in the soil metatranscriptomes. Labels and sites as described in Fig. 60 and Table 16.

To enable better taxonomic assignment and phylogenetic placement of these enigmatic soil protist groups we assembled larger SSU rRNA sequences from the short 454 reads. Three assembled SSU rRNA contigs of Foraminifera (742 to 890 bp length) were quite similar (≥ 91%) to sequences obtained in a recent focused PCR-based molecular survey targeting soil Foraminifera (Lejzerowicz et al. 2010), but showed substantial sequence dissimilarity to described species (maximum SSU rRNA sequence identity of ≤ 76%). Several unassembled SSU rRNA sequences, however, closely matched sequences typically obtained from freshwater and marine environments, such as the genera *Astrammina*, *Bathysiphon*, *Allogromia* and diverse uncultured species.

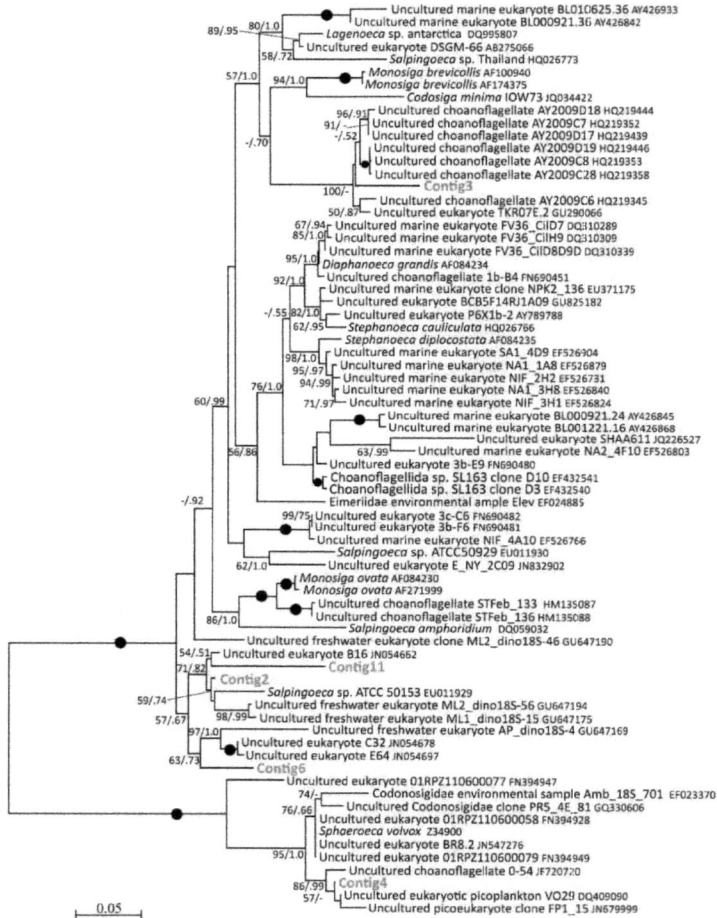

Fig. 64. Maximum likelihood tree of Choanoflagellida placing assembled long SSU rRNA contigs (red) with sequences of described and uncultivated choanoflagellates. 71 sequences with 1,232 unambiguously aligned positions were used; the tree is unrooted; values higher than 50 for maximum likelihood analyses (left) and 0.50 for Bayesian analyses (right) shown. Black circles indicate full support.

Many Choanoflagellida-affiliated SSU rRNA sequences closely matched (with ≥99% maximum identity) published sequences of uncultivated choanoflagellates (e.g. with GenBank accession numbers HQ219439 [freshwater], EF024012 [soil], JF706236 and GQ330606 [peat soil]) while others reached similarities of > 95% with uncultivated choanoflagellate sequences among typical freshwater and marine genera, such as *Monosiga, Codonosiga, Salpingoeca,* and more rarely with *Lagenoeca, Stephanoeca, Didymoeca, Diaphanoeca, Desmarella* and *Acanthoeca*. Five assembled long SSU rRNAs (763 to 1163 bp) had high sequence similarities of 96 to 99% to uncultivated freshwater choanoflagellates (Fig. 64), but the closest hit to a formally described species was < 92%.

Discussion

Metatranscriptomics-enabled census of active soil protists

Cultivation-based studies have shown that soil protists are diverse, abundant and ecologically important (Ekelund and Rønn 1994, de Ruiter et al. 1995, Bonkowski 2004, Pawlowski et al. 2012), but fail to detect a majority of taxa (Epstein and López-García 2008). In contrast, cultivation-independent PCR-based SSU rRNA gene sequencing studies have revealed a much higher diversity of soil protists as previously anticipated (Lara et al. 2007a, Lejzerowicz et al. 2010, Bates et al. 2013); however, they still fail to amplify the SSU rRNA genes of a wide range of protists and provide a highly skewed picture of protist communities (Epstein and López-García 2008, Weber and Pawlowski 2013, Stoeck et al. 2014). This is exemplified by the negative selection of amoebozoan sequences in PCR-primer based approaches (Berney et al. 2004, Amaral-Zettler et al. 2009), explaining the virtual absence of nearly the entire supergroup in HTS surveys (Baldwin et al. 2013, Bates et al. 2013), while the high rRNA gene copy numbers in ciliates (Gong et al. 2013) generally lead to their over-proportional representation.

Through the usage of metatranscriptomics we avoided some of the above-mentioned issues to gain a more accurate picture of the protist diversity in soil systems. In fact, this study and earlier metatranscriptomes have revealed a different soil protist community structure than suggested before (Urich et al. 2008, Baldwin et al. 2013). The only similarity to primer-based HTS surveys is merely the dominance of the supergroup SAR, which was due to high sequence abundances of Rhizaria (and Alveolata in arctic peat soils) (Baldwin et al. 2013, Bates et al. 2013). However, SSU rRNA sequences of Amoebozoa dominated over those of Alveolata in all non-peat soils. This is in line with cultivation-based studies, which show that small flagellates, especially cercomonads and glissomonads (Rhizaria) (Finlay et al. 2000, Ekelund et al. 2001, Howe et al. 2009) and amoebae represent the numerically dominant soil protists (Schaefer and Schauermann 1990, Finlay et al. 2000, Robinson et al. 2002). Similarly, other taxonomically highly divergent heterotrophic protist taxa especially abundant in cultivation based studies such as bodonid and euglenid euglenozoans (Finlay et al. 2000, Geisen et al. 2014a), chrysophytes (Finlay et al. 2000, Boenigk et al. 2005, Chatzinotas et al. 2013), bicosoecids (Ekelund and Patterson 1997, Lentendu et al. 2014) and heteroloboseans (Geisen et al. 2014a) were identified, but contributed relatively little to the entire protist community.

It should be noted here that metatranscriptomics also has several biases and shortcomings. For example can the reported community composition be biased by different accessibilities of SSU rRNA regions to primers and enzymes due to RNA secondary structure. Further the number of ribosomes can vary dependent of the physiological status of the cell. The higher costs related to cDNA synthesis and missing specificity as compared to PCR-amplicons make this approach rather low-throughput.

Diversity of Amoebozoa: the neglected supergroup in soil

In addition to the abovementioned negative biases of Amoebozoa in HTS approaches, a lack of molecular data and SSU rRNA reference sequences until now prevented thorough molecular diversity analyses of this supergroup. Therefore the diversity of Amoebozoa in terrestrial samples is still largely unknown (Smirnov et al. 2005, Smirnov et al. 2008). Applying the most recent taxonomy and a newly built reference database allowed us to detect a large diversity within all amoebozoan classes. The general dominance of the classes Tubulinea and Discosea confirmed former cultivation-based studies (e.g. Bass and Bischoff 2001, Finlay et al 2000) , whereas the high abundance of Variosea is entirely novel. This class was only constructed in 2004, hosting mostly large plasmodial, branching or reticulate amoebae (Cavalier-Smith et al. 2004, Smirnov et al. 2008, Smirnov et al. 2011b). To date only few variosean species have formally been described, and molecular information on those taxa is even more limited (Berney et al in press). Recent cultivation studies specifically targeting Variosea confirm the results obtained here and revealed an unprecedented diversity of these large amoebae in soils (Geisen et al. 2014a). The addition of several of these recently published variosean SSU rRNA sequences (Geisen et al. 2014a) to our reference database resulted in improved sequence assignments of variosean-specific SSU rRNAs and confirmed that this group of amoebae has been entirely overlooked in earlier studies.

Protist parasites

Terrestrial Oomycetes, distinguished for containing diverse and devastating plant-pathogens (Latijnhouwers et al. 2003) such as *Phytophtora infestans,* the causative agent of potato blight (Martin et al. 2012), also include a variety of pathogens of other organisms (Benhamou et al. 1999, Phillips et al. 2008). Oomycetes in soil have been targeted with little success (Coince et al. 2013). Our metatranscriptome approach revealed that Oomycetes were ubiquitous, abundant, and active members among soil protists, suggesting a significant role as structuring elements of natural plant communities. Unlike in Urich et al. (2008) plant parasitic plasmodiophorids were little abundant in this study.

Apicomplexa are common parasites of soil invertebrates and may potentially play a comparable role in structuring soil food webs (Altizer et al. 2003, Field and Michiels 2005). Similar to a recent DNA based study (Bates et al. 2013), we found *Apicomplexa* in all samples. However, we found higher relative abundance in the dry grassland and forest habitats, in contrast to Bates et al. (2013), who detected Apicomplexan SSU rRNA genes with higher relative abundance in the wet soils. Since the detected sequences were derived from

239

RNA, they probably did not originate from encysted Apicomplexa (Ruiz et al. 1973), but most likely from parasitized soil invertebrates.

Most ichthyosporean taxa are animal parasites and have been described as inhabitants of aquatic organisms, but have recently been found as parasites in terrestrial animals (Glockling et al. 2013). We support these recent findings as we detected Ichthosporea in significant abundances in all habitats studied.

Protist sequences closely resembling potential human pathogens were detected in all soils. Similar to cultivation-based studies (Page 1988, Geisen et al. 2014b), *Acanthamoeba* spp. were common, some of which might be causative agents of amoebic keratitis and amoeboic encephalitis. However, as the SSU rRNA sequences never showed perfect sequence matches with described species and only functional studies on cultivated species allow drawing reliably conclusions on potential pathogenicity, we can only state that these could potentially act as pathogens. Amoebic encephalitis can also be caused by *Balamuthia mandrillaris* and *Naegleria fowleri* (Excavata, Heterolobosea) (Visvesvara et al. 1993, Schuster and Visvesvara 2004, Visvesvara et al. 2007, Siddiqui and Ahmed Khan 2012). The finding of related SSU rRNA sequences support recent studies showing the presence of both *B. mandrillaris* (Lares-Jiménez et al. 2014) and *N. fowlerii* in soil (Moussa et al. 2015).

Unexpected presence of typically marine and freshwater protists

The surprising finding of several typically aquatic protist groups corroborates the notion that a plethora of soil protist taxa have formerly been missed in both, cultivation- and PCR-primer-based studies. For example, we detected Choanoflagellida in all samples; these protists are typically marine and only a few taxa are known from freshwater systems (Tong et al. 1997, Arndt et al. 2000, Stoupin et al. 2012). Mostly anecdotic evidence from few cultivation-based studies exists on the existence of the choanoflagellate genera *Monosiga*, *Codosiga* and *Salpingoeca* in soils (Ekelund and Patterson 1997, Finlay et al. 2000, Ekelund et al. 2001, Tikhonenkov et al. 2012), and only a few molecular soil surveys have reported choanoflagellate SSU rRNA sequences (Lesaulnier et al. 2008, Lara et al. 2011), several of which closely resemble SSU rRNA sequences obtained in this study. However, other sequences, among them our assembled rRNA contigs (Fig. 63), more closely resembled sequences obtained in freshwater surveys (Chen et al. 2008, Monchy et al. 2011, Stoupin et al. 2012). As most transcripts showed highest similarity with SSU rRNAs of uncultivated species, the future linkage of sequence information with morphological and functional information on the respective protist species in soil will be essential (Bachy et al. 2013).

Foraminifera are another group of typically marine protists commonly not associated with the soil environment, with *Edaphoallogromia australica* representing the only foraminiferan species reported from soil (Meisterfeld et al. 2001). Our study is the first to assess their relative abundance among active soil protists (Lejzerowicz et al. 2010). Although Foraminifera appeared to comprise a small fraction of the active protist community, they were diverse and present at all sites. This is supported by a recent targeted DNA survey that detected diverse foraminiferan SSU rRNA genes in 17 out of 20 soils samples (Lejzerowicz et al. 2010). Most sequences and SSU rRNA contigs closely matched the sequences obtained by Lejzerowicz et al. (2010), but several non-assembled transcripts more closely resembled the typically marine genera *Astrammina*, *Bathysiphon* and *Allogromia*.

The presence of rRNA transcripts of Choanoflagellida and Foraminifera in soil has important implications. First, these taxa are genuine autochthonous, active inhabitants of soils, and not simply dormant states accidentally dispersed to soil by wind or water. Second, there is a need to cultivate these protists, because this will provide information about their ecological roles, morphology, and adaptations to the terrestrial habitat, as well as enable comparative studies with their marine relatives that may provide insights in the evolutionary origin of soil protists.

Conclusions

Our study demonstrates the power of metatranscriptomics for obtaining a census of soil protist communities by circumventing major biases commonly associated with cultivation and molecular studies. Still, significant gaps in the taxonomic information prevail in all soil protist supergroups, especially in *Amoebozoa* where reliable assignment of several SSU rRNA sequences beyond order or even class level was still impeded. Cultivation based approaches (Geisen et al. 2014c, Geisen et al. 2014d) are necessary in order to fill these gaps. Furthermore, taxonomic expertise remains a crucial prerequisite to interpret protist SSU rRNA data and is as indispensable as a reliable reference database for the correct assignment of taxa. Taking the high sequence coverage into account, we are confident that our study provides the most detailed and potentially closest picture of the true composition of active protist communities in soils so far.

Acknowledgements

We thank Ave Tooming-Klunderud and colleagues at the Centre for Ecological and Evolutionary Synthesis (CEES, University of Oslo, Norway) for 454-sequencing. We thank Sarah Jennings (NIOO, Netherlands) for help in cDNA generation from Rothamsted soil and thank Christoph Bayer (Vienna) for help in sequence processing. Christa Schleper (Vienna) is thanked for her generous and continuous support. Special thanks to Fionn Clissman for his helpful suggestions on the wording. Part of this work was supported by a research grant from the EU-project 'EcoFINDERS' No. 264465.

Author Contributions

TU, SG and MB designed research. TU, AT, MS and IC collected the soil samples. Metatranscriptomes were generated by TU and AT. TU processed metatranscriptome data, analysed by SG and TU. SG and TU generated Amoebozoa-CREST. SG, TU and MB wrote the manuscript, aided by all authors.

Part 3

Ecological studies to broaden the knowledge on soil protist function other than feeding on bacteria

Part 3 – Chapter 9

Mycophagous protists in soils: diverse and widespread

Geisen Stefan[1*], Hünninghaus Maike[1], Dumack Kenneth[1], Koller Robert[1,3], Urich Tim[2], Bonkowski Michael[1]

[1] Department of Terrestrial Ecology, Institute of Zoology, University of Cologne, Germany

[2] Department of Ecogenomics and Systems Biology, University of Vienna, Austria

[3] Current address: Institute of Bio- and Geosciences, IBG-2: Plant Sciences, Forschungszentrum Jülich GmbH, 52425 Jülich, Germany.

Abstract

Protists are suggested to be bacterivores and serve together with bacterivorous nematodes as the main controllers of the bacterial energy channel in soil food webs, while the fungal energy channel is assumed to be controlled by arthropods and mycophagous nematodes. This perspective accepted by most soil biologists is, however, challenged by functional studies conducted by taxonomists that revealed range of mycophagous protists. In order to increase the knowledge on the functional importance of mycophagous protists we isolated and initiated cultures of protist taxa and tested eight for facultative feeding on diverse fungi in microcosm experiments. Two different flagellate species of the genus *Cercomonas*, the testate amoebae *Cryptodifflugia* sp. and four genera of naked amoebae (*Acanthamoeba* sp., *Leptomyxa* sp., two *Mayorella* spp. and *Thecamoeba* spp.) fed and grew on yeasts with four taxa (*Cercomonas* sp., *Leptomyxa* sp., *Mayorella* sp., and *Thecamoeba* sp.) also thriving on spores of a plant pathogenic hyphal-forming fungus.

To identify the potential importance of mycophagous protists in the environment we applied a data-mining approach on sequences obtained in two small subunit (SSU) rRNA-based high throughput sequencing studies. We focused our analyses on the distribution and relative abundances of two well-studied mycophagous protist groups, vampyrellid amoebae and grossglockneriid ciliates. Both groups were detected in all highly contrasting terrestrial samples, comprising up to 5.4 % of all protist SSU rRNA transcripts.

Taken together, this study provides evidence that mycophagy among soil protists is common and might be of substantial but hitherto overlooked ecological importance in terrestrial ecosystems. Future studies should aim at evaluating taxon-specific (facultative) mycophagy, decipher changes caused in the fungal community and quantitatively evaluate the functional importance of this trophic position in soil ecosystems.

Introduction

Soil biologists generally discriminate the nutrient flows in soil food webs into a bacterial and a fungal-based energy channel (Moore and Hunt 1988, Holtkamp et al. 2011). Heterotrophic protists are considered as major consumers of bacterial biomass supplementing higher trophic levels with nutrients bound in bacterial biomass, while they are supposed to be of marginal importance in the fungal energy channel where microarthropods and mycophagous nematodes are suggested to be the predominant consumers (Hunt et al. 1987, de Ruiter et al. 1995, Bonkowski 2004).

A major reason why soil biologists largely treat protists as bacterivores derived from traditional extraction and cultivation methods that select for bacterivorous protists (Page 1988, Berthold and Palzenberger 1995, Ekelund 1998). Protist taxonomists have, however, long realized that diverse facultative and obligate mycophagous protist taxa are common in soils (Old and Darbyshire 1978, Petz et al. 1985, Ekelund 1998). For example, all described ciliates of the family Grossglockneriidae are obligatory mycophagous (Foissner and Didier 1983, Petz et al. 1985, Petz et al. 1986). Facultative mycophagous protists (omnivores) that feed on a range of soil eukaryotes are vampyrellid amoebae (Old and Darbyshire 1978, Hess et al. 2012), *Thecamoeba* spp.

(Bamforth 2004) and few testate amoebae (Mitchell et al. 2008, Wilkinson 2008, Wilkinson and Mitchell 2010). All of the aforementioned taxa are large (often >100 µm), but mycophagy has even been reported in small flagellates (Hekman et al. 1992, Flavin et al. 2000). Still, detailed knowledge on diversity and functional importance of mycophagous protists in terrestrial ecosystems remains limited, but could be much more important than assumed, considering that mycophagous protists might reach biomasses similar to those of bacterivorous protists (Ekelund 1998).

Recent methodological advances using molecular techniques provided means of investigating the largely unknown diversity of formerly uncultivable soil protists. These sequence based studies have revealed that cultivable protists only represent a small fraction of the total protist community in soils (Foissner 1999b, Bates et al. 2013, Kamono et al. 2013). The ecological function of uncultivated protists remains, however, largely unknown, as it is usually impossible to reliably link molecular phylogenetic information with realized ecological functioning. Molecular sequence-information needs to be supplemented with functional information on ecological traits of the respective taxon, which relies mainly on in-*vitro* studies on cultivated protists. Experiments on cultivated species revealed not only a variety of mycophagous protists (op cit.), but also that even closely related protists differentially feed on bacterial prey (Böhme et al. 2009, Glücksman et al. 2010, Saleem et al. 2012). Differential feeding on fungal prey might therefore also be characteristic for mycophagous protists, thereby structuring the fungal community soils. Some evidence for this hypothesis came from former studies showing that yeasts are a preferred food source for protists, while hyphae forming fungi being less suitable as prey (Heal 1963, Bunting et al. 1979, Allen and Dawidowicz 1990).

In order to increase the knowledge on the diversity and importance of mycophagy among soil protists we (1) cultivated soil protists and tested if facultative mycophagy is common for "bacterivorous" protists. Further (2) we used a data mining approach on SSU rRNA sequences obtained from a variety of soils (Geisen et al. 2015b) to investigate the presence and relative abundance of mycophagous protists in highly diverse terrestrial samples.

Materials and Methods

Isolation of protists from soil and microscopic observation on facultative mycophagy

Soil samples were taken in Pulheim Stommeln (Germany; 51°01'N, 6°45'E); Müncheberg (52°30'N, 14°07'E), in Les Verrines (France; 46°25'N, 0°7'E) and Cologne (Germany; 50°55'N, 6°55' E). The organic soil horizon was sampled in two locations (upper 2 cm in Pulheim Stommeln, 10 cm in Müncheberg and 10 cm in Cologne). Additionally, soil from earthworm burrows (2 mm around burrows) was sampled at Les Verrines (Table 18).

Enrichment cultures were established to isolate (facultative) mycophagous protists. From each soil sample, 1 g dry weight of soil was suspended in 250 ml Neff's Modified Amoeba Salina (NMAS) according to Page (1988). After shaking on an orbital shaker (Köttermann, Uetze, Germany) at 100 rpm for 10 min and fourfold dilution with NMAS, 20 µl of the suspension was added to wells of a 24 multiwell-plate (Sarstedt, Nümbrecht, Germany). A mixed fungal inoculum of 80 µl of a 0.4 g $*$ l^{-1} NMAS solution of dried *Saccharomyces cerevisiae* (Ruf, Quakenbrück, Germany) and 160 µl of a *Fusarium culmorum* spore solution with a concentration of four spores $*$ $µl^{-1}$ were added to each well. Plates were sealed with Parafilm and stored at 15 °C in the dark. These enrichment cultures were examined microscopically for mycophagous protists, i.e. growth on fungi and ingestion of fungal material, 7 and 21 days after incubation using an inverted microscope (Nikon Eclipse TS100, Japan) at 100 x and 400 x magnification.

Subsequently, enrichment cultures with fungal growth medium were initiated using malt extract agar (MEA; 1.5 %). MEA plates were prepared by adding malt extract (1.5 %; AppliChem, Darmstadt, Germany) and agarising it by adding non-nutrient agar followed by autoclaving (122°C, 20'). MEA plates were inoculated with 100 µl suspension of *F. culmorum* spores and hyphae in H_2O_{dest}

to establish active fungal cultures. Cultures of the yeast *Cryptococcus laurentii* were incubated on potato glucose agar (1.5 %; Sigma-Aldrich, St. Louis, USA) supplemented with yeast extract (0.5 %; Oxoid Limited; Hampshire, England).

Amoebae and amoeboflagellates that showed indications of facultative mycophagy (Table 18) were cultivated monoxenically on bacteria for subsequent microcosm experiments. For that, individual protists were transferred from enrichment cultures to 60 mm Petri dishes filled with NMAS using a tapered glass pipette under an inverted phase-contrast microscope (Nikon TS-100, Japan). These monoclonal protist cultures were incubated at room temperature. Observations and microphotographs of protists were performed on a Nikon Eclipse 90i (Japan) equipped with phase contrast and Differential Interference Contrast optics at 100 – 400 x magnification.

Protists cultures obtained were tested for their feeding preferences on three fungal taxa, i.e. two yeasts *C. laurentii and S. cerevisiae* and spores of the hyphal-forming fungus *F. culmorum*. A low-density fungal suspension (100 µl fungal suspension) with a concentration of 300 cells * μl^{-1} (*S. cerevisiae* and *C. laurentii*) or 40 cells * μl^{-1} (*F. culmorum*) was directly added to protist cultures. The protist cultures were grown on accompanying bacteria in 60 mm Petri dish for one week. The cultures were microscopically investigated for uptake of fungal material 2 and 24 hours after inoculation, and microphotographs of protists ingesting fungal material were recorded.

Facultative mycophagy of the bacterivorous protist *Acanthamoeba castellanii*

Further microcosm tests were conducted using the model protist *Acanthamoeba castellanii* Neff strain to test potential for facultative mycophagy on four different fungi. Two strains of the single celled yeasts *S. cerevisiae* and two filamentous fungi, *Neurospora crassa* and *Coprinus cinerea* were presented as potential prey for *A. castellanii* grown axenically (proteose peptone-yeast extract-glucose medium, 4:2:1 mixture, respectively) (Rosenberg et al. 2009). The experiment was run in 96 well-plates (flat-bottom; Sarstedt, Nümbrecht, Germany), filled with a 150 µl sterile mixture of NMAS enriched with nutrient broth (Merck, Darmstadt, Germany) at 1:9 v/v (NB-NMAS). 100

spores of all four fungi were inoculated either alone or together with *A. castellanii*. Control treatments contained only NB-NMAS or *A. castellanii* in NB-NMAS. All treatments were replicated eight-fold. Before use, *A. castellanii* cultures were washed three times with sterile NMAS and 100 amoebae were added to each well of the *A. castellanii* treatments. Control plates received equivalent amounts of NMAS. Plates were sealed with Parafilm and directly placed in an automated microplate reader (Varioscan, Thermo Scientific, Waltham, USA) at room temperature. Optical density (OD) as an estimate of changes in fungal biomass was measured every hour for a total of four days. Plates were additionally examined microscopically for amoebae growth and contamination every second day.

Table 18. Overview on potential facultative mycophagous amoebae and amoeboflagellates Isolated in this study; Taxonomic affinities according to Smirnov et al. (2011b) and Adl et al. (2012); l = large taxon; s = small taxon

Protist taxon	Order	Class	Supergroup	Morphotype	Length [μm]	Sampling site	GPS
Acanthamoeba sp.	Centramoebida	Discosea	Amoebozoa	Naked amoeba	~40	Müncheberg	52°30′N 14°07′E
Acanthamoeba castellanii	Centramoebida	Discosea	Amoebozoa	Naked amoeba	~30	Pacific Grove	36°60′N 121°93′W
Cercomonas sp. (s)	Cercomonadida	Sarcomonadea	SAR	Amoeboflagellate	~12	Müncheberg	52°30′N 14°07′E
Cercomonas sp. (l)	Cercomonadida	Sarcomonadea	SAR	Amoeboflagellate	~30	Müncheberg	52°30′N 14°07′E
Cryptodifflugia sp.	Arcellinida	Tubulinea	SAR	Testate amoeba	~18	Pulheim Stommeln	51°01′N 6°45′E
Leptomyxa sp.	Leptomyxida	Tubulinea	Amoebozoa	Naked amoeba	>100	Cologne	50°55′N 6°55′E
Mayorella sp. (s)	Dermamoebida	Discosea	Amoebozoa	Naked amoeba	~40	Les Verrines	46°25′N 0°7′E
Mayorella sp. (l)	Dermamoebida	Discosea	Amoebozoa	Naked amoeba	~100	Cologne	50°55′N 6°55′E
Thecamoeba sp.	Thecamoebida	Discosea	Amoebozoa	Naked amoeba	~50	Cologne	50°55′N 6°55′E

Environmental sequencing approaches

Metagenetic approach on SSU rRNA

Data obtained in a high-throughput amplicon sequencing (HTAS) study targeting the small subunit (SSU) rRNA of eukaryotes (Hünninghaus et al. in prep) were analysed for transcripts of known mycophagous protist groups, vampyrellid amoebae and ciliates of the family Grossglockneriidae. In this experiment maize (*Zea mays*, Ronaldhinio) was grown in rhizoboxes containing

135 g of soil from an agricultural field side (Kramer et al. 2012). rRNA from bulk and rhizosphere soil was extracted 26 days post seed germination as described by Lueders et al. (2004). RNA extracts were subjected to cesium trifluoroacetate (CsTFA) density centrifugation with subsequent fractionation into 12-13 fractions per sample as described by Glaubitz et al. (2009). Fractions 3 and 8 were selected for 454 amplicon pyrosequencing. The primers 20f (5' - TGC CAG TAG TCA TAT GCT TGT - 3') and Euk302r+3 (5' - ATT GGA GGR CAA GTC TGG T - 3') were used to amplify a > 500 bp long fragment from a wide range of soil eukaryotes using the conditions described by Euringer and Lueders (2008). Only sequences longer than 250 bp were taxonomically assigned with CREST (Lanzén et al. 2012) and MEGAN (settings: min. bit score 330, min support 1, top percent 2) using BLASTn against the CREST-Silvamod database. All sequences assigned to the orders Vampyrellida or the family Grossglockneriidae were manually subjected to BLASTn searches (http://blast.ncbi.nlm.nih.gov/Blast.cgi) using default parameters. Vampyrellid and grossglockneriid relative sequence abundance of the entire protist community were calculated for each sample.

Screening of soil metatranscriptomes

Protist communities in diverse, replicated soil and litter samples were investigated using a metatranscriptomic approach (Geisen et al. 2015b) and mined for vampyrellid and grossglockneriid sequences.

Details about sampling, extraction of nucleic acids and 454 pyrosequencing are detailed in Geisen et al. (2015b) and references therein. Processing of raw reads is also described in Geisen et al. (2015b) and references therein. In short, sequences were filtered using LUCY (Chou and Holmes 2001) to remove short (< 150 bp) and low-quality sequences (> 0.2 % error probability). Small subunit (SSU) ribosomal RNA sequences of eukaryotes were identified by MEGAN analysis of BLASTn files against a SSU rRNA reference database (Lanzen et al. 2011; parameters: min. bit score 150, min. support 1, top percent 10; 50 best blast hits). All eukaryotic SSU rRNAs were reanalysed with CREST (Lanzén et al. 2012) using the Silvamod database with LCA parameters min bit score 250, top percent 2 for classification of protist sequences. Correct taxonomic assignment of all vampyrellid and grossglockneriid sequences was verified by manual BLASTn searches against the NCBI GenBank nt database.

Statistical analyses

Data obtained in the experiment testing facultative mycophagy of *A. castellanii*, evaluating biomass changes for *N. crassa*, *C. cinerea* and strains of *S. cerevisiae*, were analysed by repeated measures analyses of variance with factors time and amoebae. For that, data from 24 OD measurements per analysis time were combined to one mean value and were log (x + 1) transformed when required to satisfy the assumption of ANOVA. SAS 8.0 (Statistical Analysis System, SAS Institute Inc., Cary, USA) software package was used for statistical analyses.

Results

Isolation of protists from soil and microscopic observation on facultative mycophagy

Eight protist taxa of different morphology and taxonomy were examined microscopically to evaluate feeding and growth on fungi. All thrived on the yeast taxa *S. cerevisiae* and *C. laurentii*, while *Cryptodifflugia* sp., *Cercomonas* sp. (l), *Leptomyxa* sp. and *Mayorella* sp. (l) also fed on spores of *F. culmorum* (Table 19, Fig. 65). Along with ingestion of fungi, all protists reproduced in presence of the respective fungal prey.

Table 19. Feeding experiment of isolated mycophagous protists on the three different fungi (yeast species C. laurentii and S. cerevisiae and the hyphae forming F. culmorum). X = ingestion of fungal material, protist growth; - = no ingestion, no protist growth

Protist genus	Supergroup; Class; Order	*C. laurentii*	*S. cerevisiae*	*F. culmorum*
Acanthamoeba	Amoebozoa; Discosea; Centramoebida	X	X	-
Cercomonas (s)	SAR; Sarcomonadea; Cercomonadida	X	X	-
Cercomonas (l)	SAR; Sarcomonadea; Cercomonadida	X	X	X
Cryptodifflugia	Amoebozoa; Tubulinea; Arcellinida	X	X	X
Leptomyxa	Amoebozoa; Tubulinea; Leptomyxida	X	X	X
Mayorella (s)	Amoebozoa; Discosea; Dermamoebida	X	X	-
Mayorella (l)	Amoebozoa; Discosea; Dermamoebida	X	X	X
Thecamoeba	Amoebozoa; Discosea; Thecamoebida	X	X	-

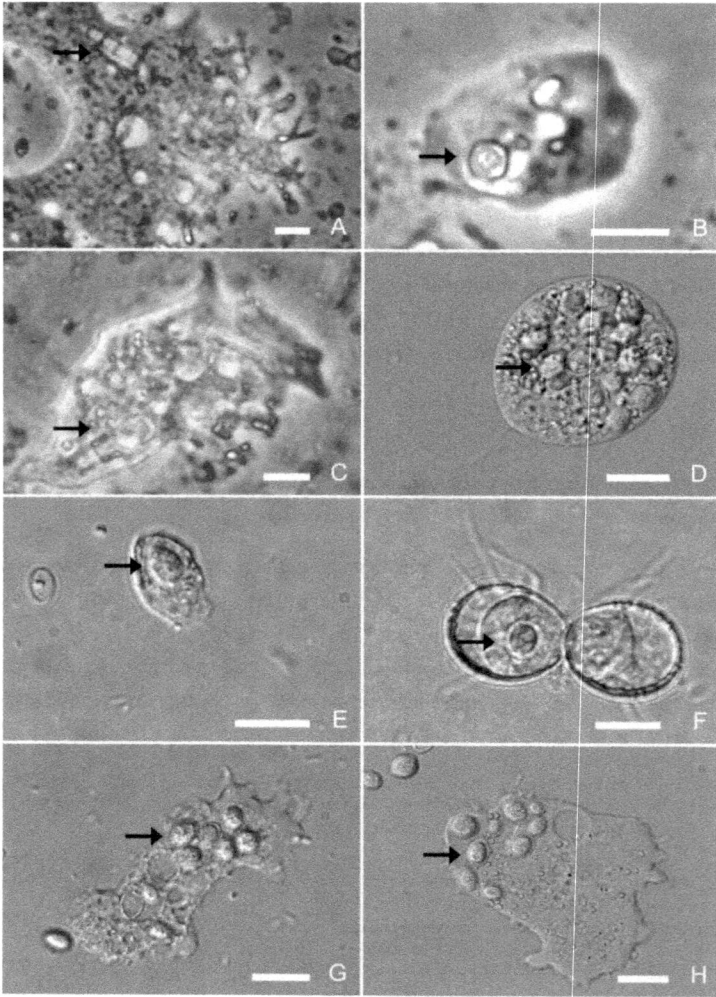

Fig. 65. Isolated soil protists with ingested fungi; (a) *Leptomyxa* sp. with ingested spores of *F. culmorum;* (b) *Thecamoeba* sp. with ingested *S. cerevisiae;* (c) *Mayorella* sp. feeding on spores of *F. culmorum;* (d) *Cercomonas* sp. (l) with ingested *S. cerevisiae;* (e) *Cercomonas* sp. (s) with ingested *C. laurentii;* (f) *Cryptodifflugia* sp. feeding on *S. cerevisiae;* (g) *Mayorella* sp. (s) with engulfed *S. cerevisiae* (h) *Acanthamoeba* sp. with ingested *S. cerevisiae;* arrows indicate ingested fungal material; scale bar = 10 μm.

Facultative mycophagy of the bacterivorous protist *Acanthamoeba castellanii*

A. castellanii inhibited growth of *C. cinerea* and both *S. cerevisiae* strains resulting in significantly reduced biomasses of those fungi compared to controls without amoebae (Fig. 66). Trophozoites and cysts of *A. castellanii* were present throughout the experiment in controls, with *C. cinerea* and both yeast strains, but a reliable quantification of amoebae could not be conducted as cells were masked by fungal material. In contrast, neither trophozoits nor cysts of *A. castellanii* could be seen in presence of *N. crassa*. In line, biomass of *N. crassa* was unaffected by amoebae (Fig. 66).

Fungal growth over time differed between strains; both yeast strains immediately started gaining biomass, while growth of *C. cinerea* and *N. crassa* until a biomass increase was observed was delayed (>24 and >40 hours, respectively; Fig. 66). Therefore, no effect of *A. castellanii* was detected on these hyphae forming fungi during the first 24 hours after inoculation, while biomasses of the *S. cerevisiae* yeast strains were reduced by 30 % on average (F = 17.5, p < 0.001). After 48 hours, amoebae inhibited growth of *C. cinerea* resulting in a biomass loss of 90 % (F = 545.1, p < 0.001, respectively). The observed effects were always highest when fungal growth started and fungi gradually compensated for feeding losses. Nevertheless amoebae strongly reduced biomasses of *S. cerevisiae* by 15 % (F = 5.2, p < 0.05) and *C. cinerea* by 82 % (F = 99.9, p < 0.001) (Fig. 66).

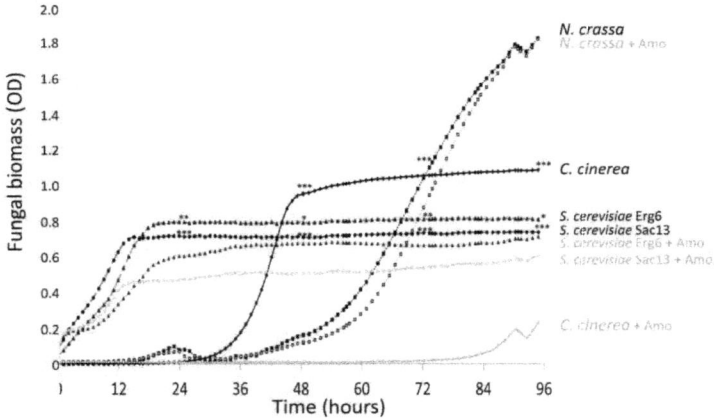

Fig. 66. Fungal biomass over a period of 95 hours determined by optical density (OD). Different fungal treatments (Neurospora crassa (squared symbols), Coprinus cinerea (diamonds), and Saccharomyces cerevisiae (triangles and circles) are shown; differences in fungal biomass in absence (filled symbols, black line) or presence of amoebae (open symbols, grey line; "+Amo") are indicated at day 1, 2, 3 and 4; * = p < 0.05, ** = p < 0.01; *** = p < 0.001.

Presence and relative abundance of metabolically active mycophagous protists in soils

Mycophagous protists revealed by amplicon sequencing targeting the SSU rRNA

We detected a substantial fraction of sequences specific for both fungal feeding protist groups, Grossglockneriidae and Vampyrellidae, in maize rhizosphere of agricultural soil. Sequences of Grossglockneriidae represented 0.4 and 0.6 % of all protist sequences, respectively, 26 days after maize planting in bulk and rhizosphere soil (Fig. 67). BLASTn searches of sequences assigned as Grossglockneriidae often yielded perfect matches to *Pseudoplatyophora nana* or *Mykophagophrya terricola*. However, several sequences showed closer affinities to uncultured species with *P. nana* or *M. terricola* as the closest described species indicating unexpected high diversity of grossglockneriids in agricultural soil samples.

Vampyrellid sequences were more abundant, representing 4.9 and 2.9 % of all protist sequences in bulk and rhizosphere soil, respectively. Vampyrellid sequence diversity appeared more diverse, resembling various members of Vampyrellida, such as *Theratromyxa weberi*, *Platyreta germanica* or *Vampyrella* sp., but often yielded highest identity to sequences of undescribed vampyrellids. Most sequences were not perfectly matching known taxa, sometimes differing by more than 5 % different to described species.

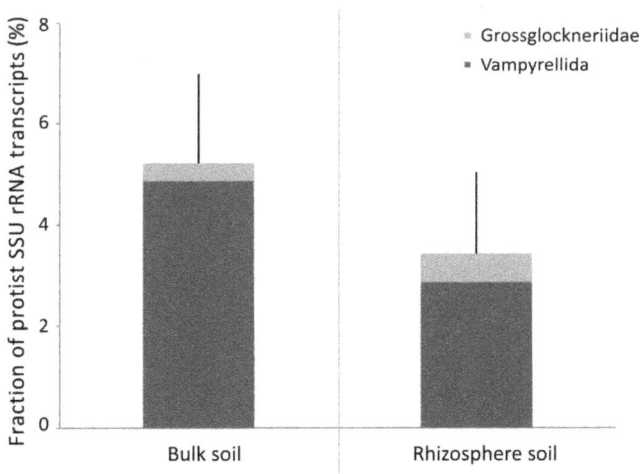

Fig. 67. rRNA sequences assigned as grossglockneriids and vampyrellids shown as relative abundance of all protist sequences in bulk (left) and rhizosphere soil (right) using rRNA as a substrate; ± SD of both mycophagous groups.

Mycophagous protists revealed by a metatranscriptomic approach

Vampyrellid and grossglockneriid rRNA transcripts were found in all investigated samples, representing 0.1 to 3.0 % of all protist-specific sequences. While Vampyrellida and Grossglockneriidae contributed little to the diversity of protists in artic peatlands (< 0.4 % of all protists), higher fractions were found in forest soil and litter (1.5 % of all protists) with highest relative abundance in grassland (3.0 %; Fig. 68).

Most grossglockneriid sequences most closely resembled *P. nana* and *M. terricola*, with especially short sequences showing identical similarities to both species. Some transcripts again resembled uncultivated grossglockneriids better than the best identified hits, *P. nana* and *M. terricola*.

The majority of sequences assigned to Vampyrellida most closely resembled *Theratromyxa* and *Arachnula*, but also sequences closely resembling uncultivated vampyrellids were common.

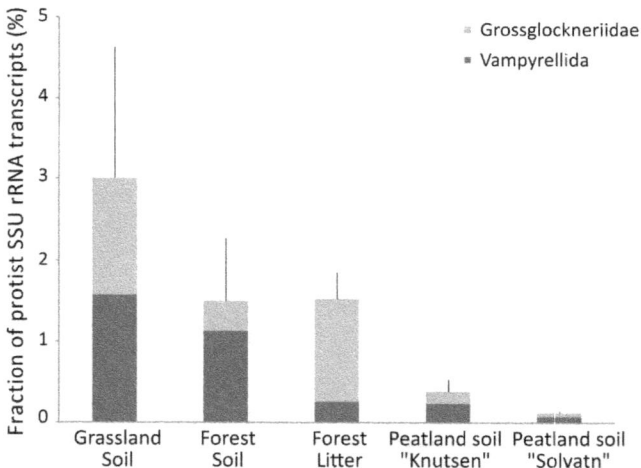

Fig. 68. Sequences assigned as grossglockneriids and vampyrellids shown as relative abundance of all protist sequences obtained in a metatranscriptomic analyses of different soils; ± SD of both mycophagous groups. We found a strong positive correlation between the relative abundances of mycophagous protists and the relative abundances of fungi, while the entity of protists showed the opposite trend (Fig. 69).

We found a strong positive correlation between the relative abundances of mycophagous protists and the relative abundances of fungi, while the entity of protists showed the opposite trend (Fig. 69).

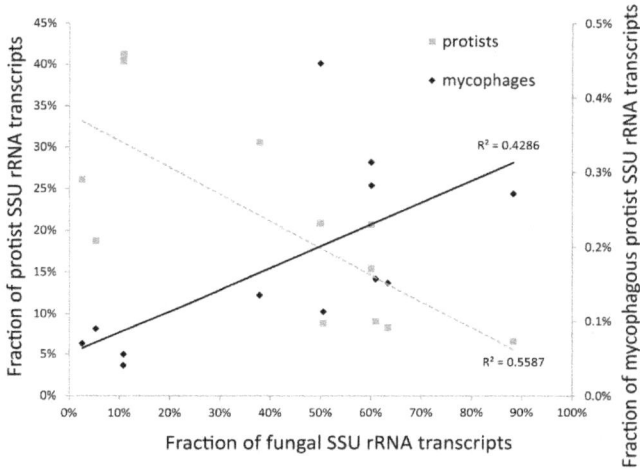

Fig. 69. Relative SSU rRNA abundance of all protists (grey squares, first x-axis) and mycophagous protists (black diamonds, second x-axis) to relative abundances of fungi (y-axis) among all eukaryote sequences. Note: positive correlation between mycophagous protists and fungi while all protists and fungi showed a negative correlation.

Discussion

In this work we highlight the hitherto neglected role of soil protists as mycophages. The functional group of mycophages among protists is almost exclusively recognized by taxonomists often characterized as highly specialized obligatory fungal feeders (e.g. Grossglockneriidae (Foissner 1980, Petz et al. 1985, Foissner 1999a)). We found that diverse soil protist taxa, previously classified as bacterivores, selectively fed on yeasts and on spores of soil fungi. Therefore, facultative mycophagy seems to be a widespread evolutionary feeding characteristic among soil protists. Unselective phagotrophy, characterized by an unselective object-uptake of suitable size, might partially explain this phenomenon. Engulfed potential prey items are then subject to extreme conditions in food vacuoles such as low pH (Laybourn-Parry 1984) and a battery of enzymes (Bowers and Korn 1973, Laybourn-Parry 1984, Khan 2009). Among the latter can be chitinases, which are encoded in the genome and can be expressed by diverse protist taxa (Tracey 1955, Anderson et al.

257

2005, Fouque et al. 2012). Therefore it is not unlikely, that many protists can access fungal-bound nutrients from engulfed prey.

In line with size-dependent uptake of potential prey, our study revealed that yeasts were commonly taken up by morphologically and taxonomically diverse protist taxa. Transport of yeasts inside amoebae without damage of the cells has been reported (Heal 1963, Chakraborty and Old 1982), but we could observe active digestion inside food vacuoles of protists leading to inhibition of fungal growth and increases in protist growth rates. All eight protists tested in this study consumed yeast cells. *C. operculata* and *Cercomonas* sp. (s) ingested single yeast cells, whereas the other, larger protists ingested several cells simultaneously. Fungal hyphae were never found inside protists. This is in line with former studies that reported protist uptake of yeasts but not hyphae (Heal 1963, Bunting et al. 1979, Allen and Dawidowicz 1990), suggesting that these fungal structures are mechanically protected against grazing of facultative mycophagous protists, that do not have specifically adapted mechanisms to perforate hyphae and spores as used by vampyrellids (Old and Darbyshire 1978, Old and Oros 1980, Chakraborty and Old 1982) or feeding structures such as those in grossglockneriids (Petz et al. 1985, Foissner 1999a). Size has also been shown to determine protist grazing on bacteria, as bacteria can avoid grazing through morphological changes, e.g. by becoming larger or smaller, producing filaments or colonies (Matz and Kjelleberg 2005, Jousset 2011).

Size of potential prey fungi does, however, not ultimately determine whether protists can feed on fungi; fungal spores and yeasts are often of similar size, but we found that *A. castellanii* almost entirely arrested growth of the filamentous *C. cinerea*, strongly reduced yeast biomass, while the filament-forming *N. crassa* was unaffected. Secondary metabolites against protist grazing might protect *N. crassa*, as genes encoding unknown, putative secondary metabolites have been found in the genome of *N. crassa* (Galagan et al. 2003). Generally, toxin production adds grazing resistance to both fungi (Stotefeld et al. 2012) and bacteria (Matz and Kjelleberg 2005, Jousset et al. 2006, Jousset et al. 2009). Nevertheless, four protist out of 8 protist species studied here, the amoeboflagellate *Cercomonas* sp. and the amoebae *Cryptodifflugia* sp.,

Leptomyxa sp. and *Mayorella* sp. ingested and grew on spores of *F. culmorum*, which is a plant pathogenic, toxin-producer (Scherm et al. 2013). This rather applied aspect has been investigated before and other plant pathogenic fungi have been found to be digested by protists (Tapilskaja 1967, Old and Oros 1980, Chakraborty and Old 1982, Chakraborty et al. 1983), suggesting that toxins only protect from grazing in a species specific manner. Taken together, no universal protection mechanism to mycophagy can be assumed and future studies need to include a wider range of fungi and mycophagous protists to investigate species-specific interactions.

Interestingly, we found that the model protist *A. castellanii* Neff strain took up and reduced biomasses of *S. cerevisiae* and *C. cinerea*. Amoebae of this ubiquitous and highly abundant genus *Acanthamoeba* (Page 1988, Rodríguez-Zaragoza 1994, Geisen et al. 2014b), especially those species identified as *A. castellanii* are commonly treated as obligatory bacterivores (Chakraborty et al. 1983, Bamforth 1988). *A. castellanii* has been used as a bacterivore model protist in a number of experiments (Weekers et al. 1993, Bonkowski and Brandt 2002, Rønn et al. 2002b, Neidig et al. 2010, Koller et al. 2013) and was consistently found to alter bacterial community composition in the plant rhizosphere (Kreuzer et al. 2006, Herdler et al. 2008, Rosenberg et al. 2009). Our results show that *A. castellanii* also exhibits selective feeding preference for fungal species, even suppressing the growth of the hyphae forming fungi *C. cinerea*. Unlike larger mycophagous amoeba such as vampyrellids that perforate hyphae (Old and Darbyshire 1978, Old and Oros 1980, Chakraborty and Old 1982), *Acanthamoeba* spp. rely on phagocytosis, which excludes long hyphae from the food spectrum. Engulfed potential prey is subjected to an immense potential enzymatic repertoire such as chitinases detected in the genome of *A. castellanii* Neff strain (Anderson et al. 2005, Clarke et al. 2013a), suggesting that *A. castellanii* has the intrinsic capacity to thrive on fungi. The potential for facultative mycophagy is also present in other *Acanthamoeba* spp. as chitinases were found in *A. culbertsoni* (Krishna Murti and Shukla 1984), *A. polyphaga* was reported to grow on *Cryptococcus neoformans* (Bunting et al. 1979) and chitinolytic activity was described in *A. glebae* (Tracey 1955). The enormous species-diversity of *Acanthamoeba* spp. (Gast et al. 1996, Stothard et al. 1998, Gast 2001, Corsaro and Venditti 2010, Qvarnstrom et al. 2013, Geisen et al. 2014b) their wide enzymatic repertoire and obvious omnivory of some taxa might explain the ubiquity, high abundance of acanthamoebae in basically all environments (Sawyer and Griffin 1975, Page 1988, Rodríguez-Zaragoza 1994). This highlights the need to carefully study ecological functions not only of one or few model taxa, but investigating ecological roles of a higher diversity of taxa.

Presence and relative abundance of metabolically active mycophagous protists in soils

Our environmental sequencing approaches based on rRNA transcripts revealed that vampyrellids and grossglockneriids as known mycophages (Old and Darbyshire 1978, Foissner 1980, Petz et al. 1986, Hess et al. 2012) were present in all soils investigated. Grossglockneriidae have been described in 1980 with the family now containing nine species in six genera, all being obligate mycophages (Foissner 1980, Petz et al. 1985, Petz et al. 1986, Foissner 1999a). Cultivation based studies indicated their presence in a wide range of soils (Foissner 1999a), but still little is known on the distribution and ecological importance of these ciliates.

Our molecular analyses of a diverse range of soils support the findings of Foissner (1999a) as sequences of grossglockneriids were found in all investigated samples. Despite an enormous diversity of soil protists, grossglockneriids represented up to 1.4 % of all protist sequences, indicating not only substantial abundances but also their potential functional importance in soil systems. Several sequences showed highest similarity to undescribed grossglockneriid species and the molecular diversity of grossglockneriids in samples generally appeared high. This indicates that grossglockneriids are undersampled and the true diversity remains unknown. Currently only *Mykophagophrys terricola* and *Pseudoplatyophrya nana* have been sequenced (Lynn et al. 1999, Dunthorn et al. 2008) urging for cultivation and sequencing efforts to supplement the remaining described (and potentially undescribed) grossglockneriid species to investigate the full extent of diversity in this mycophagous group of ciliates.

Vampyrellid amoebae have been the focus of a recent detailed investigation including data-mining of HTS datasets showing that vampyrellids are highly diverse both in sediments and soils (Berney et al. 2013). Our HTAS and metatranscriptomic data clearly support the findings of Berney et al. (2013), as vampyrellid sequences were ubiquitously retrieved, seemed diverse, and abundant (up to 7.4 % of all protist sequences) in all analysed samples. Sequences perfectly matching described species, such as *Theratromyxa weberi* and *Platyreta germanica* were detected, but sequences also often most closely resembled undescribed and uncultivated species in Vampyrellida. While only cultivation based efforts followed by targeted experiments can only reliably unravel the true ecological importance of vampyrellids as mycophages, our data indicate that this functional role could be of importance in the soil food web.

The strong correlation of the two mycophagous protist groups targeted with fungi strongly confirms that both groups are strongly interacting, likely through a direct trophic interaction as shown before. Further we show that the entity of protists showed the opposite pattern, indicating that the most abundant protists do not depend on fungi as a prey and that there are rather competitive interactions between both groups instead of trophic interactions. Further functional studies are, however needed to prove this hypothesis. As abundant soil inhabitants (Berney and Pawlowski 2006) fungi and protists inevitably co-occur in every soil environment. Therefore a variety of interactions between these two groups must have evolved. It is not unlikely that these interactions have contributed to the evolution of both groups and increasing pathogenicity in fungal taxa to resist protist predators as suggested for the "arms race" between bacteria and protists (Brüssow 2007).

Taken together, mycophagous protists could be critical in controlling fungi in soils as facultative mycophagy is known several little investigated protist groups (Pussard et al. 1979, Chakraborty et al. 1983, Old et al. 1985, Ekelund 1998, Flavin et al. 2000), which we here confirm for eight studied taxa. Sequence data from two independent studies showed that even obligate mycophagous protists are common members of the protist community. The potential importance of mycophagy among protists is reinforced in a detailed investigation of a range of soils in the Netherlands showing that protists are the major source of fungal derived nutrients (Morrien et al. in prep). Consequently protists are essential control points for carbon and nutrient flow to higher trophic levels in soil food webs (Fig. 70) and the classical separation of the bacterial and fungal energy channel seems highly artificial, especially when protists are treated as a single trophic node in these schemes.

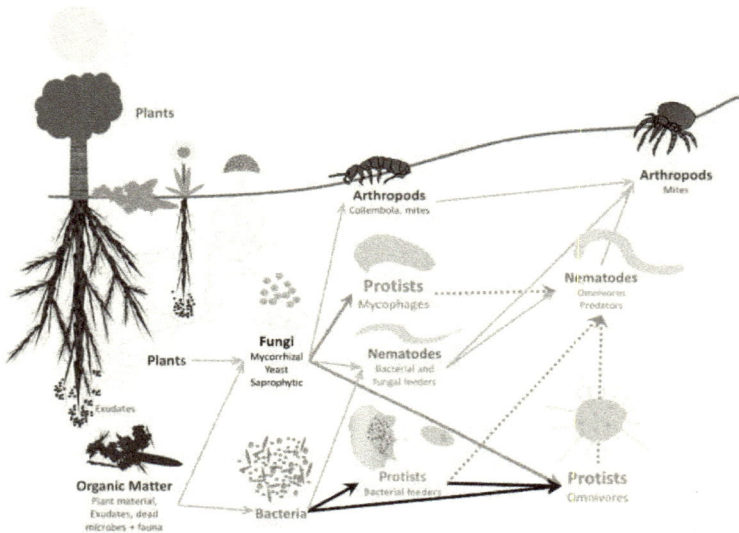

Fig. 70. The soil food web revisited showing the diverse role of free-living heterotrophic protists; bold black arrows: Known nutrient flows to protists; dotted arrows: suggested nutrient flow to higher trophic levels; red: formerly largely unconsidered trophic positions of protists.

Acknowledgements

This research was funded by the European Commission through the EcoFINDERS project (FP7-264465), with support by the DFG-Research Unit FOR 918 (Carbon flow in belowground food-webs assessed by isotope tracers), subproject BO 1907/3-1.

Part 3 – Chapter 10

Pack hunting by a common soil amoeba on nematodes

Stefan Geisen[1,2,*], Jamila Rosengarten[1], Robert Koller[1,3], Christian Mulder[4], Tim Urich[5,6], Michael Bonkowski[1]

[1] Department of Terrestrial Ecology, Institute of Zoology, University of Cologne, Germany
[2] Department of Terrestrial Ecology, Netherlands Institute of Ecology (NIOO-KNAW), Wageningen, The Netherlands
[3] Forschungszentrum Jülich, IBG-2: Plant Sciences, Jülich, Germany
[4] National Institute for Public Health and the Environment (RIVM), Bilthoven, The Netherlands
[5] Department of Ecogenomics and Systems Biology, University of Vienna, Austria
[6] Bacterial Physiology, Institute for Microbiology, Ernst Moritz Arndt University Greifswald, Germany

Abstract

Soils host the most complex communities on Earth, including the most diverse and abundant eukaryotes, i.e. heterotrophic protists. Protists are generally considered being bacterivores, but evidence for negative interactions with nematodes both from laboratory- and field-studies exist. However, direct impacts of protists on nematodes remain unknown.

We isolated the soil-borne testate amoeba *Cryptodifflugia operculata* and found a highly specialized and effective pack-hunting strategy to prey on bacterivorous nematodes. Enhanced reproduction in presence of prey nematodes suggests a beneficial predatory life-history of these omnivorous soil amoebae. *C. operculata* appears to selectively impact the nematode community composition as reductions of nematode numbers were species-specific. Furthermore, we investigated 12 soil metatranscriptomes from five distinct locations throughout Europe for 18S ribosomal RNA transcripts of *C. operculata*. The presence of *C. operculata* transcripts in all samples, representing up to 4 % of the active protist community, indicates a potential ecological importance of nematophagy performed by *C. operculata* in soil food webs.

The unique pack-hunting strategy on nematodes that was previously unknown from protists, together with molecular evidence that these pack-hunters are likely to be abundant and widespread in soils, imply a considerable importance of the hitherto neglected trophic link "nematophagous protists" in soil food webs.

Introduction

Soils provide the most essential ecosystem services for humans and soil biodiversity is suggested to be the key driver for these services (Bardgett and van der Putten 2014, Bradford et al. 2014, Wall and Six 2015). Heterotrophic protists – hereafter simplified as protists – and nematodes are both aquatic organisms, which inhabit the water-filled soil pores (Neher 2010, Geisen et al. 2014a). While nematodes are comparatively well investigated and their diverse trophic roles in soil food webs are widely acknowledged (Ferris et al. 2012), with notorious examples of common devastating plant-parasitic organisms (Back et al. 2002), knowledge on protists generally remains scarce (Caron et al. 2008). In soils, protists are predominantly suggested to be the major controllers of the soil bacterial biomass (Clarholm 1981, Bonkowski 2004). Together with bacterivorous nematodes, they represent a crucial bottleneck by transferring bacterial-bound nutrients and energy to higher trophic levels in the complex soil food web (Hunt et al. 1987, de Ruiter et al. 1995, Crotty et al. 2012a). However, both groups also act antagonistically, and evidence suggests strong intraguild predation, i.e. killing and eating of potential competitors from the same trophic guild, such as of "bacterivorous" nematodes on bacterivorous protists (Crotty et al. 2012a, Rønn et al. 2012).

The functional role of the majority of protists that spread across the entire eukaryotic tree of life (Adl et al., 2012; Pawlowski, 2013) is largely unknown.

264

For convenience, most heterotrophic protists are treated as bacterivores. However, heterotrophic protists occupy more diverse trophic positions, such as being algivorous (Smirnov et al. 2011a, Hess and Melkonian 2013), mycophagous (Old and Darbyshire 1978, Chakraborty and Old 1982, Petz et al. 1986, Ekelund 1998), parasitic (Schuster 2002, Thines 2014, Geisen et al. 2015a) or feed on other protists (Page 1977, Tapia et al. 2013). Even large-sized nematophagous protists, all exceeding 50 µm and many reaching sizes of several 100 µm, directly consume nematodes: among them are the vampyrellid amoebae *Arachnula impatiens*, *Platyreta germanica* and *Theratromyxa weberi* (Sayre 1973, Old and Darbyshire 1980, Bass et al. 2009a), the arcellinid testate amoebae *Apodera vas* and *Difflugia lanceolata* (Yeates and Foissner 1995) and the ciliates *Stylonychia pustulata* and *Urostyla* sp. (Doncaster and Hooper 1961). In contrast, direct feeding of more abundant smaller-sized protists (body < 20 µm in diameter) on larger-sized nematodes (body > 250 µm in length) is not yet known. However, evidence obtained in liquid microcosm studies similar to those used in our approach revealed that unclassified tiny protists might affect nematode behaviour (Bjørnlund and Rønn 2008, Neidig et al. 2010). All these studies suggest that the trophic positions of protists cannot be restricted into a single role like 'bacterivores' and even feeding on larger organisms such as nematodes might be common among soil protists.

However, no phylogenetic conservation of non-bacterivorous protists exists, as the above-mentioned taxa taxonomically belong to different eukaryotic supergroups. Among them, lobose testate amoebae are monophyletic in the order Arcellinida, branching deeply within naked amoebae in the class Tubulinea (Nikolaev et al. 2005, Smirnov et al. 2011b, Adl et al. 2012). Amoebae in this highly diverse order are well characterized due to distinct shell morphologies. Smaller taxa, such as the ~20 morphologically described species of the genus *Cryptodifflugia* often exhibit a wide distribution throughout soil environments and are considered as bacterivorous (Page 1966, Hedley 1977, Nicholls 2006), while two large taxa, *Apodera vas* and *Difflugia lanceolata*, have been shown to ingest entire nematodes (Yeates and Foissner 1995).

265

We here investigated potential intraguild predation of the putative bacterivorous protist *Cryptodifflugia operculata* on nematodes. Furthermore, we investigated the distribution of these small testate amoebae in a wide range of European soil habitats by analysing metatranscriptomic datasets to reveal their presence, relative abundance and potential significance in the soil food web.

Results

Interactions between *Cryptodifflugia operculata* and nematodes

Direct inverted microscopy observations revealed that *C. operculata* attacked and consumed all nematode species tested, active and dead individuals, juveniles and adults. The attack followed a specific pattern: nematode prey-escape was prevented by one amoeba attaching the tail of slowly moving nematodes by parts of their pseudopods to the substratum. Then other amoebae were attracted and together immobilized the nematode by fixing their anterior end to the substratum. Attached nematodes were subsequently disintegrated from both ends by an increasing number of attracted amoebae. This process of joint amoebae feeding dissolved and absorbed their prey within 12 hours. A sequential array of time-lapse photographs of the feeding procedure is shown in Fig. 71, with a microscopic time-lapse movie provided online as Supplementary Movie S1.

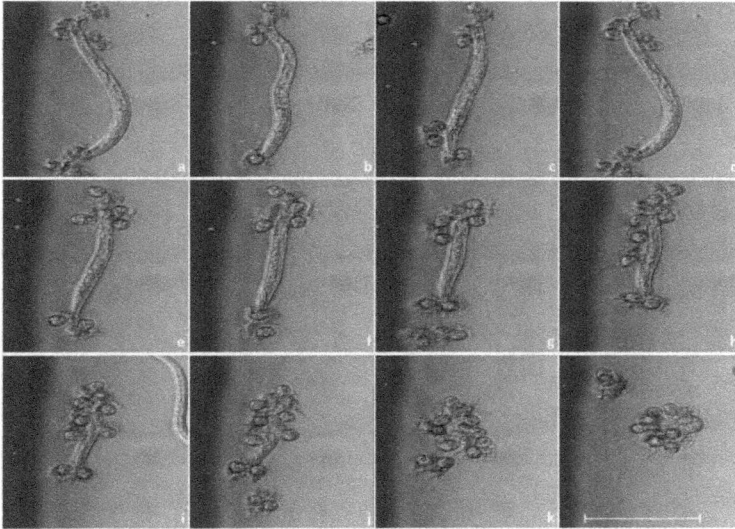

Fig. 71: Time-lapse photos showing the feeding procedure of *Cryptodifflugia operculata* on the nematode *Acrobeloides buetschlii* over a period of twelve hours at intervals of one hour; scale bar = 100 μm.

The different nematode species had strongly different effects on the reproduction of *C. operculata*. Populations of *C. operculata* decreased in control treatments containing bacteria only by 41 %, but increased by factors of 2.5, 2.4, 2.4, 1.4 and 1.4, in co-cultivation with the nematodes *Acrobeloides buetschlii*, *Rhabditis dolichura*, *R. terricola*, *R. belari* and *A. camberenensis*, respectively ($F_{[5,18]}$ = 6.34, p < 0.01; Fig. 72). However, the growth increase in *C. operculata* did not reflect the decline in nematode abundances. After 18 days, *C. operculata* had reduced the numbers of all nematode species: *R. terricola* by 89 % ($F_{[4,15]}$ = 864, p < 0.001), *A. camberenensis* by 77 % ($F_{[4,15]}$ = 99.4, p < 0.001), *R. belari* by 32 % (F = 10.7, p < 0.05), *A. buetschlii* by 31 % ($F_{[4,15]}$ = 5.12, p = 0.06) and *R. dolichura* by 25 % ($F_{[4,15]}$ = 9.3, p < 0.05; Fig. 73).

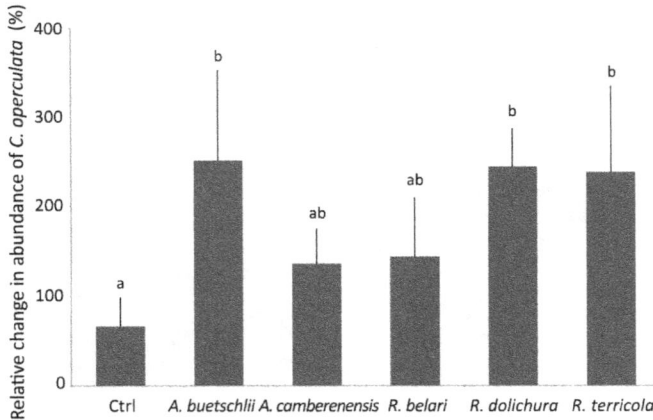

Fig. 72: Relative change in abundance of *Cryptodifflugia operculata* over the course of the experiment (18 days) in comparison to the beginning of the experiment (T0). Nematode-free controls (Ctrl) and treatments with *Acrobeloides buetschlii*, *A. camberenensis*, *Rhabditis belari*, *R. dolichura* and *R. terricola* as nematode prey species. Different letters indicate significant differences (Tukey-test, $p = 0.05$). Error bars represent standard deviation.

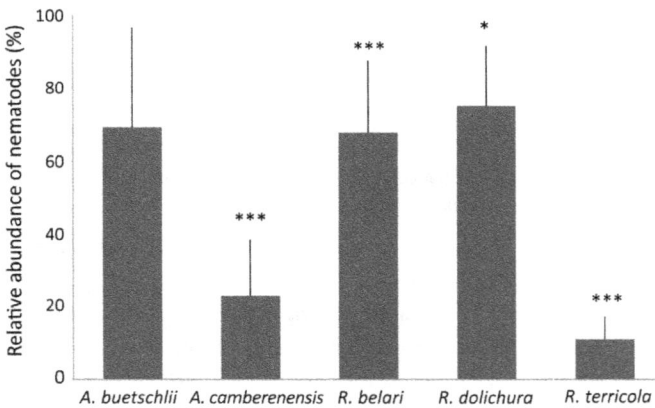

Fig. 73: Relative abundance changes to the beginning of the experiment (T0) induced by *Cryptodifflugia operculata* of nematode species *Acrobeloides buetschlii*, *A. camberenensis*, *Rhabditis belari*, *R. dolichura* or *R. terricola* in comparison to non-amoebae treatments. Stars indicate significant differences to treatments without *C. operculata* at * $p < 0.05$ and *** $p < 0.001$. Error bars represent standard deviation.

Interestingly, according to the nematode life-stages the Mantel tests point to contrasting correlations between the abundances of *C. operculata* and the abundances of either juvenile or adult nematodes. The Mantel test (9999 runs Monte Carlo simulation, $p < 0.15$) shows in the case of juvenile nematodes a negative association between matrices (8532 runs with Z > observed Z), while in the case of adult nematodes it shows the opposite, namely a positive association between matrices (8564 runs with Z < observed Z). Hence, besides the aforementioned species-specific predatory activities, the contrasting associations between the matrices of *C. operculata* and the life-stages of soil nematodes support the existence of a trait-specific feeding behaviour of these amoebae, with juvenile nematodes clearly being their preferred prey.

The relative abundance of active *Cryptodifflugia* in litter and soils

Metatranscriptomes of 12 soil and litter samples from terrestrial habitats in Europe contained 18S rRNA sequences that could be assigned to *C. operculata* with > 98 % sequence identity (Fig. 74). However, those sequences also shared a similar sequence identity to *C. oviformis*. Especially 18S rRNA sequences from forest sites often had the highest sequence identity to uncultivated species, with *C. operculata* 18S rRNA being the best match to a described species.

18S rRNA sequences resembling *C. operculata* were particularly abundant in forest litter and soil samples, comprising up to 18 % of total Amoebozoa and 4 % of all protists sequences (Fig. 74). *C. operculata* still represented 4.8 % of the amoebozoan sequences (1.1 % of all protists) in grassland soils, while both Arctic peatlands hosted only a relatively low fraction of *C. operculata* (< 5 % of Amoebozoa and < 0.7 % of all protist sequences; Fig. 74).

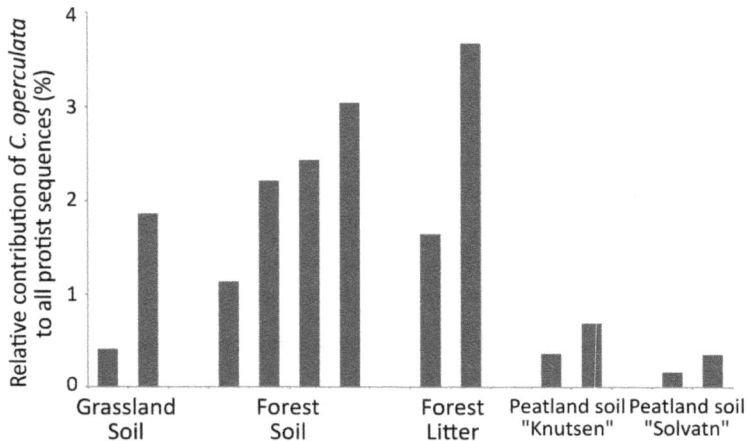

Fig. 74: 18S rRNA sequences assigned to *Cryptodifflugia operculata* in the different soil and litter metatranscriptomes summarized in Geisen et al. (2015b), shown here as relative abundance of all sequences assigned to protists. Individual bars show individual biological replicates of the five sites studied.

Discussion

Predation of testate amoebae on nematodes has been shown before (Yeates and Foissner 1995), but predation of protists on nematodes was seen as an exception confirming the rule. Our combined findings from laboratory experiments and field surveys suggest that the predation of protists on soil nematodes, especially juveniles, might be an important trophic link in the soil food web that could connect adjacent (sub)compartments as a loop within the "bacterial" pathway.

More sophisticated evidence for intraguild interactions between protists and nematodes have recently been reported (Bjørnlund and Rønn 2008, Neidig et al. 2010). Neidig et al., (2010) showed that exoproducts of amoebae repelled nematodes and exhibited marked nematostatic activity, while exoproducts of nematodes increased encystation and reduced growth of their bacterivore competitors. Likewise Bjørnlund and Rønn (2008) found flagellates that killed nematodes at high flagellate abundance, whereas nematodes consumed flagellates at lower densities; however, neither flagellates nor nematodes benefited from this trophic interaction under any of the experimental conditions. Even the metabolic scalings are highly different for protists (Makarieva et al. 2008), with amoebae showing a strong reduction of metabolic rates with body size (Mulder et al. 2012, Sechi et al. 2015). Furthermore, amoeboid locomotion is less energy-demanding than active swimming by flagellates and ciliates (Fenchel 1987). If locomotion in water is energetically easier for amoebae than for other protists, then the performance of amoebae and their trophic importance will strongly be enhanced by their superior mobility in soils and their capacity to extend into micro-sites not reached by other protists. Furthermore, amoebae, in contrast to flagellates and ciliates, become more dominant in terms of relative abundance when moisture content is reduced (Geisen et al. 2014a) and only amoebae remain capable of moving on the ultrathin water film on agar surfaces. Allometric rules contribute to explain the abundance of amoebae compared to nematodes and it lately has been shown that the ratio of abundance to body mass of amoebae compared to nematodes is even higher than expected (Sechi et al. 2015). Combined, these characteristics make amoebae the perfect predators in soils.

It has repeatedly been suggested that interactions between putative bacterivorous protists and nematodes are much more complex than depicted in current soil food webs. Evidence from a number of experiments already indicated that putative bacterivorous nematodes are important predators of naked amoebae (Anderson et al. 1977, Elliott et al. 1980b, Alphei et al. 1996, Rønn et al. 1996, Bonkowski et al. 2000b, Crotty et al. 2013). Here, we demonstrate that intraguild predation can also be an important strategy of amoebae to redirect a significant amount of food resources from bacterivorous nematodes.

So far, *C. operculata* represents the smallest nematophagous soil protist. The other taxa reported to directly feed on nematodes, the vampyrellid amoeba *T. weberi*, as well as the testate amoebae *A. vas* and *Difflugia* sp., are 5- to 10-fold larger (Sayre 1973, Yeates and Foissner 1995), resulting in an even more increased difference in the biovolume between *C. operculata* and the larger testate amoebae. Interestingly, *C. operculata* can grow entirely on bacterial cells, while the majority of the larger nematophages cannot be cultivated merely on bacteria and are suggested to be broader omnivores or predators, all needing nematodes, algae or fungi as supplementary food (Yeates and Foissner 1995, Berney et al. 2013). The major difference in terms of potential functional importance is the strong growth benefit of *C. operculata* when feeding on nematodes, while no growth benefit of the larger amoebae has yet been shown (Sayre 1973, Yeates and Foissner 1995).

Even the predation strategy of *C. operculata* profoundly differs from other known nematophagous protists and shows a high level of trait-mediated evolutionary adaptation; in contrast to *C. operculata*, the larger nematophages individually attack nematodes and ingest them entirely (Sayre 1973, Yeates and Foissner 1995). Much different and highly distinctive was the pack hunting mode employed by *C. operculata*: always one individual attacker fixed the posterior end of the nematode and then more and more conspecifics were attracted, most likely by chemical signals. We are not aware of chemical signals involved in the communication between single-celled protists, but chemical cues are secreted by social amoebae to form multinucleated plasmodia and fruiting bodies, such as in myxomycetes (Stephenson et al. 2011). Therefore it is likely that single celled protists also have the potential to communicate using chemical signals, but more detailed investigations are needed to test this hypothesis and determine potential signalling compounds.

Quickly other *C. operculata* amoebae were attracted, immobilizing the nematode at both ends. Subsequently, the amoebae gradually and completely digested the nematode by the concerted action of an increasing number of accumulating amoebae within 12 hours. Interestingly, the reproduction of *A. buetschlii*, the nematode species that most strongly favoured the reproduction of *C. operculata*, was only marginally affected, while all other nematode species were strongly reduced. *A. buetschlii* might be less prone to predation losses as this species is a clear *r*-strategist characterized by asexual reproduction and short generation times (Frey 1971), while the sexual *A. camberenensis* could likely not recover as quickly. This suggests that indirect effects of *C. operculata* affecting the composition of the nematofauna might be much more important than direct predation and the overall reduction in nematode densities *per se*.

It has to be noted that the observations of *C. operculata* feeding on nematodes were done in Petri-dishes and 24 well-plates using a liquid medium. This artificial system is of course much less complex than the soil environment. Therefore, population growth of the amoebae is likely higher in our experiments than in soils and amoebae might be more restricted to approach a trapped nematode in the more complex soil habitat. However, highly abundant bacterial populations in distinct patches (Vos et al. 2013) probably attract protist and nematode predators, bringing these bacterivores in direct proximity. Furthermore, fast replication rates of amoebae in these bacterial hotspots are likely to result in high numbers of amoebae, a condition similar to the one in our experiment. Therefore we believe that this interaction might in fact take place and be of importance in soils. Future studies should aim at investigating this interaction in more complex environments such as artificially resembling soils (Darbyshire 2005).

The metatranscriptomic data revealed that 18S rRNA sequences assigned to *C. operculata* represented a significant proportion of the entire active protist community in a broad variety of soil and litter samples across Europe. Since the biomass of protists in soil usually is several orders of magnitude greater than the biomass of nematodes (Paustian et al. 1990, Schaefer and Schauermann 1990), relative abundances of *C. operculata* of up to 4 % of all protist 18S rRNA sequences indicate a significant potential for widespread intraguild predation of protists on nematodes. It should be noted that several 18S rRNA sequences showed higher similarities to closely-related uncultivated species, which might represent other currently unknown species within the genus *Cryptodifflugia* or those that were described solely based on morphological features. Based on morphology, more than twenty *Cryptodifflugia* species have already been described (Page 1966), while sequence data are only available for *C. operculata*

(Lahr et al. 2011) and *C. oviformis* (Gomaa et al. 2012). Further taxonomic work is needed to fill the missing sequence information of species within the genus *Cryptodifflugia*, while functional studies are essential to decipher whether species closely related to *C. operculata* are also capable of preying on nematodes.

Taken together, this work is, to our knowledge, the first study identifying a widespread, omnivorous soil protist with a highly evolved strategy to prey on nematodes. Sequences assigned as *C. operculata* represented a significant proportion of the entire fraction of active protists and *C. operculata* might therefore affect community composition and total abundances of nematodes in soils. These findings suggest that protist nematophagy might be an important ecological process in soils. Due to the increasing appreciation in studies aiming at deciphering protist community structures (Pawlowski et al. 2012, Geisen et al. 2014a, Keeling et al. 2014, Geisen et al. 2015b) and investigating functional versatility (Glücksman et al. 2010, Saleem et al. 2012, Saleem et al. 2013) it is likely that different novel ecological functions performed by soil protists will be revealed. Furthermore, protists in addition to nematophagous fungi (Gray 1987) might serve as potent biocontrol agents against plant-pathogenic nematodes and future surveys should aim at investigating this promising potential.

Experimental Procedures

Isolation and description of *Cryptodifflugia operculata*

C. operculata was isolated from the mineral layer of one pasture soil in the Netherlands (52°01′ N, 5°99′ E) using random enrichment cultivation in 9 cm Petri-dishes as described in (Smirnov and Brown 2004). Prescott-James medium (Page 1988), enriched with 0.15 % wheat grass (WG; Weizengras, Sanatur, Singen, Germany) was used at room temperature to initiate a mixed protist culture in respective Petri-dishes as detailed in Geisen et al., (2014b)   . Individual protists were manually transferred to new Petri-dishes filled with WG medium to establish monoxenic protist cultures (i.e. without the presence of other eukaryotes) living only on accompanying bacteria. *C. operculata* grew well in these cultures on bacteria only. Cell sizes were determined microscopically using a Nikon Eclipse TE2000-E inverted microscope at 400 x magnification by measuring the average of 20 cells. Shell dimensions were on average 16.9 ± 2.1 SD μm in length and 14.9 ± 1.8 SD μm in width. Genomic DNA was isolated using a guanidine isothiocyanate method (Maniatis et al. 1982a) followed by amplification of the small subunit (SSU) ribosomal DNA with primers, cycling conditions and enzymatic purification as described in Geisen et al., (2014b) . A partial sequence obtained revealed a perfect match to the *C. operculata* strain deposited under GenBank accession number JF694280 (Lahr et al. 2011).

Intraguild predation of *C. operculata* on bacterivorous nematodes

The bacterial-feeding nematode species *Acrobeloides buetschlii, A. camberenensis,* and *Rhabditis belari, R. dolichura* and *R. terricola* were grown on *Escherichia coli* OP-50 on 2 % agar plates enriched with 0.1 % of 0.4 g cholesterol in 80 ml 95 % ethanol. Nematodes were suspended in 0.15 % WG medium, washed twice by centrifugation at 600 g for 60 s in fresh WG medium and their numbers were subsequently determined.

Interactions between *C. operculata* and nematode species were studied in 24 well-plates (Sarstedt, Germany). *C. operculata* were pre-inoculated in 500 μl WG medium per well in respective treatments one week before the experiment started, and enumerated at the beginning of the experiment. 500 μl sterile WG medium was added to the remaining wells. The experiments were started by inoculating 33 ± 9 SD nematodes in 200 μl WG medium to each well of the respective treatments, while 200 μl sterile WG medium was added to non-nematode treatments. In total, four replicates of each treatment (*C. operculata* alone and in combination with each of the five nematode species) were setup resulting in 24 replicates in total. The 24 well-plates were sealed with Parafilm and incubated in the dark at room temperature for 18 days. The abundance of *C. operculata* in each well was determined microscopically (Nikon Eclipse TE2000-E) at 100× 2, 4, 6, 8, 10, 14 and 18 days past inoculation, and the numbers of living nematodes were determined at day 8 and day 18. Predation success was calculated by subtracting the final from the initial nematode abundances.

To document the interaction of *C. operculata* with nematodes, *A. buetschlii* was chosen as a model nematode and inoculated to all wells of a 24 well-plate pre-inoculated with *C. operculata* three days before. Attachment and feeding of *C. operculata* on living nematodes were investigated with an inverse microscope (Nikon Eclipse TE2000-E) at 100–400× and a sequential array of time-lapse photographs over the period of 12 hours taken.

In order to investigate if *C. operculata* also consumes dead nematodes, a supportive observational experiment was conducted. For that, *A. buetschlii* in WG medium were killed by addition of boiling H_2O_{dest} (1:1 v/v). One ml of the heat-killed nematode suspension was then added to a 9 cm Petri-dish that had been pre-inoculated with *C. operculata* as described above. Attachment and feeding of *C. operculata* on the dead nematodes were investigated as described above 24 hours after inoculation.

Detection of *C. operculata* 18S rRNA in soils

Soil metatranscriptomes of soils and litter from grasslands, forests and arctic peatlands across Europe (Geisen et al. 2015b) were investigated for the relative contribution of 18S rRNA transcripts of *Cryptodifflugia* to the total abundance of protist 18S rRNA transcripts.

Sampling, extraction of nucleic acids, 454 pyrosequencing and processing of raw reads are detailed in Geisen et al., (2015b) and references therein. Sequences were filtered using LUCY (Chou and Holmes 2001) to remove short (< 150 bp) and low-quality sequences (> 0.2 % error probability). Small subunit (SSU) ribosomal RNA sequences of eukaryotes were identified by MEGAN analysis of BLASTn files against a SSU rRNA reference database (Lanzen et al., 2011; parameters: min. bit score 150, min. support 1, top percent 10; 50 best blast hits). All eukaryotic SSU rRNAs were reanalysed with CREST (Lanzén et al. 2012) using the Silvamod database with LCA parameters min bit score 250, top percent 2 for classification of protist sequences. Correct taxonomic assignment of *Cryptodifflugia* reads was verified by manual BLASTn searches against the

NCBI GenBank nucleotide database. Less than 1 % of sequences had to be removed after manual inspection because they were wrongly assigned to uncultivated *Arcella* spp.

Statistical analyses

Changes in relative density of *C. operculata* and nematode species in 24-well plates were calculated after 18 days. Percentage data were square root arcsin-transformed prior to statistical analyses and analysed by a generalized linear model (GLM) using SAS 9.1 (Cary, FL, USA). Comparison of the means for the individual treatments was done at the 5 % probability level with a Tukey-test (Tukey's honestly significant difference, HSD). Additionally, given the computational independence of the two matrices, a Mantel analysis with 9999 runs has been carried out in PC-ORD 6 to assess whether the association between the abundances of nematodes in each of their two life-stages and the abundance of *C. operculata* listed in the same order was stronger than could result from chance (Douglas and Endler 1982).

Acknowledgements

This research was largely funded by the European Commission through the EcoFINDERS project (FP7-264465), with support by the DFG-Research Unit FOR 918 (Carbon flow in belowground food webs assessed by isotope tracers), subproject BO 1907/3-1. We thank Prof. Dr. Einhard Schierenberg, Dept. of Experimental Morphology, Institute of Zoology, University of Cologne for providing the nematode cultures. Last, we thank Edward Mitchell and three unknown referees for detailed and highly constructive comments on the manuscript.

Supplementary Files

Supplementary Movie S1: Time-lapse movie showing the feeding procedure of *Cryptodifflugia operculata* on the nematode *Acrobeloides buetschlii* over a period of 12 hours (http://onlinelibrary.wiley.com/doi/10.1111/1462-2920.12949/suppinfo)

General Discussion

The work described in this thesis made significant contributions to fill the gaps of knowledge on the extremely abundant, diverse, and functionally important group of soil protists (Clarholm 1985, Foissner 1987, Ekelund and Rønn 1994, Finlay et al. 2000, Bonkowski 2004, Bonkowski and Clarholm 2012). It has been shown that specific protist taxa, and in particular amoebae, can exhibit important functions, but knowledge on the diversity and community composition of protists in soils was scarce. Those studies investigating and showing important ecological functions exhibited by soil protists largely focused on one or few taxa (Bonkowski and Brandt 2002, Krome et al. 2009, Pedersen et al. 2009, Koller et al. 2013, Saleem et al. 2013). Taken into account the enormous diversity of (soil) protists (Foissner 1987, Cavalier-Smith 1998, Adl et al. 2012, Bates et al. 2013), generalizations on protists functioning derived from these simplistic studies more than likely provide an artificially oversimplified picture. Further, methodological drawbacks prevent assessing the real soil protist community composition with many species remaining undiscovered (Foissner 1999b, Moreira and López-García 2002, Epstein and López-García 2008).

Within this thesis, several hundred amoebae were cultivated, classified into morphogroups and more specifically described using sequence information in order to increase the knowledge of soil protists. Using this information a total of 16 new species and 7 new genera have been formally described (Part 1, Chapters 1 - 4). These basal cultivation efforts targeting amoebae emphasized the enormous diversity of soil protist. To get a more exhaustive knowledge on the protists community, Part 2 of this thesis aimed to describe the community composition of soil protists by applying and optimizing a range of cultivation-based, and cultivation-independent molecular methods (Part 2, Chapters 5 - 8). These studies further emphasise the enormous diversity of the soil protist community revealing that protist communities differ between soils and that the results strongly depend on the method being used. The last part aimed at increasing the knowledge of species-specific ecological functions performed by soil protists (Part 3, Chapter 9 - 10) and reveals that protists are far more than

278

bacterial feeders. Distinct taxa were shown to feed on fungi and nematodes and environmental sequencing revealed that protists with these ecological functions are widely distributed and abundant in soils.

Cultivation efforts remain an essential component in the study of soil protists (Part 1)

As the first major part of this thesis, the cultivation and description of 16 new species and even 7 genera of soil amoebae reveals that soil protists and especially amoebae are largely undersampled and reinforces the notion that a plethora of currently unknown protists inhabit soils (Moreira and López-García 2002, Epstein and López-García 2008).

The description of two new *Stenamoeba* spp. shows that this genus is species-rich, despite the short history of the genus *Stenamoeba* (Smirnov et al. 2007) with currently only three described species (Dyková et al. 2010b). The presence of MTOCs in *Stenamoeba* further demonstrates the scarce knowledge about taxonomic and morphological characteristics specific for or shared between amoebae within the supergroup Amoebozoa. Therefore, morphological features still add pivotal information to taxonomic affinities of unresolved groups based on phylogenetic information, such as between members of the class Discosea (Cavalier-Smith et al. 2004, Smirnov et al. 2005, Kudryavtsev and Pawlowski 2013), here shown by the ultrastructural feature of MTOCs (Chapter 1).

The description of another discosean amoeba, *Cochliopodium plurinucleolum* reveals that species-specific morphological characters and phylogenetic affinities even within a well-investigated amoebozoan genus such as *Cochliopodium* are far from being deciphered. Several *Cochliopodium* spp. with nearly identical sequences are further shown to exhibit inconsistent morphological characters, indicating that only a combination of molecular and morphological tools enables reliable identification of *Cochliopodium* spp. These discrepancies need to be considered when identifying amoebae based on morphological characters. Even the differentiation between higher taxonomic

279

ranks often are impossible, such as between vahlkampfiids (Brown and De Jonckheere 1999, De Jonckheere and Brown 2005), tubulinids (Page 1985, Smirnov et al. 2011b), or the focus group of Chapter 3, i.e. "Variosea-like" amoebae (Smirnov et al. 2008, Lahr et al. 2012, Berney et al. 2013). Therefore, it is hardly surprising that amoebae described without providing molecular information have commonly been misidentified and later transferred to other positions in the eukaryotic tree (Brown and De Jonckheere 1999, Smirnov et al. 2007, Smirnov et al. 2008, Smirnov et al. 2011b, Lahr et al. 2012). But as indicated by the results obtained within this chapter, species-level identification based on purely molecular information need careful interpretation. Intra-specific differences are often higher than inter-specific differences (Smirnov et al. 2007, Qvarnstrom et al. 2013) and morphologically clearly distinguishable species or genera sometimes share identical sequences, especially when partial sequences are being used (De Jonckheere and Brown 2005, Smirnov et al. 2009, Anderson and Tekle 2013). Further, annotated sequences are commonly mislabelled and some taxonomic affinities of and between groups remain uncertain, preventing reliable sequence assignments without knowledge on morphology (Berney et al. 2004, Smirnov et al. 2008, De Jonckheere et al. 2012). Therefore, identification up to species level should only be made by combining morphological with sequence information.

The description of six new genera of amoebae in the class Variosea helped at strongly increasing the knowledge on the diversity of and phylogenetic affinities within Variosea. Morphologically species-differentiation within Variosea remains hardly impossible, due to profound intra-clonal morphological plasticity that often surpasses differences even between genera. The variosean morphology, i.e. cells more or less branching and extended with filose pseudopodia, generally seems widespread among amoebae and is adopted by amoebae in the orders Leptomyxida, Centramoebida and Varipodida (Amoebozoa) and Vampyrellida (Cercozoa) (Adl et al. 2005, Bass et al. 2009a, Smirnov et al. 2011b, Hess et al. 2012). Therefore it is hardly surprising that many taxa have taken a long way until finding their current taxonomic affinity. One striking example is Leptomyxida that long combined all variosean-like amoebae but was later divided based on molecular information;

280

Leptomyxa and *Gephyramoeba* are now placed in Leptomyxida, *Balamuthia* in Centramoebida, while *Acramoeba dendroida* (initially mislabelled as "*Gephyramoeba*") (Smirnov et al. 2008), *Grellamoeba robusta* (Dyková et al. 2010a) and *Telaepolella tubasferens* (initially mislabelled "*Arachnula impatiens*") (Lahr et al. 2012) found their home in Varipodida (Amaral-Zettler et al. 2000, Smirnov et al. 2011b). Morphological and molecular information on the new cultivated genera provided and by reliable placing many sequences from uncultivated species inside Variosea an enormous diversity of the class Variosea is shown.

Similar to those species descriptions of amoebae in the supergroup Amoebozoa, seven new vahlkampfiid species including the new genus *Parafumarolamoeba* are described, revealing a high unknown diversity of heterolobosean amoebae (supergroup Excavata). Six new species placed in the recently erected genus *Allovahlkampfia* (Walochnik and Mulec 2009) were isolated from all geographically distant soils indicating a wide distribution of *Allovahlkampfia* spp. in soils. "*Solumitrus*" *palustris* is included as *A. palustris* as it reliably placed inside this genus, confirming previous studies (Brown et al. 2012, Harding et al. 2013) and shows that taxonomic affinities of other protists and amoebae might still change in the future due to increased taxon-sampling and multi-gene approaches. *Parafumarolamoeba alta* as the type species of a new genus branched with only uncultivated taxa in phylogenetic analyses pointing out that cultivation efforts are necessary in assigning sequences obtained in cultivation-independent soil surveys such as by using high-throughput sequencing (HTS).

The formal description of all these new species and genera of amoebae is in line with other recent descriptions of soil amoebae, which, however, remain rare (De Jonckheere et al. 2011b, Atlan et al. 2012). Cultures of several other new species and genera were obtained as part of this PhD work. An overview of sequenced amoebae of the supergroup Amoebozoa is shown in Fig. 75, demonstrating that several species only known from environmental sequencing approaches have been successfully cultivated, such as Vannellidae (e.g. strains Nl174, Sar32 and Tib97), Dermamoebidae (strains Sar17, Nl179 and Tib196),

Angulamoeba sp. F2 and *Variosea* sp. (strains G5, Tib48, Tib90). Most other clones branch inside known genera but often sharing low identity to described species, so that they are likely to represent new species, such as *Hartmannella* sp. (strains Nl117, Sar7 and Tib2), *Cochliopodium* sp. (strains Tib64 and Tib174), *Vannella* sp. (strains Nl7, Nl176 and Sar88) and *Filamoeba* sp. (strain Tib69). Among the cultivated amoebae are also many known species, such as *Saccamoeba limax* (strain Nl46), *Vermamoeba vermiformis* (strains Sar34 and Tib103), *Vannella simplex* (Sar36) and *Vexillifera bacilipedes* (strain Nl6). Several more examples for each of those three categories are illustrated in Fig. 75. Despite less diverse and lower abundant in cultures, several heterolobosean amoebae were also cultivated that can be placed in these three categories and also an amoeba showing unique morphology and molecular patterns that could not reliably be assigned to any eukaryotic supergroup (data not shown). All these examples prove that many species remain unknown or only known from environmental sequencing approaches (Lara et al. 2007a, Lejzerowicz et al. 2010, Berney et al. 2013). However, the cultivation of entirely unknown amoebae or known only from environmental sequencing approaches reveals that at least a subset of the so-called uncultivable amoebae are in fact cultivable. Molecular techniques such as HTS are now rapidly being improved, replacing traditional cultivation-based studies (Lara et al. 2007a, Urich et al. 2008, Medinger et al. 2010, Bates et al. 2013). Much, if not major information is currently lost as many sequences cannot reliably be assigned to known species, reinforcing the notion that traditional cultivation-based methods can- and should not entirely be replaced by molecular tools. In addition to the benefit of being able to assign molecular sequences to morphological information, cultivation-based efforts allow detailed ecological investigations on distinct species to eventually allow functional assignment to respective sequences.

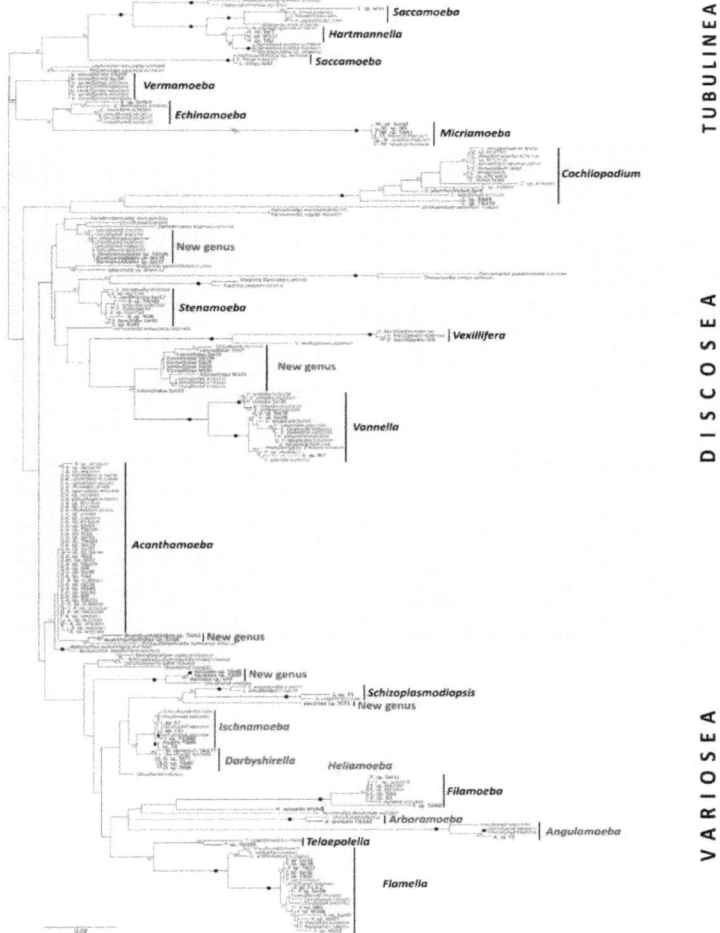

Fig. 75. Maximum likelihood phylogenetic tree of amoebozoan amoebae (classes Tubulinea, Discosea and Variosea) cultivated in this thesis with closest BLASTn hits (default parameters) obtained of each respective sequence. Amoebae most likely resembling known species are shown in red, new species and genera in pink; Genera of amoebae cultivated in this study highlighted in bold.

RaxML analysis (version 7.3.2, Stamatakis 2006), 1,113 nucleotide positions of the SSU-rDNA gene, GTR+γ+I model of nucleotide substitution, rooted with Tubulinea; bootstrap values shown > 60, solid circles = 100. Branches with a break cut in half; scale bar = 0.08 substitutions / site.

283

Diversity, distribution and community structure (Part 2)

The second major part of this thesis aimed at increasing the knowledge on the diversity and community structure of soil protists by applying a battery of different methods. Among those were a traditional cultivation based enumeration study combined with morphological protist identification (Chapter 5), a combination of cultivation and subsequent molecular identification of *Acanthamoeba* clones (Chapter 6), a HTAS study (Chapter 7) and a metatranscriptomic approach (Chapter 8).

Traditional cultivation based methods have the longest history and used to provide the only possibility of studying the entity of soil protists (Darbyshire et al. 1974, Foissner 1987, Smirnov 2003). As a first method to study soil protist communities a modified liquid aliquot method (LAM) (Butler and Rogerson 1995), decreasing workload, allowing deeper taxonomic identification of protist clades and enabling to obtain information on abundances of respective taxa. The high taxonomic resolution according to the most recent taxonomic classification (Smirnov et al. 2011b, Jeuck and Arndt 2013) is the first in this depth and provides evidence, that global climate changes might impact distinct protist clades differentially and that soil moisture has a profound impact on the abundance of soil protists. Therefore, cultivation based studies allow answering specific questions such as in ecological studies and remain unique in providing biomass estimations of a range of soil protists. A major advantage of the LAM is that protist taxa are usually growing in monoclonal cultures enabling downstream deep taxonomic classifications using molecular sequencing (e.g. Chapter 6) and functional investigations (Chapters 9 and 10).

An example for an approach using morphologically classified cultivated taxa from geographically distant locations with subsequent high-resolution molecular identification was applied in Chapter 6 as a second example of studying soil protist communities. Interestingly, all sequences obtained from morphologically often indistinguishable *Acanthamoeba* spp. are different to previously published sequences. Further, none of the sequences obtained at

284

one location is identical to sequences recovered at another location and also the community composition was found to differ between sites. The fairly low number of sequenced clones might have missed known and strains of *Acanthamoeba* identical between sites. Nevertheless, the observed difference in the *Acanthamoeba* community structure with only previously unknown strains sequenced strongly suggests that a huge diversity of *Acanthamoeba* strains remain to be discovered. Analysing a wider range of soils by deep-sequencing are needed to decipher the dispersal of sequence-identical *Acanthamoeba* strains and evaluate the factors that determine the community composition of *Acanthamoeba* spp. The observed patterns are even more profound taken into account that *Acanthamoeba* is among the best studied protists due to the presence of several facultative human pathogenicity strains (Schuster 2002, Schuster and Visvesvara 2004). One of the strains isolated in Dutch soils showed dramatic cytopathogenic characteristics, indicating that soils serve as a reservoir for pathogenic protists. Further functional investigations are highly desirable to investigate whether sequence differences between strains can be used to derive ecological functions and / or pathogenicity.

A high-throughput amplicon sequencing (HTAS) approach using cercozoan specific primers is used in Chapter 7 as a third technique to investigate the diversity and community composition of soil protists. Sample throughput and depth of community analyses are highly increased and the need to cultivate protists is circumvented by directly targeting DNA (Dawson and Hagen 2009, Creer et al. 2010, Medinger et al. 2010). Fundamental differences in the cercozoan community are detected that differ depending on geographic location and soil treatment. Interestingly, cercozoan communities differ strongly even between comparable soil environments strongly opposing the famous hypothesis that "everything is everywhere, the environment selects" (Baas-Becking 1934, Finlay 2002) and support recent findings that challenged this concept (Foissner 2006, Bass et al. 2007, Fontaneto and Hortal 2013, Heger et al. 2013). Further, this study shows that differences between soils become much more evident when increasing the level of taxonomic resolution suggesting that HTS approaches allow deciphering even minor differences in soil protist community structures.

Using metatranscriptomics as a forth method to study the entire diversity and community structure of soil protists allows deep analyses and comparisons of soil and litter samples from distinct locations in Europe. The deepest resolution of the soil protist community obtained to date reveals that amoebae represent a high proportion of the protist community contrasting previous HTAS approaches (Baldwin et al. 2013, Bates et al. 2013), that protist communities differ between locations on taxonomic ranks from supergroup to genus level and that protist groups basically unknown from soils such as choanoflagellates and foraminifera comprise a significant part of the protist community. Therefore it seems that research on soil protist diversity and community composition up to now only scratched the tip of the iceberg, with presumably only a small fraction of the entire protist community being known and a virtual absence of reliable information on protist community compositions.

Biases for all of the applied methods remain and need to be considered when studying protists; despite applying a modified LAM (Chapter 5), cultivation based techniques suffer from low sample-throughput, the prerequisite of protist cultivability (Berthold and Palzenberger 1995, Foissner 1999b) and the need of expert skills to morphologically identify many protist groups even to shallow taxonomic levels, e.g. family or genus level (Foissner 1999b, De Jonckheere and Brown 2005, Smirnov et al. 2008, Howe et al. 2009). Downstream sequencing of cultivated protists increases the taxonomic resolution and this combination of morphological and molecular tools remains the only reliable method to identify the majority of protists up to species level (De Jonckheere 1998, Brown and De Jonckheere 1999, Pawlowski et al. 1999, De Jonckheere and Brown 2005, Pawlowski and Burki 2009, Brabender et al. 2012). However, the work- and time-load synchronously increases dramatically, lowering the sample throughput to a minimum (Chapter 6). As the name suggests, "high-throughput" sequencing methods in form of HTAS or metatranscriptomics solve this problem. Additionally, HTS methods enable identification of uncultivable protists and avoid expert knowledge in protist identification. On the downside, high start-up costs, advanced bioinformatic capabilities and skills, taxonomic expertise in sequence interpretation by dealing with wrongly annotated sequences need to be considered (Berney et al. 2004, Epstein and López-García 2008, Medinger et al. 2010, Pawlowski et al. 2011). Not only do all those methods for studying protist diversity necessitate different prerequisites and skills, while all being affected by distinct sources of error, each method reveals a different picture on protist communities. This suggests that no single method allows deciphering the entire soil protist community in all aspects (Fig. 76) and that the method of choice depends on the question being addressed.

It has to be noted, however, that the methods applied here targeted different questions and were used to study dissimilar soils. Differences in observed protist communities are affected at least in part by abiotic factors such as differences in soil texture, organic matter content, plant communities, land-use and moisture (Foissner 1997, 1999b, Anderson 2002, Bates et al. 2013). These differences render direct method comparisons based on our results impossible, but it remains unquestionable that each method reveals a different part of the protist community. Future direct comparisons of the entire battery of methods to study the soil protist community are essential to decipher inter-methodological differences that will ultimately allow to reliable identify information that can be extrapolated between methods. Only this knowledge will finally allow a reliable estimation of protist abundances, dispersal, diversity and community composition in soils, which until then remains highly speculative.

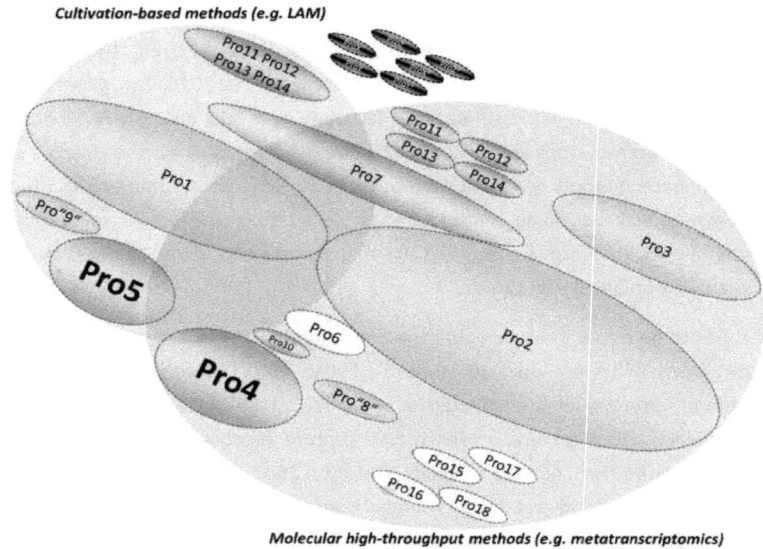

Fig. 76. Simplified schematics comparing two major methodological approaches for studying soil protists, i.e. cultivation based (smaller blue circle; top left hand corner) and cultivation-independent (larger grey circle; bottom right hand corner). Differences between approaches are illustrating showing methodological artefacts that need to be considered when studying protist communities. Sphere sizes indicate (relative) abundances of respective protist clades; Equal coloured spheres: comparable errors between methods

Pro1 - 18: protist taxa. Pro1: abundant, well cultivable; Pro2: abundant, not-well cultivable; Pro3: abundant, uncultivable; Pro4: overrepresented when applying HTS technologies (e.g. multiple copies of target molecules, amplification of extracellular DNA as well as PCR-artefacts in HTAS or high numbers of rRNA transcripts in metatranscriptomics); Pro5: overrepresented in cultivation-based studies, e.g. mainly present inactively; Pro6: underrepresented due to low abundances close to the detection limit in cultivation efforts; Pro7: equally represented between methods; Pro„8": wrongly assigned (e.g. sequencing errors or mislabelled published sequence); Pro„9": morphologically misidentified; Pro10: adversely affected in sequencing (e.g. extraction of target molecules prevented by incomplete cell lyses or mismatches of primers in HTAS); Pro11 - Pro14: morphologically (nearly) identical, distinguishable only using cultivation-independent techniques; Pro15 - 18: low abundant taxa recovered only by HTS; ProX-ProZ[2]: uncultivable protists, sequences obtained but removed in quality filtering or very low abundant taxa, i.e. „missing diversity".

Ecological importance (Part 3)

The third major part of this thesis reveals that protists are more than bacterivores as is usually been suggested in soil food web models (Hunt et al. 1987, de Ruiter et al. 1995, Crotty et al. 2011). Mycophagous (Chapter 9) and nematophagous protists (Chapter 10) are described, both functional groups showing high relative abundances in diverse soil and litter samples.

In the first study (Chapter 9) several protists previously suggested as bacterivorous are shown to feed on distinct fungi. True obligate mycophagous protists have rarely been studied, but few are known, such as ciliates in the family Grossglockneriidae (Foissner 1980, Petz et al. 1986, Foissner 1999a) and some facultative mycophagous groups, such as vampyrellid amoebae (Old and Darbyshire 1980, Old and Oros 1980, Hess et al. 2012). Sequences specific for these two protist groups were discovered in targeted sequence-mining of HTAS and metatranscriptomic (Chapter 8) datasets, being present in all samples investigated and represent a substantial fraction of the entire protist community. Therefore, soil protists are likely more than bacterivores and mycophagous protists should be considered as an important trophic node in soil systems (Fig. 77).

A small and common testate amoeba, *Cryptodifflugia operculata*, is shown in a second study to interact with a range of other soil organisms, most profoundly feeding and proliferating on a range of nematodes. *C. operculata* grows in monoclonal cultures on bacteria only, but in presence of nematodes, abundances strongly increased suggesting that this facultative nematophagy suits as an important feeding strategy in soils. The high representation of *Cryptodifflugia*-like sequences among all protist sequences in the metatranscriptomic analyses of several soils (Chapter 8) provides further evidence that the trophic level of protists is not identical for all of these highly diverse taxa. Nematophagous or omnivorous protists are likely to deserve an own node in soil food webs and potentially constituting an important alternative link to other trophic levels (Fig. 77). Also conceivable is that the microbial loop in soil and eventually plant growth (Clarholm 1985, Bonkowski 2004) is further stimulated by omnivorous protists as they supplement plants and bacteria with nutrients released from presumably higher trophic levels.

Due to their enormous diversity, vastly outreaching that of all multicellular eukaryotes (Cavalier-Smith 1993, Cavalier-Smith 1998, Adl et al. 2012) it is little surprising that protists occupy distinct environmental niches and perform alternate ecological functions, despite the contradictory view of treating protists in a single functional unit as bacterivores (Hunt et al. 1987, de Ruiter et al. 1995). The studies in this chapter provide further evidence for feeding differences between protist taxa, confirming studies revealing strongly diverging feeding strategies, such as by feeding on other protists (Page 1977, Smirnov et al. 2007, Berney et al. 2013), fungi (Old and Oros 1980, Petz et al. 1985, Ekelund 1998) and nematodes (Doncaster and Hooper 1961, Sayre 1973, Yeates and Foissner 1995). Peculiar, however, is the small size and high abundance of the newly found nematophagous amoeba *C. operculata* suggesting that food web anomalies are common and that higher turnover and abundances of these smaller organisms are of major ecological importance in controlling nematode numbers. The sequence information obtained from both mycophagous and nematophagous protists further shows their high representation in soils and litter indicating that traditional soil food webs need to be complemented by several nodes of functionally distinct protist clades, where e.g. nematophagous and mycophagous protists should find their home (Fig. 77). Taken together, soil protists are a diverse assembly of organisms that host diverse functionally different groups most likely functioning as key nodes in soil food webs.

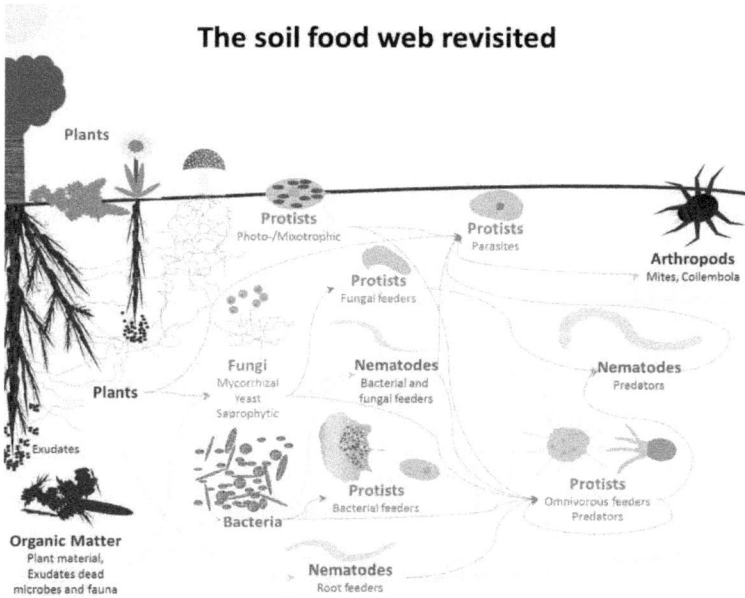

Fig. 77. The soil food web focusing on protists; direct interactions indicated with arrows (red: interactions of protists with other soil organisms, grey: interactions between non-protists); different organisms encoded by different colours.

References

Abu Kwaik, Y. 1996. The phagosome containing *Legionella pleumophila* within the protozoan *Hartmannella vermiformis* is surrounded by the rough endoplasmic reticulum. Appl. Environ. Microbiol. **62**:2022-2028.

Adl, M. S., and V. V. S. R. Gupta. 2006. Protists in soil ecology and forest nutrient cycling. Can. J. For. Res. **36**:1805–1817.

Adl, S. M., A. Habura, and Y. Eglit. 2014. Amplification primers of SSU rDNA for soil protists. Soil Biol. Biochem. **69**:328–342.

Adl, S. M., A. G. B. Simpson, M. A. Farmer, R. A. Andersen, O. R. Anderson, J. R. Barta, S. S. Bowser, G. Brugerolle, R. A. Fensome, S. Fredericq, T. Y. James, S. Karpov, P. Kugrens, J. Krug, C. E. Lane, L. A. Lewis, J. Lodge, D. H. Lynn, D. G. Mann, R. M. McCourt, L. Mendoza, O. Moestrup, S. E. Mozley-Standridge, T. A. Nerad, C. A. Shearer, A. V. Smirnov, F. W. Spiegel, and M. Taylor. 2005. The new higher level classification of eukaryotes with emphasis on the taxonomy of protists. J. Eukaryot. Microbiol. **52**:399-451.

Adl, S. M., A. G. B. Simpson, C. E. Lane, J. Lukeš, D. Bass, S. S. Bowser, M. W. Brown, F. Burki, M. Dunthorn, V. Hampl, A. Heiss, M. Hoppenrath, E. Lara, L. le Gall, D. H. Lynn, H. McManus, E. A. D. Mitchell, S. E. Mozley-Stanridge, L. W. Parfrey, J. Pawlowski, S. Rueckert, L. Shadwick, C. L. Schoch, A. Smirnov, and F. W. Spiegel. 2012. The revised classification of eukaryotes. J. Eukaryot. Microbiol. **59**:429-514.

Ahmad, A. F., P. W. Andrew, and S. Kilvington. 2011. Development of a nested PCR for environmental detection of the pathogenic free-living amoeba *Balamuthia mandrillaris*. J. Eukaryot. Microbiol. **58**:269-271.

Alabouvette, C., M. M. Couteaux, K. M. Old, M. Pussard, O. Reisinger, and F. Toutain. 1981. Les protozoaires du sol: aspects écologiques et méthodologiques. Annales Biologique **3**:255-303.

Allen, P. G., and E. A. Dawidowicz. 1990. Phagocytosis in *Acanthamoeba*: I. A mannose receptor is responsible for the binding and phagocytosis of yeast. J. Cell. Physiol. **145**:508-513.

Alphei, J., M. Bonkowski, and S. Scheu. 1996. Protozoa, Nematoda and Lumbricidae in the rhizosphere of *Hordelymus europeaus* (Poaceae): faunal interactions, response of microorganisms and effects on plant growth Oecologia **106**:111-126.

Altizer, S., D. Harvell, and E. Friedle. 2003. Rapid evolutionary dynamics and disease threats to biodiversity. Trends Ecol. Evol. **18**:589-596.

Altschul, S. F., W. Gish, W. Miller, E. W. Myers, and D. J. Lipman. 1990. Basic local alignment search tool. J. Mol. Biol. **215**:403-410.

Amaral-Zettler, L. A., F. Gómez, E. Zettler, B. G. Keenan, R. Amils, and M. L. Sogin. 2002. Microbiology: eukaryotic diversity in Spain's River of Fire. Nature **417**:137-137.

Amaral-Zettler, L. A., E. A. McCliment, H. W. Ducklow, and S. M. Huse. 2009. A method for studying protistan diversity using massively parallel sequencing of V9 hypervariable regions of small-subunit ribosomal RNA genes. PLoS One **4**:e6372.

Amaral-Zettler, L. A., T. A. Nerad, C. J. O'Kelly, M. T. Peglar, P. M. Gillevet, J. D. Silberman, and M. L. Sogin. 2000. A molecular reassessment of the leptomyxid amoebae. Protist **151**:275-282.

Amaral Zettler, L. A., T. A. Nerad, C. J. O'Kelly, M. T. Peglar, P. M. Gillevet, J. D. Silberman, and M. L. Sogin. 2000. A molecular reassessment of the Leptomyxid amoebae. Protist **151**:275-282.

Anand, C., A. Skinner, A. Malic, and J. Kurtz. 1983. Interaction of *L. pneumophilia* and a free living amoeba (*Acanthamoeba palestinensis*). J. Hyg. Camb. **91**:167.

Anderson, I. J., R. F. Watkins, J. Samuelson, D. F. Spencer, W. H. Majoros, M. W. Gray, and B. J. Loftus. 2005. Gene discovery in the *Acanthamoeba castellanii* genome. Protist **156**:203-214.

Anderson, M. J. 2001. A new method for non-parametric multivariate analysis of variance. Austral Ecol. **26**:32-46.

Anderson, M. J., and T. J. Willis. 2003. Canonical analysis of principal coordinates: a useful method of constrained ordination for ecology. Ecology **84**:511-525.

Anderson, O. R. 2000. Abundance of terrestrial gymnamoebae at a northeastern U. S. site: a four-year study, including the El Nino winter of 1997–1998. J. Eukaryot. Microbiol. **47**:148-155.

Anderson, O. R. 2002. Laboratory and field-based studies of abundances, small-scale patchiness, and diversity of gymnamoebae in soils of varying porosity and organic content: evidence of microbiocoenoses. J. Eukaryot. Microbiol. **49**:17-23.

Anderson, O. R. 2008. The role of amoeboid protists and the microbial community in moss-rich terrestrial ecosystems: biogeochemical implications for the carbon budget and carbon cycle, especially at higher latitudes. J. Eukaryot. Microbiol. **55**:145-150.

Anderson, O. R., and Y. I. Tekle. 2013. A description of *Cochliopodium megatetrastylus* n. sp. isolated from a freshwater habitat. Acta Protozool. **52**:55-64.

Anderson, O. R., W. E. N. Wang, S. P. Faucher, K. Bi, and H. A. Shuman. 2011. A new heterolobosean amoeba *Solumitrus palustris* n. g., n. sp. isolated from freshwater marsh soil. J. Eukaryot. Microbiol. **58**:60-67.

Anderson, R. V., E. T. Elliott, J. F. McClellan, D. C. Coleman, C. V. Cole, and H. W. Hunt. 1977. Trophic interactions in soils as they affect energy and nutrient dynamics. III. Biotic interactions of bacteria, amoebae, and nematodes. Microb. Ecol. **4**:361-371.

Angell, R. W. 1975. Structure of *Trichosphaerium micrum* sp. n. J. Eukaryot. Microbiol. **22**:18-22.

Angell, R. W. 1976. Observations on *Trichosphaerium platyxyrum* sp. n. J. Eukaryot. Microbiol. **23**:357-364.

Arndt, H., D. Dietrich, B. Auer, E.-J. Cleven, T. Gräfenhan, M. Weitere, and A. P. Mylnikov. 2000. Functional diversity of heterotrophic flagellates in aquatic ecosystems. The flagellates: unity, diversity and evolution:240-268.

Atlan, D., B. Coupat-Goutaland, A. Risler, M. Reyrolle, M. Souchon, J. Briolay, S. Jarraud, P. Doublet, and M. Pélandakis. 2012. *Micriamoeba tesseris* nov. gen. nov. sp.: a new taxon of free-living small-sized amoebae non-permissive to virulent Legionellae. Protist **163**:888-902.

Baas-Becking, L. G. M. 1934. Geobiologie, of inleiding tot de milieukunde. Van Stockum.

Bachy, C., J. R. Dolan, P. Lopez-Garcia, P. Deschamps, and D. Moreira. 2013. Accuracy of protist diversity assessments: morphology compared with cloning and direct pyrosequencing of 18S rRNA genes and ITS regions using the conspicuous tintinnid ciliates as a case study. ISME J. **7**:244-255.

Back, M. A., P. P. J. Haydock, and P. Jenkinson. 2002. Disease complexes involving plant parasitic nematodes and soilborne pathogens. Plant Pathol. **51**:683-697.

Bagella, S., L. Salis, G. M. Marrosu, I. Rossetti, S. Fanni, M. C. Caria, and P. P. Roggero. 2013. Effects of long-term management practices on grassland plant assemblages in Mediterranean cork oak silvo-pastoral systems. Plant Ecol. **214**:621-631.

Baldwin, D. S., M. J. Colloff, G. N. Rees, A. A. Chariton, G. O. Watson, L. N. Court, D. M. Hartley, M. j. Morgan, A. J. King, J. S. Wilson, M. Hodda, and C. M. Hardy. 2013. Impacts of inundation and drought on eukaryote biodiversity in semi-arid floodplain soils. Mol. Ecol. **22**:1746-1758.

Bamforth, S. S. 1971. The numbers and proportions of Testacea and ciliates in litters and soils. J. Protozool. **18**:24-28.

Bamforth, S. S. 1980. Terrestrial Protozoa. J. Eukaryot. Microbiol. **27**:33-36.

Bamforth, S. S. 1988. Interactions between protozoa and other organisms. Agr. Ecosyst. Environ. **24**:229-234.

Bamforth, S. S. 2004. Water film fauna of microbiotic crusts of a warm desert. J. Arid Environ. **56**:413-423.

Bamforth, S. S. 2007. Protozoa from aboveground and ground soils of a tropical rain forest in Puerto Rico. Pedobiologia **50**:515-525.

Bamforth, S. S. 2010. Distribution of and insights from soil protozoa of the Olympic coniferous rain forest. Pedobiologia **53**:361-367.

Bardgett, R. D., and W. H. van der Putten. 2014. Belowground biodiversity and ecosystem functioning. Nature **515**:505-511.

Bark, A. 1973. A study of the genus *Cochliopodium* Hertwig and Lesser 1874. Protistologica **9**:119-138.

Bass, D., and T. Cavalier-Smith. 2004. Phylum-specific environmental DNA analysis reveals remarkably high global biodiversity of Cercozoa (Protozoa). Int. J. Syst. Evol. Microbiol. **54**:2393-2404.

Bass, D., E. E. Y. Chao, S. Nikolaev, A. Yabuki, K.-i. Ishida, C. Berney, U. Pakzad, C. Wylezich, and T. Cavalier-Smith. 2009a. Phylogeny of novel naked filose and reticulose Cercozoa: Granofilosea cl. n. and Proteomyxidea revised. Protist **160**:75-109.

Bass, D., A. T. Howe, A. P. Mylnikov, K. Vickerman, E. E. Chao, J. Edwards Smallbone, J. Snell, C. Cabral Jr, and T. Cavalier-Smith. 2009b. Phylogeny and classification of Cercomonadida (Protozoa, Cercozoa): *Cercomonas*, *Eocercomonas*, *Paracercomonas*, and *Cavernomonas* gen. nov. Protist **160**:483-521.

Bass, D., T. Richards, L. Matthai, V. Marsh, and T. Cavalier-Smith. 2007. DNA evidence for global dispersal and probable endemicity of protozoa. BMC Evol. Biol. **7**:162.

Bass, D., A. Yabuki, S. Santini, S. Romac, and C. Berney. 2012. *Reticulamoeba* is a long-branched granofilosean (Cercozoa) that is missing from sequence databases. PLoS One **7**:e49090.

Bass, P., and P. J. Bischoff. 2001. Seasonal variability in abundance and diversity of soil gymnamoebae along a short transect in southeastern USA. J. Eukaryot. Microbiol. **48**:475-479.

Bates, S. T., J. C. Clemente, G. E. Flores, W. A. Walters, L. W. Parfrey, R. Knight, and N. Fierer. 2013. Global biogeography of highly diverse protistan communities in soil. ISME J. **7**:652-659.

Baumgartner, M., S. Eberhardt, J. F. De Jonckheere, and K. O. Stetter. 2009. *Tetramitus thermacidophilus* n. sp., an amoeboflagellate from acidic hot springs. J. Eukaryot. Microbiol. **56**:201-206.

Behnke, A., M. Engel, R. Christen, M. Nebel, R. R. Klein, and T. Stoeck. 2011. Depicting more accurate pictures of protistan community complexity using pyrosequencing of hypervariable SSU rRNA gene regions. Environ. Microbiol. **13**:340-349.

Benhamou, N., P. Rey, K. Picard, and Y. Tirilly. 1999. Ultrastructural and cytochemical aspects of the interaction between the mycoparasite *Pythium oligandrum* and soilborne plant pathogens. Phytopathol. **89**:506-517.

Bennett, W. 1986. Fine structure of the trophic stage of *Endostelium* Olive, Bennett & Deasey, 1984 (Eumycetozoea, Protosteliia). Protistologica **22**:205-212.

Benwitz, G., and K. G. Grell. 1971a. Ultrastruktur mariner Amöben II. *Stereomyxa ramosa*. Arch. Protistenkd. **113**:51-67.

Benwitz, G., and K. G. Grell. 1971b. Ultrastruktur mariner Amöben III. *Stereomyxa angulosa*. Arch. Protistenkd. **113**:68-79.

Berney, C., J. Fahrni, and J. Pawlowski. 2004. How many novel eukaryotic 'kingdoms'? Pitfalls and limitations of environmental DNA surveys. BMC Biol. **2**:13.

Berney, C., S. Geisen, J. Van Wichelen, F. Nitsche, P. Vanormelingen, M. Bonkowski, and D. Bass. 2015. Expansion of the 'Reticulosphere': diversity of novel branching and network-forming amoebae helps to define Variosea (Amoebozoa). Protist **166**:271-295.

Berney, C., and J. Pawlowski. 2006. A molecular time-scale for eukaryote evolution recalibrated with the continuous microfossil record. Proc. Biol. Sci. **273**:1867-1872.

Berney, C., S. Romac, F. Mahé, S. Santini, R. Siano, and D. Bass. 2013. Vampires in the oceans: predatory cercozoan amoebae in marine habitats. ISME J. **7**:2387-2399.

Berthold, A., and M. Palzenberger. 1995. Comparison between direct counts of active soil ciliates (Protozoa) and most probable number estimates obtained by Singh's dilution culture method. Biol. Fertil. Soils **19**:348-356.

Bischoff, P. J. 2002. An analysis of the abundance, diversity and patchiness of terrestrial gymnamoebae in relation to soil depth and precipitation events following a drought in southeastern U.S.A. Acta Protozool. **41**:183-189.

Bjørnlund, L., M. Liu, R. Rønn, S. Christensen, and F. Ekelund. 2012. Nematodes and protozoa affect plants differently, depending on soil nutrient status. Eur. J. Soil Biol. **50**:28-31.

Bjørnlund, L., and R. Rønn. 2008. 'David and Goliath' of the soil food web – Flagellates that kill nematodes. Soil Biol. Biochem. **40**:2032-2039.

Bobrov, A. A. 2005. Testate amoebas and the regularities of their distribution in soil. Eurasian Soil Sci. **38**:1001-1008.

Boenigk, J., and H. Arndt. 2002. Bacterivory by heterotrophic flagellates: community structure and feeding strategies. Anton. Leeuw. **81**:465-480.

Boenigk, J., M. Ereshefsky, K. Hoef-Emden, J. Mallet, and D. Bass. 2012. Concepts in protistology: species definitions and boundaries. Eur. J. Protistol. **48**:96-102.

Boenigk, J., S. Jost, T. Stoeck, and T. Garstecki. 2007. Differential thermal adaptation of clonal strains of a protist morphospecies originating from different climatic zones. Environ. Microbiol. **9**:593-602.

Boenigk, J., K. Pfandl, P. Stadler, and A. Chatzinotas. 2005. High diversity of the '*Spumella*-like' flagellates: an investigation based on the SSU rRNA gene sequences of isolates from habitats located in six different geographic regions. Environ. Microbiol. **7**:685-697.

Böhme, A., U. Risse-Buhl, and K. Küsel. 2009. Protists with different feeding modes change biofilm morphology. FEMS Microbiol. Ecol. **69**:158-169.

Bonkowski, M. 2004. Protozoa and plant growth: the microbial loop in soil revisited. New Phytol. **162**:617-631.

Bonkowski, M., and F. Brandt. 2002. Do soil protozoa enhance plant growth by hormonal effects? Soil Biol. Biochem. **34**:1709-1715.

Bonkowski, M., and M. Clarholm. 2012. Stimulation of plant growth through interactions of bacteria and protozoa: testing the auxiliary microbial loop hypothesis. Acta Protozool. **51**:237-247.

Bonkowski, M., B. Griffiths, and C. Scrimgeour. 2000a. Substrate heterogeneity and microfauna in soil organic 'hotspots' as determinants of nitrogen capture and growth of ryegrass. Appl. Soil Ecol. **14**:37-53.

Bonkowski, M., B. Griffiths, and C. Scrimgeour. 2000b. Substrate heterogeneity and microfauna in soil organic 'hotspots' as determinants of nitrogen capture and growth of ryegrass. Appl. Soil Ecol. **14**:37-53.

Bonkowski, M., and M. Schaefer. 1997. Interactions between earthworms and soil protozoa: a trophic component in the soil food web. Soil Biol. Biochem. **29**:499-502.

Booton, G. C., D. J. Kelly, Y.-W. Chu, D. V. Seal, E. Houang, D. S. C. Lam, T. J. Byers, and P. A. Fuerst. 2002. 18S ribosomal DNA typing and tracking of *Acanthamoeba* species isolates from corneal scrape specimens, contact lenses, lens cases, and home water supplies of *Acanthamoeba* keratitis patients in Hong Kong. J. Clin. Microbiol. **40**:1621-1625.

Borken, W., and E. Matzner. 2009. Reappraisal of drying and wetting effects on C and N mineralization and fluxes in soils. Global Change Biology **15**:808-824.

Bouwman, L. A., and K. B. Zwart. 1994. The ecology of bacterivorous protozoans and nematodes in arable soil. Agriculture, Ecosystems and Environment **51**:145-160.

Bowers, B., and E. D. Korn. 1968. The fine structure of *Acanthamoeba castellanii*: I. The Trophozoite. J. Cell Biol. **39**:95-111.

Bowers, B., and E. D. Korn. 1973. Cytochemical identification of phosphatase activity in the contractile vacuole of *Acanthamoeba castellanii*. J. Cell Biol. **59**:784-791.

Brabender, M., Á. K. Kiss, A. Domonell, F. Nitsche, and H. Arndt. 2012. Phylogenetic and morphological diversity of novel soil cercomonad species with a description of two new genera (*Nucleocercomonas* and *Metabolomonas*). Protist **163**:495-528.

Bradford, M. A. 2013. Thermal adaptation of decomposer communities in warming soils. Front. Microbiol. **4**.

Bradford, M. A., S. A. Wood, R. D. Bardgett, H. I. Black, M. Bonkowski, T. Eggers, S. J. Grayston, E. Kandeler, P. Manning, H. Setala, and T. H. Jones. 2014. Discontinuity in the responses of ecosystem processes and multifunctionality to altered soil community composition. Proc. Natl. Acad. Sci. USA **111**:14478-14483.

Brown, M. W., J. D. Silberman, and F. W. Spiegel. 2012. A contemporary evaluation of the acrasids (Acrasidae, Heterolobosea, Excavata). Eur. J. Protistol. **48**:103-123.

Brown, M. W., F. W. Spiegel, and J. D. Silberman. 2007. Amoeba at attention: phylogenetic affinity of *Sappinia pedata*. J. Eukaryot. Microbiol. **54**:511-519.

Brown, S., and J. F. De Jonckheere. 1999. A reevaluation of the amoeba genus *Vahlkampfia* based on SSUrDNA sequences. Eur. J. Protistol. **35**:49-54.

Brown, S., and A. V. Smirnov. 2004. Diversity of gymnamoebae in grassland soil in southern Scotland. Protistology **3**:191-195.

Brüssow, H. 2007. Bacteria between protists and phages: from antipredation strategies to the evolution of pathogenicity. Mol. Microbiol. **65**:583-589.

Bunting, L. A., J. B. Neilson, and G. S. Bulmer. 1979. *Cryptococcus neoformans*: gastronomic delight of a soil ameba. Sabouraudia **17**:225-232.

Burki, F., K. Shalchian-Tabrizi, M. Minge, Å. Skjæveland, S. I. Nikolaev, K. S. Jakobsen, and J. Pawlowski. 2007. Phylogenomics reshuffles the eukaryotic supergroups. PLoS One **2**:e790.

Butler, H., and A. Rogerson. 1995. Temporal and spatial abundance of naked amoebae (Gymnamoebae) in marine benthic sediments of the Clyde Sea Area, Scotland. J. Eukaryot. Microbiol. **42**:724-730.

Butler, H., and A. Rogerson. 1997. Consumption rates of six species of marine benthic naked amoebae (Gymnamoebia) from sediments in the Clyde Sea Area. J. Mar. Biol. Assoc. U. K. **77**:989-997.

Canter-Lund, H., and J. W. Lund. 1995. Freshwater algae: their microscopic world explored. Pages 272-273 Freshwater algae: their microscopic world explored. Biopress Ltd., Bristol.

Carlson, M. L., L. A. Flagstad, F. Gillet, and E. A. D. Mitchell. 2010. Community development along a proglacial chronosequence: Are above-ground and below-ground community structure controlled more by biotic than abiotic factors? J. Ecol. **98**:1084-1095.

Caron, D. A., A. Z. Worden, P. D. Countway, E. Demir, and K. B. Heidelberg. 2008. Protists are microbes too: a perspective. ISME J. **3**:4-12.

Carson, J. K., V. Gonzalez-Quiñones, D. V. Murphy, C. Hinz, J. A. Shaw, and D. B. Gleeson. 2010. Low pore connectivity increases bacterial diversity in soil. Appl. Environ. Microbiol. **76**:3936-3942.

Cavalier-Smith, T. 1993. Kingdom protozoa and its 18 phyla. Microbiol. Rev. **57**:953-994.

Cavalier-Smith, T. 1998. A revised six-kingdom system of life. Biol. Rev. **73**:203-266.

Cavalier-Smith, T. 2003. Protist phylogeny and the high-level classification of Protozoa. Eur. J. Protistol. **39**:338-348.

Cavalier-Smith, T., and E. E. Chao. 1995. The opalozoan *Apusomonas* is related to the common ancestor of animals, fungi, and choanoflagellates. Proc. R. Soc. Lond. B **261**:1-6.

Cavalier-Smith, T., E. E. Y. Chao, and B. Oates. 2004. Molecular phylogeny of Amoebozoa and the evolutionary significance of the unikont *Phalansterium*. Eur. J. Protistol. **40**:21-48.

Chakraborty, S., and K. Old. 1982. Mycophagous soil amoeba: interactions with three plant pathogenic fungi. Soil Biol. Biochem. **14**:247-255.

Chakraborty, S., K. M. Old, and J. H. Warcup. 1983. Amoebae from a take-all suppressive soil which feed on *Gaeumannomyces graminis tritici* and other soil fungi. Soil Biol. Biochem. **15**:17-24.

Chatzinotas, A., S. Schellenberger, K. Glaser, and S. Kolb. 2013. Assimilation of cellulose-derived carbon by microeukaryotes in oxic and anoxic slurries of an aerated soil. Appl. Environ. Microbiol. **79**:5777-5781.

Chen, M., F. Chen, Y. Yu, J. Ji, and F. Kong. 2008. Genetic diversity of eukaryotic microorganisms in Lake Taihu, a large shallow subtropical lake in China. Microb. Ecol. **56**:572-583.

Chiellini, C., R. Iannelli, L. Modeo, V. Bianchi, and G. Petroni. 2012. Biofouling of reverse osmosis membranes used in river water purification for drinking purposes: analysis of microbial populations. Biofouling **28**:969-984.

Chou, H.-H., and M. H. Holmes. 2001. DNA sequence quality trimming and vector removal. Bioinformatics **17**:1093-1104.

Christensen, S., L. Björnlund, and M. Vestergard. 2007. Decomposer biomass in the rhizosphere to assess rhizodeposition. Oikos **116** 65-74.

Clarholm, M. 1981. Protozoan grazing of bacteria in soil—impact and importance. Microb. Ecol. **7**:343-350.

Clarholm, M. 1985. Interactions of bacteria, protozoa and plants leading to mineralization of soil nitrogen. Soil Biol. Biochem. **17**:181-187.

Clarholm, M., M. Bonkowski, and B. Griffiths. 2007. Protozoa and other protista in soil. Pages 148-175 *in* J. Van Elsas, J. Jansson, and J. Trevors, editors. Modern Soil Microbiology, 2nd edn.CRC, Boca Raton London New York. CRC Press, Boca Raton, FL, USA.

Clarke, K. R. 1993. Non-parametric multivariate analyses of changes in community structure. Aust. J. Ecol. **18**:117-143.

Clarke, M., A. Lohan, B. Liu, I. Lagkouvardos, S. Roy, N. Zafar, C. Bertelli, C. Schilde, A. Kianianmomeni, T. Burglin, C. Frech, B. Turcotte, K. Kopec, J. Synnott, C. Choo, I. Paponov, A. Finkler, C. Heng Tan, A. Hutchins, T. Weinmeier, T. Rattei, J. Chu, G. Gimenez, M. Irimia, D. Rigden, D. Fitzpatrick, J. Lorenzo-Morales, A. Bateman, C.-H. Chiu, and P. Tang. 2013a. Genome of *Acanthamoeba castellanii* highlights extensive lateral gene transfer and early evolution of tyrosine kinase signaling. Genome Biol. **14**:R11.

Clarke, M., A. J. Lohan, B. Liu, I. Lagkouvardos, S. Roy, N. Zafar, C. Bertelli, C. Schilde, A. Kianianmomeni, and T. R. Bürglin. 2013b. Genome of *Acanthamoeba castellanii* highlights extensive lateral gene transfer and early evolution of tyrosine kinase signaling. Genome Biol. **14**:R11.

Coince, A., O. Caël, C. Bach, J. Lengellé, C. Cruaud, F. Gavory, E. Morin, C. Murat, B. Marçais, and M. Buée. 2013. Below-ground fine-scale distribution and soil versus fine root detection of fungal and soil oomycete communities in a French beech forest. Fungal Ecol. **6**:223-235.

Corsaro, D., and D. Venditti. 2010. Phylogenetic evidence for a new genotype of *Acanthamoeba* (Amoebozoa, Acanthamoebida). Parasitol. Res. **107**:233-238.

Corsaro, D., and D. Venditti. 2011. More *Acanthamoeba* genotypes: limits to the use rDNA fragments to describe new genotypes. Acta Protozool. **50**:49.

Couradeau, E., K. Benzerara, D. Moreira, E. Gérard, J. Kaźmierczak, R. Tavera, and P. López-García. 2011. Prokaryotic and eukaryotic community structure in field and cultured microbialites from the alkaline lake Alchichica (Mexico). PLoS One **6**:e28767.

Creer, S., V. G. Fonseca, D. L. Porazinska, R. M. Giblin-Davis, W. Sung, D. M. Power, M. Packer, G. R. Carvalho, M. L. Blaxter, P. J. D. Lambshead, and W. K. Thomas. 2010. Ultrasequencing of the meiofaunal biosphere: practice, pitfalls and promises. Mol. Ecol. **19**:4-20.

Crotty, F., S. Adl, R. Blackshaw, and P. Murray. 2012a. Protozoan pulses unveil their pivotal position within the soil food web. Microb. Ecol. **63**:905-918.

Crotty, F., R. Blackshaw, and P. Murray. 2011. Tracking the flow of bacterially derived 13C and 15N through soil faunal feeding channels. Rapid Commun. Mass Sp. **25**:1503-1513.

Crotty, F. V., S. M. Adl, R. P. Blackshaw, and P. J. Murray. 2012b. Using stable isotopes to differentiate trophic feeding channels within soil food webs. J. Eukaryot. Microbiol. **59**:520-526.

Crotty, F. V., S. M. Adl, R. P. Blackshaw, and P. J. Murray. 2013. Measuring soil protist respiration and ingestion rates using stable isotopes. Soil Biol. Biochem. **57**:919-921.

Darbyshire, J. 1994. Soil Protozoa. CAB International, Wallingford.

Darbyshire, J. F. 1976. Effect of water suctions on the growth in soil of the ciliate *Colpoda steini,* and the bacterium *Azotobacter chroococcum.* J. Soil Sci. **27**:369-376.

Darbyshire, J. F. 2005. The use of soil biofilms for observing protozoan movement and feeding. FEMS Microbiol. Lett. **244**:329-333.

Darbyshire, J. F., B. S. Griffiths, D. M. S., and W. J. McHardy. 1989. Ciliate distribution amongst soil aggregates. Rev. Ecol. Biol. Sol **26**:47-56.

Darbyshire, J. F., R. E. Whitley, M. P. Graebes, and R. H. E. Inkson. 1974. A rapid micromethod for estimating bacterial and protozoan populations in soil. Rev. Ecol. Biol. Sol **11**:465-475.

Darriba, D., G. L. Taboada, R. Doallo, and D. Posada. 2012. jModelTest 2: more models, new heuristics and parallel computing. Nat. Methods **9**:772-772.

Dawson, S., and K. Hagen. 2009. Mapping the protistan 'rare biosphere'. J. Biol. **8**:1-3.

De Jonckheere, J. F. 1994. Comparison of partial SSUrDNA sequences suggests revisions of species names in the genus *Naegleria*. Eur. J. Protistol. **30**:333-341.

De Jonckheere, J. F. 1998. Sequence variation in the ribosomal internal transcribed spacers, including the 5.8 S rDNA, of *Naegleria* spp. Protist **149**:221-228.

De Jonckheere, J. F. 2002. A century of research on the amoeboflagellate genus *Naegleria*. Acta Protozool. **41**:309-342.

De Jonckheere, J. F. 2004. Molecular definition and the ubiquity of species in the genus *Naegleria*. Protist **155**:89-103.

De Jonckheere, J. F. 2006. Isolation and molecular identification of free-living amoebae of the genus *Naegleria* from Arctic and sub-Antarctic regions. Eur. J. Protistol. **42**:115-123.

De Jonckheere, J. F. 2012. The impact of man on the occurrence of the pathogenic free-living amoeboflagellate *Naegleria fowleri*. Future Microbiol. **7**:5-7.

De Jonckheere, J. F., M. Baumgartner, S. Eberhardt, F. R. Opperdoes, and K. O. Stetter. 2011a. *Oramoeba fumarolia* gen. nov., sp. nov., a new marine heterolobosean amoeboflagellate growing at 54°C. Eur. J. Protistol. **47**:16-23.

De Jonckheere, J. F., and S. Brown. 2005. The identification of vahlkampfiid amoebae by ITS sequencing. Protist **156**:89-96.

De Jonckheere, J. F., S. Gryseels, and M. Eddyani. 2012. Knowledge of morphology is still required when identifying new amoeba isolates by molecular techniques. Eur. J. Protistol. **48**:178-184.

De Jonckheere, J. F., J. Murase, and F. R. Opperdoes. 2011b. A new thermophilic heterolobosean amoeba, *Fumarolamoeba ceborucoi*, gen. nov., sp. nov., isolated near a fumarole at a volcano in Mexico. Acta Protozool. **50**:41-48.

De Rijk, P., J. Wuyts, and R. De Wachter. 2003. RnaViz 2: an improved representation of RNA secondary structure. Bioinformatics **19**:299-300.

de Ruiter, P. C., J. C. Moore, K. B. Zwart, L. A. Bouwman, J. Hassink, J. Bloem, J. A. de Vos, J. C. Y. Marinissen, W. A. M. Didden, G. Lebrink, and L. Brussaard. 1993. Simulation of nitrogen mineralization in the below-ground food webs of two winter wheat fields. J. Appl. Ecol. **30**:95-106.

de Ruiter, P. C., A. M. Neutel, and J. C. Moore. 1995. Energetics, patterns of interaction strengths, and stability in real ecosystems. Science **269**:1257-1260.

de Vries, F., M. Liiri, L. Bjornlund, H. Setälä, S. Christensen, and R. Bardgett. 2012a. Legacy effects of drought on plant growth and the soil food web. Oecologia **170**:821-833.

de Vries, F. T., M. E. Liiri, L. Bjørnlund, M. A. Bowker, S. Christensen, H. M. Setala, and R. D. Bardgett. 2012b. Land use alters the resistance and resilience of soil food webs to drought. Nat. Clim. Change **2**:276-280.

Domonell, A., M. Brabender, F. Nitsche, M. Bonkowski, and H. Arndt. 2013. Community structure of cultivable protists in different grassland and forest soils of Thuringia. Pedobiologia **56**:1-7.

Doncaster, C. C., and D. J. Hooper. 1961. Nematodes attacked by protozoa and tardigrades. Nematologica **6**:333-335.

Douglas, M. E., and J. A. Endler. 1982. Quantitative matrix comparisons in ecological and evolutionary investigations. J. Theor. Biol. **99**:777-795.

Dunnebacke, T. H., F. L. Schuster, S. Yagi, and G. C. Booton. 2004. *Balamuthia mandrillaris* from soil samples. Microbiology **150**:2837-2842.

Dunthorn, M., W. Foissner, and L. A. Katz. 2008. Molecular phylogenetic analysis of class Colpodea (phylum Ciliophora) using broad taxon sampling. Mol. Phylogenet. Evol. **46**:316-327.

Dyková, I., M. Kostka, and H. Pecková. 2010a. *Grellamoeba robusta* gen. n., sp. n., a possible member of the family Acramoebidae Smirnov, Nassonova et Cavalier-Smith, 2008. Eur. J. Protistol. **46**:77-85.

Dyková, I., M. Kostka, and H. Pecková. 2010b. Two new species of the genus *Stenamoeba* Smirnov, Nassonova, Chao et Cavalier-Smith, 2007. Acta Protozool. **49**:245-251.

Dyková, I., J. Lom, and B. Machaèkova. 1998. *Cochliopodium minus*, isolated from organs of perch *Perca fluviatilis*. Dis Aquat Org **34**:205-210.

Dyková, I., J. Lom, J. M. Schroeder-Diedrich, G. C. Booton, and T. J. Byers. 1999. *Acanthamoeba* strains isolated from organs of freshwater fishes. J. Parasitol. **85**:1106-1113.

Edgar, R. C. 2013. UPARSE: highly accurate OTU sequences from microbial amplicon reads. Nat. Methods **10**:996-998.

Efron, N., G. Young, and N. A. Brennan. 1989. Ocular surface temperature. Curr. Eye Res. **8**:901-906.

Ekelund, F. 1998. Enumeration and abundance of mycophagous protozoa in soil, with special emphasis on heterotrophic flagellates. Soil Biol. Biochem. **30**:1343-1347.

Ekelund, F., and D. J. Patterson. 1997. Some heterotrophic flagellates from a cultivated garden soil in Australia. Arch. Protistenkd. **148**:461-478.

Ekelund, F., and R. Rønn. 1994. Notes on protozoa in agricultural soil with emphasis on heterotrophic flagellates and naked amoebae and their ecology. FEMS Microbiol. Rev. **15**:321-353.

Ekelund, F., R. Rønn, and B. S. Griffiths. 2001. Quantitative estimation of flagellate community structure and diversity in soil samples. Protist **152**:301-314.

Elliott, E. T., R. V. Anderson, D. C. Coleman, and C. V. Cole. 1980a. Habitable pore space and microbial trophic interactions. Oikos **35**:327-335.

Elliott, E. T., R. V. Anderson, D. C. Coleman, and C. V. Cole. 1980b. Habitable pore space and microbial trophic interactions. Oikos **35**:327-335.

Elliott, E. T., and D. C. Coleman. 1977. Soil protozoan dynamics in a shortgrass prairie. Soil Biol. Biochem. **9**:113-118.

Epstein, S., and P. López-García. 2008. "Missing" protists: a molecular prospective. Biodivers. Conserv. **17**:261-276.

Esteban, G. F., K. J. Clarke, J. L. Olmo, and B. J. Finlay. 2006. Soil protozoa — an intensive study of population dynamics and community structure in an upland grassland. Appl. Soil Ecol. **33**:137-151.

Euringer, K., and T. Lueders. 2008. An optimised PCR/T-RFLP fingerprinting approach for the investigation of protistan communities in groundwater environments. J. Microbiol. Meth. **75**:262-268.

Evans, S., and M. Wallenstein. 2012. Soil microbial community response to drying and rewetting stress: does historical precipitation regime matter? Biogeochemistry **109**:101-116.

Evans, S. E., and M. D. Wallenstein. 2014. Climate change alters ecological strategies of soil bacteria. Ecol. Lett. **17**:155-164.

Farhat, M., M. Moletta-Denat, J. Frère, S. Onillon, M.-C. Trouilhé, and E. Robine. 2012. Effects of disinfection on *Legionella* spp., eukarya, and biofilms in a hot water system. Appl. Environ. Microbiol. **78**:6850-6858.

Feest, A., and R. Campbell. 1986. The microbiology of soils under successive wheat crops in relation to take-all disease. FEMS Microbiol. Lett. **38**:99-111.

Fell, J. W., G. Scorzetti, L. Connell, and S. Craig. 2006. Biodiversity of micro-eukaryotes in Antarctic Dry Valley soils with <5% soil moisture. Soil Biol. Biochem. **38**:3107-3119.

Felsenstein, J. 1981. Evolutionary trees from DNA sequences: a maximum likelihood approach. J. Mol. Evol. **17**:368-376.

Felsenstein, J. 1985. Confidence limits on phylogenies: an approach using the bootstrap. Evolution **39**:783-791.

Fenchel, T. 1987. Ecology of protozoa: the biology of free-living phagotrophic protists. Science Tech Publishers, Madison, WI.

Fenchel, T. 2010. The life history of *Flabellula baltica* Smirnov (Gymnamoebae, Rhizopoda): Adaptations to a spatially and temporally heterogeneous environment. Protist **161**:279-287.

Ferris, H., B. S. Griffiths, D. L. Porazinska, T. O. Powers, K.-H. Wang, and M. Tenuta. 2012. Reflections on plant and soil nematode ecology: past, present and future. J. Nematol. **44**:115-126.

Field, S. G., and N. K. Michiels. 2005. Parasitism and growth in the earthworm *Lumbricus terrestris*: fitness costs of the gregarine parasite *Monocystis* sp. Parasitology **130**:397-403.

Fierer, N., J. Schimel, and P. Holden. 2003. Influence of drying–rewetting frequency on soil bacterial community structure. Microb. Ecol. **45**:63-71.

Fierer, N., and J. P. Schimel. 2003. A proposed mechanism for the pulse in carbon dioxide production commonly observed following the rapid rewetting of a dry soil. Soil Sci. Soc. Am. J. **67**:798-805.

Finlay, B. J. 1998. The global diversity of protozoa and other small species. Int. J. Parasitol. **28**:29-48.

Finlay, B. J. 2002. Global dispersal of free-living microbial eukaryote species. Science **296**:1061-1063.

Finlay, B. J., H. I. J. Black, S. Brown, K. J. Clarke, G. F. Esteban, R. M. Hindle, J. L. Olmo, A. Rollett, and K. Vickerman. 2000. Estimating the growth potential of the soil protozoan community. Protist **151**:69-80.

Fiore-Donno, A. M., A. Kamono, E. E. Chao, M. Fukui, and T. Cavalier-Smith. 2010. Invalidation of *Hyperamoeba* by transferring its species to other genera of Myxogastria. J. Eukaryot. Microbiol. **57**:189-196.

Fischer, M., O. Bossdorf, S. Gockel, F. Hansel, A. Hemp, D. Hessenmoller, G. Korte, J. Nieschulze, S. Pfeiffer, D. Prati, S. Renner, I. Schoning, U. Schumacher, K. Wells, F. Buscot, E. K. V. Kalko, K. E. Linsenmair, E. D. Schulze, and W. W. Weisser. 2010. Implementing large-scale and long-term functional biodiversity research: The Biodiversity Exploratories. Basic and Applied Ecology **11**:473-485.

Flavin, M., C. J. O'Kelly, T. A. Nerad, and G. Wilkinson. 2000. *Cholamonas cyrtodiopsidis* gen. n., sp. n.(Cercomonadida), an endocommensal, mycophagous heterotrophic flagellate with a doubled kinetid. Acta Protozool. **39**:51-60.

Foissner, W. 1980. Colpodide Ciliaten (Protozoa: Ciliophora) aus alpinen Böden. Zool. Jb. Syst **107**:391-432.

Foissner, W. 1987. Soil protozoa: fundamental problems, ecological significance, adaptations in ciliates and testaceans, bioindicators, and guide to the literature. Prog. Protistol. **2**:69-212.

Foissner, W. 1997. Protozoa as bioindicators in agroecosystems, with emphasis on farming practices, biocides, and biodiversity. Agr. Ecosyst. Environ. **62**:93-103.

Foissner, W. 1998. An updated compilation of world soil ciliates (Protozoa, Ciliophora), with ecological notes, new records, and descriptions of new species. Eur. J. Protistol. **34**:195-235.

Foissner, W. 1999a. Description of two new, mycophagous soil ciliates (Ciliophora, Colpodea): *Fungiphrya strobli* n. g., n. sp. and *Grossglockneria ovata* n. sp. J. Eukaryot. Microbiol. **46**:34-42.

Foissner, W. 1999b. Soil protozoa as bioindicators: pros and cons, methods, diversity, representative examples. Agr. Ecosyst. Environ. **74**:95-112.

Foissner, W. 2006. Biogeography and dispersal of micro-organisms: a review emphasizing protists. Acta Protozool. **45**:111-136.

Foissner, W. 2009. Protist diversity and distribution: Some basic considerations

Protist Diversity and Geographical Distribution. Pages 1-8 *in* W. Foissner and D. L. Hawksworth, editors. Springer Netherlands.

Foissner, W., and P. Didier. 1983. Nahrungsaufnahme, Lebenszyklus und Morphogenese von *Pseudoplatyophrya nana* (KAHL, 1926)(Ciliophora, Colpodida). Protistologica **19**:103-109.

Fontaneto, D., and J. Hortal. 2013. At least some protist species are not ubiquitous. Mol. Ecol. **22**:5053-5055.

Fouque, E., M. C. Trouilhe, V. Thomas, P. Hartemann, M. H. Rodier, and Y. Héchard. 2012. Cellular, biochemical, and molecular changes during encystment of free-living amoebae. Eukaryotic Cell **11**:382-387.

Frey, F. 1971. Über die Eignung von *Acrobeloides buetschlii* (Cephalobidae) für nematologische Laboruntersuchungen. Nematologica **17**:474-477.

Fuchslueger, L., M. Bahn, K. Fritz, R. Hasibeder, and A. Richter. 2014. Experimental drought reduces the transfer of recently fixed plant carbon to soil microbes and alters the bacterial community composition in a mountain meadow. New Phytol. **201**:916-927.

Gabilondo, R., and E. Bécares. 2014. The effects of natural carbon dioxide seepage on edaphic protozoan communities in Campo de Calatrava, Ciudad Real, Spain. Soil Biol. Biochem. **68**:133-139.

Galagan, J. E., S. E. Calvo, K. A. Borkovich, E. U. Selker, N. D. Read, D. Jaffe, W. FitzHugh, L.-J. Ma, S. Smirnov, and S. Purcell. 2003. The genome sequence of the filamentous fungus *Neurospora crassa*. Nature **422**:859-868.

Galiana, E., A. Marais, C. Mura, B. Industri, G. Arbiol, and M. Ponchet. 2011. Ecosystem screening approach for pathogen-associated microorganisms affecting host disease. Appl. Environ. Microbiol. **77**:6069-6075.

Gast, R. J. 2001. Development of an *Acanthamoeba*-specific reverse dot-blot and the discovery of a new ribotype. J. Eukaryot. Microbiol. **48**:609-615.

Gast, R. J., D. R. Ledee, P. A. Fuerst, and T. J. Byers. 1996. Subgenus systematics of *Acanthamoeba*: four nuclear 18S rDNA sequence types. J. Eukaryot. Microbiol. **43**:498-504.

Geisen, S., C. Bandow, J. Römbke, and M. Bonkowski. 2014a. Soil water availability strongly alters the community composition of soil protists. Pedobiologia **57**:205–213.

Geisen, S., A. M. Fiore-Donno, J. Walochnik, and M. Bonkowski. 2014b. *Acanthamoeba* everywhere: high diversity of *Acanthamoeba* in soils. Parasitol. Res. **113**:3151-3158.

Geisen, S., A. Kudryavtsev, M. Bonkowski, and A. Smirnov. 2014c. Discrepancy between species borders at morphological and molecular levels in the genus *Cochliopodium* (Amoebozoa, Himatismenida), with the description of *Cochliopodium plurinucleolum* n. sp. Protist **165**:364-383.

Geisen, S., I. Laros, A. Vizcaíno, M. Bonkowski, and G. A. de Groot. 2015a. Not all are free-living: high-throughput DNA metabarcoding reveals a diverse community of protists parasitizing soil metazoa. Mol. Ecol.

Geisen, S., A. T. Tveit, I. M. Clark, A. Richter, M. M. Svenning, M. Bonkowski, and T. Urich. 2015b. Metatranscriptomic census of active protists in soils. ISME J.

Geisen, S., J. Weinert, A. Kudryavtsev, A. Glotova, M. Bonkowski, and A. V. Smirnov. 2014d. Two new species of the genus *Stenamoeba* (Discosea, Longamoebia): cytoplasmic MTOC is present in one more amoebae lineage. Eur. J. Protistol. **50**:153-165.

Glaubitz, S., T. Lueders, W.-R. Abraham, G. Jost, K. Jürgens, and M. Labrenz. 2009. 13C-isotope analyses reveal that chemolithoautotrophic Gamma- and Epsilonproteobacteria feed a microbial food web in a pelagic redoxcline of the central Baltic Sea. Environ. Microbiol. **11**:326-337.

Glockling, S. L., W. L. Marshall, and F. H. Gleason. 2013. Phylogenetic interpretations and ecological potentials of the Mesomycetozoea (Ichthyosporea). Fungal Ecol. **6**:237-247.

Glücksman, E., T. Bell, R. I. Griffiths, and D. Bass. 2010. Closely related protist strains have different grazing impacts on natural bacterial communities. Environ. Microbiol. **12**:3105-3113.

Gomaa, F., M. Todorov, T. J. Heger, E. A. D. Mitchell, and E. Lara. 2012. SSU rRNA phylogeny of Arcellinida (Amoebozoa) reveals that the largest arcellinid genus, *Difflugia* Leclerc 1815, is not monophyletic. Protist **163**:389-399.

Gong, J., J. Dong, X. Liu, and R. Massana. 2013. Extremely high copy numbers and polymorphisms of the rDNA operon estimated from single cell analysis of oligotrich and peritrich ciliates. Protist **164**:369-379.

Goodey, T. 1915a. Note on the remarkable retention of vitality by protozoa from old stored soils. Ann. Appl. Biol. **1**:395-399.

Goodey, T. 1915b. A preliminary communication on three new proteomyxan rhizopods from soil. Arch. Protistenkd. **35**:80-102.

Gouy, M., S. Guindon, and O. Gascuel. 2010. SeaView version 4: a multiplatform graphical user interface for sequence alignment and phylogenetic tree building. Mol. Biol. Evol. **27**:221-224.

Gower, J. C. 2005. Principal coordinates analysis. Encyclopedia of Biostatistics. John Wiley & Sons, Ltd.

Grant, J., Y. I. Tekle, O. R. Anderson, D. J. Patterson, and L. A. Katz. 2009. Multigene evidence for the placement of a heterotrophic amoeboid lineage *Leukarachnion* sp. among photosynthetic stramenopiles. Protist **160**:376-385.

Gray, N. F. 1987. Nematophagous fungi with particular reference to their ecology. Biol. Rev. **62**:245-304.

Grębecki, A. 1994. Membrane and cytoskeleton flow in motile cells with emphasis on the contribution of free-living amoebae. Pages 37-80 *in* W. J. Kwang and J. Jonathan, editors. Int. Rev. Cytol. Academic Press.

Grell, K. G. 1994. The feeding community of *Synamoeba arenaria* n. gen., n. sp. Arch. Protistenkd. **144**:143-146.

Grell, K. G. 1995. *Reticulamoeba minor* n. sp. and its reticulopodia. Arch. Protistenkd. **145**:3-9.

Grell, K. G., and G. Benwitz. 1978. Ultrastruktur mariner Amöben IV. *Corallomyxa chattoni* n. sp. Arch. Protistenkd. **120**:287-300.

Greub, G., and D. Raoult. 2004. Microorganisms resistant to free-living amoebae. Clin. Microbiol. Rev. **17**:413-433.

Griffiths, B. S., M. Bonkowski, G. Dobson, and S. Caul. 1999. Changes in soil microbial community structure in the presence of microbial-feeding nematodes and protozoa. Pedobiologia **43**:297-304.

Griffiths, B. S., M. Bonkowski, J. Roy, and K. Ritz. 2001. Functional stability, substrate utilisation and biological indicators of soils following environmental impacts. Appl. Soil Ecol. **16**:49-61.

Guillard, R. R. L., and C. J. Lorenzen. 1972. Yellow-green algae with chlorophyllide C. J. Phycol. **8**:10-14.

Guillou, L., D. Bachar, S. Audic, D. Bass, C. Berney, L. Bittner, C. Boutte, G. Burgaud, C. de Vargas, J. Decelle, J. del Campo, J. R. Dolan, M. Dunthorn, B. Edvardsen, M. Holzmann, W. H. C. F. Kooistra, E. Lara, N. Le Bescot, R. Logares, F. Mahé, R. Massana, M. Montresor, R. Morard, F. Not, J. Pawlowski, I. Probert, A.-L. Sauvadet, R. Siano, T. Stoeck, D. Vaulot, P. Zimmermann, and R. Christen. 2013. The Protist Ribosomal Reference database (PR2): a catalog of unicellular eukaryote small sub-unit rRNA sequences with curated taxonomy. Nucleic Acids Res. **41**:D597-D604.

Guindon, S., and O. Gascuel. 2003. A simple, fast, and accurate algorithm to estimate large phylogenies by maximum likelihood. Syst. Biol. **52**:696-704.

Hall, T. A. 1999. BioEdit: a user-friendly biological sequence alignment editor and analysis program for Windows 95/98/NT. Pages 95-98 *in* Nucleic Acids Symp. Ser.

Hammer, Ø., D. Harper, and P. Ryan. 2001. PAST: paleontological statistics software package for education and data analysis. Palaeontol. Electronica **4**:1-9.

Harding, T., M. W. Brown, A. Plotnikov, E. Selivanova, J. S. Park, J. H. Gunderson, M. Baumgartner, J. D. Silberman, A. J. Roger, and A. G. Simpson. 2013. Amoeba stages in the deepest branching heteroloboseans, including *Pharyngomonas*: evolutionary and systematic implications. Protist **164**:272-286.

Heal, O. 1965. Observations on testate amoebae (Protozoa: Rhizopoda) from Signy Island, South Orkney Islands. Brit. Antarct. Surv. Bull **6**:43-47.

Heal, O. W. 1963. Soil fungi as food for amoebae. Pages 289-297 *in* J. Doeksen and J. Van der Drift, editors. Soil Organisms. North-Holland Publishing Company, Amsterdam.

Hedley, R. H., Ogden, C. G. & Mordan, N. J. 1977. Biology and fine structure of *Cryptodifflugia oviformis* (Rhizopodea: Protozoa). Bull. Br. Mus. Nat. Hist. **30**:311–328.

Hedlund, K., I. Santa Regina, W. H. Van der Putten, J. Lepš, T. Díaz, G. W. Korthals, S. Lavorel, V. K. Brown, D. Gormsen, S. R. Mortimer, C. Rodríguez Barrueco, J. Roy, P. Smilauer, M. Smilauerová, and C. Van Dijk. 2003. Plant species diversity, plant biomass and responses of the soil community on abandoned land across Europe: idiosyncracy or above-belowground time lags. Oikos **103**:45-58.

Heger, T. J., E. A. D. Mitchell, and B. S. Leander. 2013. Holarctic phylogeography of the testate amoeba *Hyalosphenia papilio* (Amoebozoa: Arcellinida) reveals extensive genetic diversity explained more by environment than dispersal limitation. Mol. Ecol. **22**:5172–5184.

Hekman, W. E., P. J. H. F. van den Boogert, and K. B. Zwart. 1992. The physiology and ecology of a novel, obligate mycophagous flagellate. FEMS Microbiol. Lett. **86**:255-265.

Herdler, S., K. Kreuzer, S. Scheu, and M. Bonkowski. 2008. Interactions between arbuscular mycorrhizal fungi (*Glomus intraradices*, Glomeromycota) and amoebae (*Acanthamoeba castellanii*, Protozoa) in the rhizosphere of rice (*Oryza sativa*). Soil Biol. Biochem. **40**:660-668.

Hess, S., and M. Melkonian. 2013. The mystery of clade X: *Orciraptor* gen. nov. and *Viridiraptor* gen. nov. are highly specialised, algivorous amoeboflagellates (Glissomonadida, Cercozoa). Protist **164**:706-747.

Hess, S., N. Sausen, and M. Melkonian. 2012. Shedding light on vampires: the phylogeny of vampyrellid amoebae revisited. PLoS One 7:e31165.

Hewett, M. K., B. S. Robinson, P. T. Monis, and C. P. Saint. 2003. Identification of a new Acanthamoeba 18S rRNA gene sequence type, corresponding to the species Acanthamoeba jacobsi Sawyer, Nerad and Visvesvara, 1992 (Lobosea: Acanthamoebidae). Acta Protozool. 42:325-329.

Hollande, A., G. Nicolas, and J. Escaig. 1981. Veture glycostylaire et ultrastructure d'une Amibe marine libre (Mayorella pussardi nov. sp.: Paramoebidae) observee apres congelation ultrarapide suivie de cryosubstitution. Protistologica 17:147-154.

Holtkamp, R., A. van der Wal, P. Kardol, W. H. van der Putten, P. C. de Ruiter, and S. C. Dekker. 2011. Modelling C and N mineralisation in soil food webs during secondary succession on ex-arable land. Soil Biol. Biochem. 43:251-260.

Hong, S., J. Bunge, C. Leslin, S. Jeon, and S. S. Epstein. 2009. Polymerase chain reaction primers miss half of rRNA microbial diversity. ISME J. 3:1365-1373.

Hordijk, W., and O. Gascuel. 2005. Improving the efficiency of SPR moves in phylogenetic tree search methods based on maximum likelihood. Bioinformatics 21:4338-4347.

Horn, M., T. R. Fritsche, R. K. Gautom, K.-H. Schleifer, and M. Wagner. 1999. Novel bacterial endosymbionts of Acanthamoeba spp. related to the Paramecium caudatum symbiont Caedibacter caryophilus. Environ. Microbiol. 1:357-367.

Horn, M., and M. Wagner. 2004. Bacterial endosymbionts of free-living amoebae. J. Eukaryot. Microbiol. 51:509-514.

Howe, A. T., D. Bass, E. E. Chao, and T. Cavalier-Smith. 2011a. New genera, species, and improved phylogeny of Glissomonadida (Cercozoa). Protist 162:710-722.

Howe, A. T., D. Bass, J. M. Scoble, R. Lewis, K. Vickerman, H. Arndt, and T. Cavalier-Smith. 2011b. Novel cultured protists identify deep-branching environmental DNA clades of Cercozoa: new genera Tremula, Micrometopion, Minimassisteria, Nudifila, Peregrinia. Protist 162:332-372.

Howe, A. T., D. Bass, K. Vickerman, E. E. Chao, and T. Cavalier-Smith. 2009. Phylogeny, taxonomy, and astounding genetic diversity of Glissomonadida ord. nov., the dominant gliding zooflagellates in soil (Protozoa: Cercozoa). Protist 160:159-189.

Huang, X., and A. Madan. 1999. CAP3: A DNA sequence assembly program. Genome Res. 9:868-877.

Huelsenbeck, J. P., and F. Ronquist. 2001. MRBAYES: bayesian inference of phylogenetic trees. Bioinformatics 17:754-755.

Hunt, H. W., D. C. Coleman, E. R. Ingham, R. E. Ingham, E. T. Elliott, J. C. Moore, S. L. Rose, C. P. P. Reid, and C. R. Morley. 1987. The detrital food web in a shortgrass prairie. Biol. Fertil. Soils 3:57-68.

IPCC. 2012. Managing the Risks of Extreme Events and Disasters to Advance Climate Change Adaptation. A Special Report of Working Groups I and II of the Intergovernmental Panel on Climate Change.in C. B. Field, V. Barros, T.F. Stocker, D. Qin, D. J. Dokken, K. L. Ebi, M. D. Mastrandrea, K. J. Mach, G.-K. Plattner, S. K. Allen, M. Tignor, and P. M. Midgley, editors. Cambridge University Press, Cambridge, UK and New York, NY, USA.

Jeon, S., J. Bunge, C. Leslin, T. Stoeck, S. Hong, and S. Epstein. 2008. Environmental rRNA inventories miss over half of protistan diversity. BMC Microbiol. Biol. 8:222.

Jeuck, A., and H. Arndt. 2013. A short guide to common heterotrophic flagellates of freshwater habitats based on the morphology of living organisms. Protist 164:842-860.

Jousset, A. 2011. Ecological and evolutive implications of bacterial defences against predators. Environmental microbiology.

Jousset, A. 2012. Ecological and evolutive implications of bacterial defences against predators. Environ. Microbiol. 14:1830–1843.

Jousset, A., E. Lara, L. Wall, and C. Valverde. 2006. Secondary metabolites help biocontrol strain Pseudomonas fluorescens CHA0 to escape protozoan grazing. Applied and environmental microbiology 72:7083.

Jousset, A., L. Rochat, M. Pechy-Tarr, C. Keel, S. Scheu, and M. Bonkowski. 2009. Predators promote defence of rhizosphere bacterial populations by selective feeding on non-toxic cheaters. ISME Journal **3**:666-674.

Kaiser, C., M. Koranda, B. Kitzler, L. Fuchslueger, J. Schnecker, P. Schweiger, F. Rasche, S. Zechmeister-Boltenstern, A. Sessitsch, and A. Richter. 2010. Belowground carbon allocation by trees drives seasonal patterns of extracellular enzyme activities by altering microbial community composition in a beech forest soil. New Phytol. **187**:843-858.

Kamono, A., M. Meyer, T. Cavalier-Smith, M. Fukui, and A. M. Fiore-Donno. 2013. Exploring slime mould diversity in high-altitude forests and grasslands by environmental RNA analysis. FEMS Microbiol. Ecol. **84**:98-109.

Keeling, P. J., F. Burki, H. M. Wilcox, B. Allam, E. E. Allen, L. A. Amaral-Zettler, E. V. Armbrust, J. M. Archibald, A. K. Bharti, C. J. Bell, B. Beszteri, K. D. Bidle, C. T. Cameron, L. Campbell, D. A. Caron, R. A. Cattolico, J. L. Collier, K. Coyne, S. K. Davy, P. Deschamps, S. T. Dyhrman, B. Edvardsen, R. D. Gates, C. J. Gobler, S. J. Greenwood, S. M. Guida, J. L. Jacobi, K. S. Jakobsen, E. R. James, B. Jenkins, U. John, M. D. Johnson, A. R. Juhl, A. Kamp, L. A. Katz, R. Kiene, A. Kudryavtsev, B. S. Leander, S. Lin, C. Lovejoy, D. Lynn, A. Marchetti, G. McManus, A. M. Nedelcu, S. Menden-Deuer, C. Miceli, T. Mock, M. Montresor, M. A. Moran, S. Murray, G. Nadathur, S. Nagai, P. B. Ngam, B. Palenik, J. Pawlowski, G. Petroni, G. Piganeau, M. C. Posewitz, K. Rengefors, G. Romano, M. E. Rumpho, T. Rynearson, K. B. Schilling, D. C. Schroeder, A. G. B. Simpson, C. H. Slamovits, D. R. Smith, G. J. Smith, S. R. Smith, H. M. Sosik, P. Stief, E. Theriot, S. N. Twary, P. E. Umale, D. Vaulot, B. Wawrik, G. L. Wheeler, W. H. Wilson, Y. Xu, A. Zingone, and A. Z. Worden. 2014. The marine microbial eukaryote transcriptome sequencing project (MMETSP): illuminating the functional diversity of eukaryotic life in the oceans through transcriptome sequencing. PLoS Biol. **12**:e1001889.

Khan, N. A. 2006. *Acanthamoeba*: biology and increasing importance in human health. FEMS Microbiol. Rev. **30**:564-595.

Khan, N. A. 2009. *Acanthamoeba*: biology and pathogenesis. Caister Academic Press, Norfolk, United Kingdom.

Knacker, T., C. M. van Gestel, S. Jones, A. V. M. Soares, H.-J. Schallnaß, B. Förster, and C. Edwards. 2004. Ring-testing and field-validation of a Terrestrial Model Ecosystem (TME) – an instrument for testing potentially harmful substances: conceptual approach and study design. Ecotoxicology **13**:9-27.

Koller, R., A. Rodriguez, C. Robin, S. Scheu, and M. Bonkowski. 2013. Protozoa enhance foraging efficiency of arbuscular mycorrhizal fungi for mineral nitrogen from organic matter in soil to the benefit of host plants. New Phytol. **199**:203-211.

Kramer, P. J. 1983. Water relations of plants. Academic Press, New York.

Kramer, S., S. Marhan, L. Ruess, W. Armbruster, O. Butenschoen, H. Haslwimmer, Y. Kuzyakov, J. Pausch, N. Scheunemann, J. Schoene, A. Schmalwasser, K. U. Totsche, F. Walker, S. Scheu, and E. Kandeler. 2012. Carbon flow into microbial and fungal biomass as a basis for the belowground food web of agroecosystems. Pedobiologia **55**:111-119.

Krashevska, V., M. Bonkowski, M. Maraun, and S. Scheu. 2007. Testate amoebae (Protista) of an elevational gradient in the tropical mountain rain forest of Ecuador. Pedobiologia **51**:319-331.

Kreuzer, K., J. Adamczyk, M. Iijima, M. Wagner, S. Scheu, and M. Bonkowski. 2006. Grazing of a common species of soil protozoa (*Acanthamoeba castellanii*) affects rhizosphere bacterial community composition and root architecture of rice (*Oryza sativa* L.). Soil Biol. Biochem. **38**:1665-1672.

Krishna Murti, C. R., and O. P. Shukla. 1984. Differentiation of pathogenic amoebae: encystation and excystation of *Acanthamoeba culbertsoni* — a model. J. Bioscience **6**:475-489.

Krome, K., K. Rosenberg, M. Bonkowski, and S. Scheu. 2009. Grazing of protozoa on rhizosphere bacteria alters growth and reproduction of *Arabidopsis thaliana*. Soil Biol. Biochem. **41**:1866-1873.

Kudryavtsev, A., D. Bernhard, M. Schlegel, E. E-Y Chao, and T. Cavalier-Smith. 2005. 18S ribosomal RNA gene sequences of *Cochliopodium* (Himatismenida) and the phylogeny of Amoebozoa. Protist **156**:215-224.

Kudryavtsev, A. A. 1999. Description of *Cochliopodium larifeili* n. sp.(Lobosea, Himatismenida), an amoeba with peculiar scale structure, and notes on the diagnosis of the genus *Cochliopodium* (Hertwig and Lesser, 1874) Bark, 1973. Protistology **1**:66-71.

Kudryavtsev, A. A. 2000. The first isolation of *Cochliopodium gulosum* Schaeffer, 1926 (Lobosea, Himatismenida) since its initial description. II. Electron-microscopical study and redescription. Protistology **1**:110-112.

Kudryavtsev, A. A. 2004. Description of *Cochliopodium spiniferum* sp. n., with notes on the species identification within the genus *Cochliopodium*. Acta Protozool. **43**:345-349.

Kudryavtsev, A. A. 2005. Redescription of *Cochliopodium vestitum* (Archer, 1871), a freshwater spine-bearing *Cochliopodium*. Acta Protozool. **44**:123-128.

Kudryavtsev, A. A. 2006. "Minute" species of *Cochliopodium* (Himatismenida): description of three new fresh- and brackish-water species with a new diagnosis for *Cochliopodium minus* Page, 1976. Eur. J. Protistol. **42**:77-89.

Kudryavtsev, A. A., S. Brown, and A. V. Smirnov. 2004. *Cochliopodium barki* n. sp. (Rhizopoda, Himatismenida) re-isolated from soil 30 years after its initial description. Eur. J. Protistol. **40**:283-287.

Kudryavtsev, A. A., and J. Pawlowski. 2013. *Squamamoeba japonica* n. g. n. sp. (Amoebozoa): a deep-sea amoeba from the sea of Japan with a novel cell coat structure. Protist **164**:13-23.

Kudryavtsev, A. A., J. Pawlowski, and K. Hausmann. 2009a. Description and phylogenetic relationships of *Spumochlamys perforata* n. sp. and *Spumochlamys bryora* n. sp. (Amoebozoa, Arcellinida). J. Eukaryot. Microbiol. **56**:495-503.

Kudryavtsev, A. A., and A. V. Smirnov. 2006. *Cochliopodium gallicum* n. sp. (Himatismenida), an amoeba bearing unique scales, from cyanobacterial mats in the Camargue (France). Eur. J. Protistol. **42**:3-7.

Kudryavtsev, A. A., C. Wylezich, and J. Pawlowski. 2011. *Ovalopodium desertum* n. sp. and the phylogenetic relationships of Cochliopodiidae (Amoebozoa). Protist **162**:571–589.

Kudryavtsev, A. A., C. Wylezich, M. Schlegel, J. Walochnik, and R. Michel. 2009b. Ultrastructure, SSU rRNA gene sequences and phylogenetic relationships of *Flamella* Schaeffer, 1926 (Amoebozoa), with description of three new species. Protist **160**:21-40.

Kuikman, P. J., A. G. Jansen, and J. A. van Veen. 1991. 15N-nitrogen mineralization from bacteria by protozoan grazing at different soil moisture regimes. Soil Biol. Biochem. **23**:193-200.

Kumar, S., A. Skjaeveland, R. Orr, P. Enger, T. Ruden, B.-H. Mevik, F. Burki, A. Botnen, and K. Shalchian-Tabrizi. 2009. AIR: a batch-oriented web program package for construction of supermatrices ready for phylogenomic analyses. BMC Bioinformatics **10**:357.

Lagkouvardos, I., J. Shen, and M. Horn. 2014. Improved axenization method reveals complexity of symbiotic associations between bacteria and acanthamoebae. Environ. Microbiol. Rep. **6**:383-388.

Lagomarsino, A., S. Grego, and E. Kandeler. 2012. Soil organic carbon distribution drives microbial activity and functional diversity in particle and aggregate-size fractions. Pedobiologia **55**:101-110.

Lahr, D., G. Kubik, A. Gant, J. Grant, and R. Anderson. 2012. Morphological description of *Telaepolella tubasferens* ngn sp., isolate ATCC© 50593™, a filose amoeba in the Gracilipodida, Amoebozoa. Acta Protozool. **51**:305-318.

Lahr, D. J. G., J. Grant, and L. A. Katz. 2013. Multigene phylogenetic reconstruction of the Tubulinea (Amoebozoa) corroborates four of the six major lineages, while additionally revealing that shell composition does not predict phylogeny in the Arcellinida. Protist **164**:323-339.

Lahr, D. J. G., J. Grant, T. Nguyen, J. H. Lin, and L. A. Katz. 2011. Comprehensive phylogenetic reconstruction of Amoebozoa based on concatenated analyses of SSU-rDNA and actin genes. PLoS One **6**:e22780.

Lamoth, F., and G. Greub. 2010. Amoebal pathogens as emerging causal agents of pneumonia. FEMS Microbiol. Rev. **34**:260-280.

Lanave, C., G. Preparata, C. Sacone, and G. Serio. 1984. A new method for calculating evolutionary substitution rates. J. Mol. Evol. **20**:86-93.

Łanocha, N., D. Kosik-Bogacka, A. Maciejewska, M. Sawczuk, A. Wilk, and W. Kuźna-Grygiel. 2009. The occurrence *Acanthamoeba* (free living amoeba) in environmental and respiratory samples in Poland. Acta Protozool. **48**:271-279.

Lanzén, A., S. L. Jørgensen, M. M. Bengtsson, I. Jonassen, L. Øvreås, and T. Urich. 2011. Exploring the composition and diversity of microbial communities at the Jan Mayen hydrothermal vent field using RNA and DNA. FEMS Microbiol. Ecol. **77**:577-589.

Lanzén, A., S. L. Jørgensen, D. H. Huson, M. Gorfer, S. H. Grindhaug, I. Jonassen, L. Øvreås, and T. Urich. 2012. CREST – Classification Resources for Environmental Sequence Tags. PLoS One **7**:e49334.

Lara, E., C. Berney, F. Ekelund, H. Harms, and A. Chatzinotas. 2007a. Molecular comparison of cultivable protozoa from a pristine and a polycyclic aromatic hydrocarbon polluted site. Soil Biol. Biochem. **39**:139-148.

Lara, E., C. Berney, H. Harms, and A. Chatzinotas. 2007b. Cultivation-independent analysis reveals a shift in ciliate 18S rRNA gene diversity in a polycyclic aromatic hydrocarbon-polluted soil. FEMS Microbiol. Ecol. **62**:365-373.

Lara, E., E. A. Mitchell, D. Moreira, and P. López García. 2011. Highly diverse and seasonally dynamic protist community in a pristine peat bog. Protist **162**:14-32.

Lares-Jiménez, L. F., G. C. Booton, F. Lares-Villa, C. A. Velázquez-Contreras, and P. A. Fuerst. 2014. Genetic analysis among environmental strains of *Balamuthia mandrillaris* recovered from an artificial lagoon and from soil in Sonora, Mexico. Exp. Parasitol. **145, Supplement**:S57-S61.

Latijnhouwers, M., P. J. G. M. de Wit, and F. Govers. 2003. Oomycetes and fungi: similar weaponry to attack plants. Trends Microbiol. **11**:462-469.

Laybourn-Parry, J. 1984. A functional biology of free-living protozoa. Univ of California Press.

Lee, J., G. Leedale, and P. Bradbury. 2000. An illustrated guide to the protozoa: organisms traditionally referred to as protozoa, or newly discovered groups, 2nd ed. Society of Protozoologists, Lawrence, Kansas.

Lejzerowicz, F., J. Pawlowski, L. Fraissinet-Tachet, and R. Marmeisse. 2010. Molecular evidence for widespread occurrence of Foraminifera in soils. Environ. Microbiol. **12**:2518-2526.

Lentendu, G., T. Wubet, A. Chatzinotas, C. Wilhelm, F. Buscot, and M. Schlegel. 2014. Effects of long-term differential fertilization on eukaryotic microbial communities in an arable soil: a multiple barcoding approach. Mol. Ecol. **23**:3341-3355.

Lesaulnier, C., D. Papamichail, S. McCorkle, B. Ollivier, S. Skiena, S. Taghavi, D. Zak, and D. van der Lelie. 2008. Elevated atmospheric CO2 affects soil microbial diversity associated with trembling aspen. Environ. Microbiol. **10**:926-941.

Levine, N. D., J. O. Corliss, F. E. G. Cox, G. Deroux, J. Grain, B. M. Honigberg, G. F. Leedale, A. R. Loeblich, I. J. Lom, D. Lynn, E. G. Merinfeld, F. C. Page, G. Poljansky, V. Sprague, J. Vavra, and F. G. Wallace. 1980. A newly revised classification of the Protozoa. J Protozool. **27**:37-58.

Lueders, T., M. Manefield, and M. W. Friedrich. 2004. Enhanced sensitivity of DNA- and rRNA-based stable isotope probing by fractionation and quantitative analysis of isopycnic centrifugation gradients. Environ. Microbiol. **6**:73-78.

Lynn, D. H., A. D. Wright, M. Schlegel, and W. Foissner. 1999. Phylogenetic relationships of orders within the class Colpodea (phylum Ciliophora) inferred from small subunit rRNA gene sequences. J. Mol. Evol. **48**:605-614.

Maciver, S. K., M. Asif, M. W. Simmen, and J. Lorenzo-Morales. 2013. A systematic analysis of *Acanthamoeba* genotype frequency correlated with source and pathogenicity: T4 is confirmed as a pathogen-rich genotype. Eur. J. Protistol. **49**:217-221.

Makarieva, A. M., V. G. Gorshkov, B. L. Li, S. L. Chown, P. B. Reich, and V. M. Gavrilov. 2008. Mean mass-specific metabolic rates are strikingly similar across life's major domains: evidence for life's metabolic optimum. Proc. Natl. Acad. Sci. USA **105**:16994-16999

Maniatis, T., E. F. Fritsch, and J. Sambrook. 1982a. Molecular Cloning, A Laboratory Manual. Cold Spring Habor Laboratory, Cold Spring Habor, New York.

Maniatis, T., E. F. Fritsch, and J. Sambrook. 1982b. Molecular Cloning, A Laboratory Manual. Cold Spring Habor Laboratory, Cold Spring Habor, New York.

Martin, F. N., Z. G. Abad, Y. Balci, and K. Ivors. 2012. Identification and detection of *Phytophthora*: reviewing our progress, identifying our needs. Plant Dis. **96**:1080-1103.

Martiny, J. B. H., B. J. M. Bohannan, J. H. Brown, R. K. Colwell, J. A. Fuhrman, J. L. Green, M. C. Horner-Devine, M. Kane, J. A. Krumins, and C. R. Kuske. 2006. Microbial biogeography: Putting microorganisms on the map. Nat. Rev. Microbiol. **4**:102-112.

Matz, C., and S. Kjelleberg. 2005. Off the hook-how bacteria survive protozoan grazing. Trends Microbiol. **13**:302-307.

Medinger, R., V. Nolte, R. V. Pandey, S. Jost, B. Ottenwälder, C. Schlötterer, and J. Boenigk. 2010. Diversity in a hidden world: potential and limitation of next generation sequencing for surveys of molecular diversity of eukaryotic microorganisms. Mol. Ecol. **19**:32-40.

Medlin, L., H. J. Elwood, S. Stickel, and M. L. Sogin. 1988. The characterization of enzymatically amplified eukaryotic 16S-like rRNA-coding regions. Gene **71**:491-499.

Meisner, A., E. Bååth, and J. Rousk. 2013. Microbial growth responses upon rewetting soil dried for four days or one year. Soil Biol. Biochem. **66**:188-192.

Meisterfeld, R., M. Holzmann, and J. Pawlowski. 2001. Morphological and molecular characterization of a new terrestrial allogromiid species: *Edaphoallogromia australica* gen. et spec. nov. (Foraminifera) from northern Queensland (Australia). Protist **152**:185-192.

Miller, M. A., W. Pfeiffer, and T. Schwartz. 2010. Creating the CIPRES Science Gateway for inference of large phylogenetic trees. Pages 1-8 *in* Gateway Computing Environments Workshop (GCE), 2010.

Mitchell, E. A. D., D. J. Charman, and B. G. Warner. 2008. Testate amoebae analysis in ecological and paleoecological studies of wetlands: past, present and future. Biodivers. Conserv. **17**:2115-2137.

Monchy, S., G. Sanciu, M. Jobard, S. Rasconi, M. Gerphagnon, M. Chabé, A. Cian, D. Meloni, N. Niquil, and U. Christaki. 2011. Exploring and quantifying fungal diversity in freshwater lake ecosystems using rDNA cloning/sequencing and SSU tag pyrosequencing. Environ. Microbiol. **13**:1433-1453.

Moon-van der Staay, S. Y., R. De Wachter, and D. Vaulot. 2001. Oceanic 18S rDNA sequences from picoplankton reveal unsuspected eukaryotic diversity. Nature **409**:607-610.

Moon-van der Staay, S. Y., V. A. Tzeneva, G. W. M. Van Der Staay, W. M. De Vos, H. Smidt, and J. H. P. Hackstein. 2006. Eukaryotic diversity in historical soil samples. FEMS Microbiol. Ecol. **57**:420-428.

Moore, J. C., and H. W. Hunt. 1988. Resource compartmentation and the stability of real ecosystems. Nature **333**:261-263.

Moreira, D., and P. López-García. 2002. The molecular ecology of microbial eukaryotes unveils a hidden world. Trends Microbiol. **10**:31-38.

Moussa, M., O. Tissot, J. Guerlotté, J. De Jonckheere, and A. Talarmin. 2015. Soil is the origin for the presence of *Naegleria fowleri* in the thermal recreational waters. Parasitol. Res. **114**:311-315.

Mrva, M. 2010. Morphological studies on two rare soil amoebae *Deuteramoeba algonquinensis* and *D. mycophaga* (Gymnamoebia, Amoebidae). Protistology **6**:284-289.

Mulder, C., A. Boit, S. Mori, J. A. Vonk, S. D. Dyer, L. Faggiano, S. Geisen, A. L. González, M. Kaspari, S. Lavorel, P. A. Marquet, A. G. Rossberg, R. W. Sterner, W. Voigt, and D. H. Wall. 2012. Distributional (in)congruence of biodiversity–ecosystem functioning. Adv. Ecol. Res. **Volume 46**:1-88.

Murase, J., and P. Frenzel. 2008. Selective grazing of methanotrophs by protozoa in a rice field soil. FEMS Microbiol. Ecol. **65**:408-414.

Nakai, R., T. Abe, T. Baba, S. Imura, H. Kagoshima, H. Kanda, Y. Kohara, A. Koi, H. Niki, K. Yanagihara, and T. Naganuma. 2012. Eukaryotic phylotypes in aquatic moss pillars inhabiting a freshwater lake in East Antarctica, based on 18S rRNA gene analysis. Polar Biol. **35**:1495-1504.

Nassonova, E., A. Smirnov, J. Fahrni, and J. Pawlowski. 2010. Barcoding amoebae: comparison of SSU, ITS and COI genes as tools for molecular identification of naked lobose amoebae. Protist **161**:102-115.

Neher, D. A. 2010. Ecology of plant and free-living nematodes in natural and agricultural soil. Annu. Rev. Phytopathol. **48**:371-394.

Neidig, N., A. Jousset, F. Nunes, M. Bonkowski, R. J. Paul, and S. Scheu. 2010. Interference between bacterial feeding nematodes and amoebae relies on innate and inducible mutual toxicity. Funct. Ecol. **24**:1133-1138.

Neuhauser, S., M. Kirchmair, S. Bulman, and D. Bass. 2014. Cross-kingdom host shifts of phytomyxid parasites. BMC Evol. Biol. **14**:33.

Nicholls, K. H. 2006. *Cryptodifflugia leachi* n. sp., a minute new testate rhizopod species (Rhizopoda: Phryganellina). Acta Protozool. **45**:295-299.

Nielsen, U. N., E. Ayres, D. H. Wall, and R. D. Bardgett. 2011. Soil biodiversity and carbon cycling: a review and synthesis of studies examining diversity–function relationships. Eur. J. Soil Sci. **62**:105-116.

Nikolaev, S. I., C. Berney, N. B. Petrov, A. P. Mylnikov, J. F. Fahrni, and J. Pawlowski. 2006. Phylogenetic position of *Multicilia marina* and the evolution of Amoebozoa. Int. J. Syst. Evol. Micr. **56**:1449-1458.

Nikolaev, S. I., E. A. D. Mitchell, N. B. Petrov, C. Berney, J. Fahrni, and J. Pawlowski. 2005. The testate lobose amoebae (order Arcellinida Kent, 1880) finally find their home within Amoebozoa. Protist **156**:191-202.

Nuprasert, W., C. Putaporntip, L. Pariyakanok, and S. Jongwutiwes. 2010. Identification of a novel T17 genotype of *Acanthamoeba* from environmental isolates and T10 genotype causing keratitis in Thailand. J. Clin. Microbiol. **48**:4636-4640.

O'Brien, H. E., J. L. Parrent, J. A. Jackson, J.-M. Moncalvo, and R. Vilgalys. 2005. Fungal community analysis by large-scale sequencing of environmental samples. Appl. Environ. Microbiol. **71**:5544-5550.

Old, K. M., S. Chakraborty, and R. Gibbs. 1985. Fine structure of a new mycophagous amoeba and its feeding on *Cochliobolus sativus*. Soil Biol. Biochem. **17**:645-655.

Old, K. M., and J. F. Darbyshire. 1978. Soil fungi as food for giant amoebae. Soil Biol. Biochem. **10**:93-100.

Old, K. M., and J. F. Darbyshire. 1980. *Arachnula impatiens* Cienk., a mycophagous giant amoeba from soil. Protistologica **16**:277-287.

Old, K. M., and J. M. Oros. 1980. Mycophagous amoebae in Australian forest soils. Soil Biol. Biochem. **12**:169-175.

Olive, L. S. 1967. The Protostelida: a new order of the Mycetozoa. Mycologia **59**:1-29.

Olive, L. S. 1970. The Mycetozoa: a revised classification. Bot. Rev. **36**:59-89.

Page, F. C. 1966. *Cryptodifflugia operculata* n. sp. (Rhizopodea: Arcellinida, Cryptodifflugiidae) and the status of the genus *Cryptodifflugia*. T. Am. Microsc. Soc. **85**:506-515.

Page, F. C., editor. 1976a. An Illustrated Key to Freshwater and Soil Amoebae. Freshwater Biological Association, The Ferry House, Ableside, Cumbria.

Page, F. C. 1976b. A revised classification of the Gymnamoebia (Protozoa: Sarcodina). Zool. J. Linn. Soc. **58**:61-77.

Page, F. C. 1977. The genus *Thecamoeba* (Protozoa, Gymnamoebia) species distinctions, locomotive morphology, and protozoan prey. J. Nat. Hist. **11**:25-63.

Page, F. C. 1983a. Marine gymnamoebae. Institute of Terrestrial Ecology, Cambridge, England.

Page, F. C. 1983b. Three freshwater species of *Mayorella* (Amoebida) with a cuticle. Arch. Protistenkd. **127**:201-221.

Page, F. C. 1985. The limax amoebae: comparative fine structure of the Hartmannellidae (Lobosea) and further comparisons with the Vahlkampfiidae (Heterolobosea). Protistologica **21**:361-383.

Page, F. C. 1987. The classification of 'naked' amoebae (phylum Rhizopoda). Arch. Protistenkd. **133**:199-217.

Page, F. C. 1988. A new key to freshwater and soil gymnamoebae with instructions for culture. Freshwater Biological Association, Ambleside, Cumbria.

Page, F. C. 1991. Nackte Rhizopoda. Pages 3-187 *in* F. C. Page and F. J. Siemensma, editors. Nackte Rhizopoda und Heliozoea (Protozoenfauna Band 2). Gustav Fischer Verlag, Stuttgart, New York.

Page, F. C., and R. L. Blanton. 1985. The Heterolobosea (Sarcodina: Rhizopoda), a new class uniting the Schizopyrenida and the Acrasidae (Acrasida). Protistologica **21**:121-132.

Page, F. C., and F. Siemensma. 1991. Nackte Rhizopoda und Heliozoea (protozoenfauna band 2). Gustav Fischer Verlag, Stuttgart, New York, Stuttgart, New York.

Pánek, T., J. D. Silberman, N. Yubuki, B. S. Leander, and I. Cepicka. 2012. Diversity, evolution and molecular systematics of the Psalteriomonadidae, the main lineage of anaerobic/microaerophilic heteroloboseans (Excavata: Discoba). Protist **163**:807-831.

Park, J. S., and A. G. B. Simpson. 2011. Characterization of *Pharyngomonas kirbyi* (= "*Macropharyngomonas halophila*" nomen nudum), a very deep-branching, obligately halophilic heterolobosean flagellate. Protist **162**:691-709.

Parry, J. D., J. W. B. Allen I. Laskin, and M. G. Geoffrey. 2004. Protozoan grazing of freshwater biofilms. Adv. Appl. Microbiol. **54**:167-196.

Paustian, K., O. Andren, M. Clarholm, A.-C. Hansson, G. Johansson, J. Lagerlof, T. Lindberg, R. Pettersson, and B. Sohlenius. 1990. Carbon and nitrogen budgets of four agro-ecosystems with annual and perennial crops, with and without N fertilization. J. Appl. Ecol. **27**:60-84.

Pawlowski, J. 2000. Introduction to the molecular systematics of Foraminifera. Micropaleontol. **46**:1-12.

Pawlowski, J. 2013. The new micro-kingdoms of eukaryotes. BMC Biol. **11**:40.

Pawlowski, J., S. Audic, S. Adl, D. Bass, L. Belbahri, S. S. Bowser, I. Cepicka, J. Decelle, M. Dunthorn, A. M. Fiore-Donno, G. H. Gile, M. Holzmann, R. Jahn, M. Jirků, P. J. Keeling, M. Kostka, A. A. Kudryavtsev, E. Lara, J. Lukeš, D. G. Mann, E. A. D. Mitchell, F. Nitsche, M. Romeralo, G. W. Saunders, A. G. B. Simpson, A. V. Smirnov, J. L. Spouge, R. F. Stern, T. Stoeck, J. Zimmermann, D. Schindel, and C. de Vargas. 2012. CBOL protist working group: barcoding eukaryotic richness beyond the animal, plant, and fungal kingdoms. PLoS Biol. **10**:e1001419.

Pawlowski, J., I. Bolivar, J. F. Fahrni, C. D. Vargas, and S. S. Bowser. 1999. Molecular evidence that *Reticulomyxa filosa* is a freshwater naked foraminifer. J. Eukaryot. Microbiol. **46**:612-617.

Pawlowski, J., and F. Burki. 2009. Untangling the phylogeny of amoeboid protists. J. Eukaryot. Microbiol. **56**:16-25.

Pawlowski, J., R. Christen, B. Lecroq, D. Bachar, H. R. Shahbazkia, L. Amaral-Zettler, and L. Guillou. 2011. Eukaryotic richness in the abyss: insights from pyrotag sequencing. PLoS One **6**:e18169.

Pedersen, A. L., O. Nybroe, A. Winding, F. Ekelund, and L. Bjørnlund. 2009. Bacterial feeders, the nematode *Caenorhabditis elegans* and the flagellate *Cercomonas longicauda*, have different effects on outcome of competition among the *Pseudomonas* biocontrol strains CHA0 and DSS73. Microb. Ecol. **57**:501-509.

Peglar, M. T., L. A. A. Zettler, O. R. Anderson, T. A. Nerad, P. M. Gillevet, T. E. Mullen, S. Frasca, J. D. Silberman, C. J. O'Kelly, and M. L. Sogin. 2003. Two new small-subunit ribosomal RNA gene lineages within the subclass Gymnamoebia. J. Eukaryot. Microbiol. **50**:224-232.

Petz, W., W. Foissner, and H. Adam. 1985. Culture, food selection and growth rate in the mycophagous ciliate *Grossglockneria acuta* Foissner, 1980: first evidence of autochthonous soil ciliates. Soil Biol. Biochem. **17**:871-875.

Petz, W., W. Foissner, E. Wirnsberger, W. D. Krautgartner, and H. Adam. 1986. Mycophagy, a new feeding strategy in autochthonous soil ciliates. Naturwissenschaften **73**:560-562.

Phillips, A. J., V. L. Anderson, E. J. Robertson, C. J. Secombes, and P. van West. 2008. New insights into animal pathogenic oomycetes. Trends Microbiol. **16**:13-19.

Plassart, P., S. Terrat, B. Thomson, R. Griffiths, S. Dequiedt, M. Lelievre, T. Regnier, V. Nowak, M. Bailey, P. Lemanceau, A. Bispo, A. Chabbi, P.-A. Maron, C. Mougel, and L. Ranjard. 2012. Evaluation of the ISO standard 11063 DNA extraction procedure for assessing soil microbial abundance and community structure. PLoS One **7**:e44279.

Pombert, J.-F., and J. Janouškovec. 2013. The complete mitochondrial genome from an unidentified *Phalansterium* species. Protist Genomics **1**:25-32.

Prescott, D. M., and T. W. James. 1955a. Culturing of Amoeba-Proteus on Tetrahymena. Experimental Cell Research **8**:256-258.

Prescott, D. M., and T. W. James. 1955b. Culturing of *Amoeba proteus* on *Tetrahymena*. Exp. Cell Res. **8**:256.

Pussard, M., C. Allabouvette, and R. Pons. 1979. Étude preliminaire d'une amibe mycophage *Thecamoeba granifera* ssp. minor (Thecamoebidae, amoebida). . Protistologica **15**:139-149.

Pussard, M., and R. Pons. 1977. Morphologie de la paroi kystique et taxonomie du genre *Acanthamoeba* (Protozoa, Amoebida). Protistologica **13**:557-598.

Pussard, M., J. Senaud, and R. Pons. 1977. Observations ultrastructurales sur *Gocevia fonbrunei* Pussard 1965 (Protozoa, Rhizopodea). Protistologica **13**:265-285.

Qvarnstrom, Y., T. A. Nerad, and G. S. Visvesvara. 2013. Characterization of a new pathogenic *Acanthamoeba* species, *A. byersi* n. sp., isolated from a human with fatal amoebic encephalitis. J. Eukaryot. Microbiol. **60**:626-633.

Radax, R., T. Rattei, A. Lanzén, C. Bayer, H. T. Rapp, T. Urich, and C. Schleper. 2012. Metatranscriptomics of the marine sponge *Geodia barretti*: tackling phylogeny and function of its microbial community. Environ. Microbiol. **14**:1308-1324.

Reich, P. B., D. Tilman, F. Isbell, K. Mueller, S. E. Hobbie, D. F. B. Flynn, and N. Eisenhauer. 2012. Impacts of biodiversity loss escalate through time as redundancy fades. Science **336**:589-592.

Reynolds, C. S., G. H. M. Jaworski, H. A. Cmiech, and G. F. Leedale. 1981. On the annual cycle of the blue-green alga *microcystis aeruginosa* Kutz. emend. Elenkin.

Rhumbler, L. 1904. Systematische Zusammenstellung der rezenten Reticulosa (Nuda et Foraminifera). Arch. Protistenkd. **3**:181–294.

Risler, A., B. Coupat-Goutaland, and M. Pélandakis. 2013. Genotyping and phylogenetic analysis of *Acanthamoeba* isolates associated with keratitis. Parasitol. Res. **112**:3807-3816.

Risse-Buhl, U., M. Herrmann, P. Lange, D. M. Akob, N. Pizani, W. Schönborn, K. U. Totsche, and K. Küsel. 2013. Phagotrophic protist diversity in the groundwater of a karstified aquifer - morphological and molecular analysis. J. Eukaryot. Microbiol. **60**:467-479.

Ritz, K., and I. Young. 2011. The architecture and biology of soils: life in inner space. Cabi, Wallingford, UK.

Robinson, B. S., S. S. Bamforth, and P. J. Dobson. 2002. Density and diversity of protozoa in some arid Australian soils. J. Eukaryot. Microbiol. **49**:449-453.

Robinson, B. S., J. F. De Jonckheere, and P. J. Dobson. 2007. Two new *Tetramitus* species (Heterolobosea, Vahlkampfiidae) from cold aquatic environments. Eur. J. Protistol. **43**:1-7.

Rodríguez-Zaragoza, S. 1994. Ecology of free-living amoebae. Crit. Rev. Microbiol. **20**:225-241.

Rodriguez-Zaragoza, S., E. Mayzlish, and Y. Steinberger. 2005. Vertical distribution of the free-living amoeba population in soil under desert shrubs in the Negev desert, Israel. Appl. Environ. Microbiol. **71**:2053-2060.

Rogerson, A., F. Hannah, and G. Gothe. 1996. The grazing potential of some unusual marine benthic amoebae feeding on bacteria. Eur. J. Protistol. **32**:271-279.

Rogerson, A., A. G. Williams, and P. C. Wilson. 1998. Utilization of macroalgal carbohydrates by the marine amoeba *Trichosphaerium sieboldi*. J. Mar. Biol. Assoc. U. K. **78**:733-744.

Rønn, R., F. Ekelund, and S. Christensen. 1995. Optimizing soil extract and broth media for MPN-enumeration of naked amoebae and heterotrophic flagellates in soil. Pedobiologia **39**:10-19.

Rønn, R., M. Gavito, J. Larsen, I. Jakobsen, H. Frederiksen, and S. Christensen. 2002a. Response of free-living soil protozoa and microorganisms to elevated atmospheric CO2 and presence of mycorrhiza. Soil Biol. Biochem. **34**:923-932.

Rønn, R., B. S. Griffiths, F. Ekelund, and S. Christensen. 1996. Spatial distribution and successional pattern of microbial activity and micro-faunal populations on decomposing barley roots. J. Appl. Ecol. **33**:662-672.

Rønn, R., A. E. McCaig, B. S. Griffiths, and J. I. Prosser. 2002b. Impact of protozoan grazing on bacterial community structure in soil microcosms. Appl. Environ. Microbiol. **68**:6094-6105.

Rønn, R., M. Vestergård, and F. Ekelund. 2012. Interactions between bacteria, protozoa and nematodes in soil. Acta Protozool. **51**:223-235.

Ronquist, F., and J. P. Huelsenbeck. 2003. MrBayes 3: bayesian phylogenetic inference under mixed models. Bioinformatics **19**:1572-1574.

Rosenberg, K., J. Bertaux, K. Krome, A. Hartmann, S. Scheu, and M. Bonkowski. 2009. Soil amoebae rapidly change bacterial community composition in the rhizosphere of *Arabidopsis thaliana*. ISME J. **3**:675-684.

Ruiz, A., J. K. Frenkel, and L. Cerdas. 1973. Isolation of *Toxoplasma* from soil. J. Parasitol. **59**:204-206.

Saetre, P., and J. Stark. 2005. Microbial dynamics and carbon and nitrogen cycling following re-wetting of soils beneath two semi-arid plant species. Oecologia **142**:247-260.

Saleem, M., I. Fetzer, C. F. Dormann, H. Harms, and A. Chatzinotas. 2012. Predator richness increases the effect of prey diversity on prey yield. Nat. Commun. **3**:1305.

Saleem, M., I. Fetzer, H. Harms, and A. Chatzinotas. 2013. Diversity of protists and bacteria determines predation performance and stability. ISME J. **7**:1912-1921.

Sambrook, J., E. F. Fritsch, and T. Maniatis. 1989. Molecular cloning: a laboratory manual. Cold spring harbor laboratory press New York.

Sawyer, T. K. 1975. *Clydonella* n. g. (Amoebida: Thecamoebidae), proposed to provide an appropriate generic home for Schaeffer's marine species of rugipes, *C. vivax* (Schaeffer, 1926) N. comb. T. Am. Microsc. Soc. **94**:395-400.

Sawyer, T. K., and J. L. Griffin. 1975. A proposed new family, Acanthamoebidae n. fam. (Order Amoebida), for certain cyst-forming filose amoebae. T. Am. Microsc. Soc. **94**:93-98.

Saxton, K. E., and W. J. Rawls. 2006. Soil water characteristic estimates by texture and organic matter for hydrologic solutions. Soil Sci. Soc. Am. J. **70**:1569-1578.

Sayre, R. M. 1973. *Theratromyxa weberi*, an amoeba predatory on plant-parasitic nematodes. J. Nematol. **5**:258-264.

Schaefer, M., and J. Schauermann. 1990. The soil fauna of beech forests: comparison between a mull and a moder soil. Pedobiologia **34**:299-314.

Schaeffer, A. A. 1926. Taxonomy of the amebas with descriptions of thirty-nine new marine and freshwater species. Pap. Dept. Mar. Biol., Carnegie Inst. of Washington.

Scharroba, A., D. Dibbern, M. Hünninghaus, S. Kramer, J. Moll, O. Butenschoen, M. Bonkowski, F. Buscot, E. Kandeler, R. Koller, D. Krüger, T. Lueders, S. Scheu, and L. Ruess. 2012. Effects of resource availability and quality on the structure of the micro-food web of an arable soil across depth. Soil Biol. Biochem. **50**:1-11.

Scheiner, S. M., and J. Gurevitch. 2001. The design and analysis of ecological experiments. Oxford University Press, New York.

Scherber, C., N. Eisenhauer, W. W. Weisser, B. Schmid, W. Voigt, M. Fischer, E.-D. Schulze, C. Roscher, A. Weigelt, E. Allan, H. Beszler, M. Bonkowski, N. Buchmann, F. Buscot, L. W. Clement, A. Ebeling, C. Engels, S. Halle, I. Kertscher, A.-M. Klein, R. Koller, S. Konig, E. Kowalski, V. Kummer, A. Kuu, M. Lange, D. Lauterbach, C. Middelhoff, V. D. Migunova, A. Milcu, R. Muller, S. Partsch, J. S. Petermann, C. Renker, T. Rottstock, A. Sabais, S. Scheu, J. Schumacher, V. M. Temperton, and T. Tscharntke. 2010. Bottom-up effects of plant diversity on multitrophic interactions in a biodiversity experiment. Nature **468**:553-556.

Scherm, B., V. Balmas, F. Spanu, G. Pani, G. Delogu, M. Pasquali, and Q. Migheli. 2013. *Fusarium culmorum*: causal agent of foot and root rot and head blight on wheat. Mol. Plant Pathol. **14**:323-341.

Schloss, P. D., S. L. Westcott, T. Ryabin, J. R. Hall, M. Hartmann, E. B. Hollister, R. A. Lesniewski, B. B. Oakley, D. H. Parks, C. J. Robinson, J. W. Sahl, B. Stres, G. G. Thallinger, D. J. van Horn, and C. F. Weber. 2009. Introducing mothur: open-source, platform-independent, community-supported software for describing and comparing microbial communities. Appl. Environ. Microbiol. **75**:7537-7541.

Schröter, D., V. Wolters, and P. C. de Ruiter. 2003. C and N mineralisation in the decomposer food webs of a European forest transect. Oikos **102**:294-308.

Schuster, F. L. 1976. Fine structure of the schizont stage of the testate marine ameba, *Trichosphaerium* sp. J. Eukaryot. Microbiol. **23**:86-93.

Schuster, F. L. 2002. Cultivation of pathogenic and opportunistic free-living amebas. Clin. Microbiol. Rev. **15**:342-354.

Schuster, F. L., and G. S. Visvesvara. 2004. Free-living amoebae as opportunistic and non-opportunistic pathogens of humans and animals. Int. J. Parasitol. **34**:1001-1027.

Sechi, V., L. Brussaard, R. G. De Goede, M. Rutgers, and C. Mulder. 2015. Choice of resolution by functional trait or taxonomy affects allometric scaling in soil food webs. Am. Nat. **185**:142-149.

Shadwick, L. L., F. W. Spiegel, J. D. L. Shadwick, M. W. Brown, and J. D. Silberman. 2009. Eumycetozoa= Amoebozoa?: SSUrDNA phylogeny of protosteloid slime molds and its significance for the amoebozoan supergroup. PLoS One **4**:e6754.

Sherwood, S. C., S. Bony, and J.-L. Dufresne. 2013. Spread in model climate sensitivity traced to atmospheric convective mixing. Nature **505**:37-42.

Siddiqui, R., and N. Ahmed Khan. 2012. Biology and pathogenesis of *Acanthamoeba*. Parasit. Vectors **5**:1-13.

Smirnov, A. V. 1996. *Stygamoeba regulata* n. s p. (Rhizopoda) — a marine amoeba with an unusual combination of light-microscopical and ultrastructural features. Arch. Protistenkd. **146**:299-307.

Smirnov, A. V. 2003. Optimizing methods of the recovery of gymnamoebae from environmental samples: a test of ten popular enrichment media, with some observations on the development of cultures. Protistology **3**:47-57.

Smirnov, A. V., O. M. Bedjagina, and A. V. Goodkov. 2011a. *Dermamoeba algensis* n. sp.(Amoebozoa, Dermamoebidae)–An algivorous lobose amoeba with complex cell coat and unusual feeding mode. Eur. J. Protistol. **47**:67-78.

Smirnov, A. V., and S. Brown. 2004. Guide to the methods of study and identification of soil gymnamoebae. Protistology **3**:148-190.

Smirnov, A. V., E. Chao, E. S. Nassonova, and T. Cavalier-Smith. 2011b. A revised classification of naked lobose amoebae (Amoebozoa: Lobosa). Protist **162**:545-570.

Smirnov, A. V., and A. Goodkov. 2004. Ultrastructure and geographic distribution of the genus *Paradermamoeba* (Gymnamoebia, Thecamoebidae). Eur. J. Protistol. **40**:113-118.

Smirnov, A. V., and A. V. Goodkov. 1999. An illustrated list of basic morphotypes of Gymnamoebia (Rhizopoda, Lobosea). Protistology **1**:20-29.

Smirnov, A. V., E. Nassonova, C. Berney, J. Fahrni, I. Bolivar, and J. Pawlowski. 2005. Molecular phylogeny and classification of the lobose amoebae. Protist **156**:129.

Smirnov, A. V., E. Nassonova, J. Fahrni, and J. Pawlowski. 2009. *Rhizamoeba neglecta* n. sp. (Amoebozoa, Tubulinea) from the bottom sediments of freshwater Lake Leshevoe (Valamo Island, North-Western Russia), with notes on the phylogeny of the order Leptomyxida. Eur. J. Protistol. **45**:251-259.

Smirnov, A. V., E. Nassonova, M. Holzmann, and J. Pawlowski. 2002. Morphological, ecological and molecular studies of *Vannella simplex* Wohlfarth-Bottermann 1960 (Lobosea, Gymnamoebia), with a new diagnosis of this species. Protist **153**:367-377.

Smirnov, A. V., E. S. Nassonova, and T. Cavalier-Smith. 2008. Correct identification of species makes the amoebozoan rRNA tree congruent with morphology for the order Leptomyxida Page 1987; with description of *Acramoeba dendroida* n. g., n. sp., originally misidentified as '*Gephyramoeba* sp.'. Eur. J. Protistol. **44**:35-44.

Smirnov, A. V., E. S. Nassonova, E. Chao, and T. Cavalier-Smith. 2007. Phylogeny, evolution, and taxonomy of vannellid amoebae. Protist **158**:295-324.

Sokal, R. R. 1961. Distance as a measure of taxonomic similarity. Syst. Zool. **10**:70-79.

Stamatakis, A. 2006. RAxML-VI-HPC: maximum likelihood-based phylogenetic analyses with thousands of taxa and mixed models. Bioinformatics **22**:2688-2690.

Stephenson, S. L., A. M. Fiore-Donno, and M. Schnittler. 2011. Myxomycetes in soil. Soil Biol. Biochem. **43**:2237-2242.

Stoeck, T., H.-W. Breiner, S. Filker, V. Ostermaier, B. Kammerlander, and B. Sonntag. 2014. A morphogenetic survey on ciliate plankton from a mountain lake pinpoints the necessity of lineage-specific barcode markers in microbial ecology. Environ. Microbiol. **16**:430-444.

Stotefeld, L., S. Scheu, and M. Rohlfs. 2012. Fungal chemical defence alters density-dependent foraging behaviour and success in a fungivorous soil arthropod. Ecological Entomology **37**:323-329.

Stothard, D. R., J. M. Schroeder-Diedrich, M. H. Awwad, R. J. Gast, D. R. Ledee, S. Rodriguez-Zaragoza, C. L. Dean, P. A. Fuerst, and T. J. Byers. 1998. The evolutionary history of the genus *Acanthamoeba* and the identification of eight new 18S rRNA gene sequence types. J. Eukaryot. Microbiol. **45**:45-54.

Stoupin, D., A. K. Kiss, H. Arndt, A. V. Shatilovich, D. A. Gilichinsky, and F. Nitsche. 2012. Cryptic diversity within the choanoflagellate morphospecies complex *Codosiga botrytis* - phylogeny and morphology of ancient and modern isolates. Eur. J. Protistol. **48**:263-273.

Stout, J. D. 1984. The protozoan fauna of a seasonally inundated soil under grassland. Soil Biol. Biochem. **16**:121-125.

Tang, Y., H. Nielsen, B. Masquida, P. P. Gardner, and S. D. Johansen. 2014. Molecular characterization of a new member of the lariat capping twin-ribozyme introns. Mob. DNA **5**:25.

Tapia, J. L., B. N. Torres, and G. S. Visvesvara. 2013. *Balamuthia mandrillaris*: in vitro interactions with selected protozoa and algae. J. Eukaryot. Microbiol. **60**:448-454.

Tapilskaja, N. 1967. *Amoeba albida* Nägler und ihre Beziehungen zu dem Pilz *Verticillium dahliae* Kleb, dem Erreger der Welkekrankheit von Baumwollpflanzen. Pedobiologia **7**:156-165.

Tekle, Y. I., O. R. Anderson, A. F. Lecky, and S. D. Kelly. 2013. A new freshwater amoeba: *Cochliopodium pentatrifurcatum* n. sp. (Amoebozoa, Amorphea). J. Eukaryot. Microbiol. **60**:342-349.

Tekle, Y. I., J. Grant, O. R. Anderson, T. A. Nerad, J. C. Cole, D. J. Patterson, and L. A. Katz. 2008. Phylogenetic placement of diverse amoebae inferred from multigene analyses and assessment of clade stability within 'Amoebozoa' upon removal of varying rate classes of SSU-rDNA. Mol. Phylogenet. Evol. **47**:339-352.

Tekle, Y. I., J. Grant, J. C. Cole, T. A. Nerad, O. R. Anderson, D. J. Patterson, and L. A. Katz. 2007. A multigene analysis of *Corallomyxa tenera* sp. nov. suggests its membership in a clade that includes *Gromia*, Haplosporidia and Foraminifera. Protist **158**:457-472.

Thines, M. 2014. Phylogeny and evolution of plant pathogenic oomycetes—a global overview. Eur. J. Plant Pathol. **138**:431-447.

Thomas, V., K. Herrera-Rimann, D. S. Blanc, and G. Greub. 2006. Biodiversity of amoebae and amoeba-resisting bacteria in a hospital water network. Appl. Environ. Microbiol. **72**:2428-2438.

Tikhonenkov, D. V., A. P. Mylnikov, Y. C. Gong, W. S. Feng, and Y. Mazei. 2012. Heterotrophic flagellates from freshwater and soil habitats in subtropical China (Wuhan Area, Hubei province). Acta Protozool. **51**:65.

Tiunov, A. V., M. Bonkowski, J. A. Tiunov, and S. Scheu. 2001. Microflora, protozoa and nematoda in *Lumbricus terrestris* burrow walls: a laboratory experiment. Pedobiologia **45**:46-60.

Tkach, V., and J. Pawlowski. 1999. A new method of DNA extraction from the ethanol-fixed parasitic worms. Acta Parasitol. **2**:147-148.

Tong, S., N. Vørs, and D. J. Patterson. 1997. Heterotrophic flagellates, centrohelid heliozoa and filose amoebae from marine and freshwater sites in the Antarctic. Polar Biol. **18**:91-106.

Tracey, M. V. 1955. Cellulase and chitinase in soil amoebae. Nature **175**:815-815.

Turner, T. R., K. Ramakrishnan, J. Walshaw, D. Heavens, M. Alston, D. Swarbreck, A. Osbourn, A. Grant, and P. S. Poole. 2013. Comparative metatranscriptomics reveals kingdom level changes in the rhizosphere microbiome of plants. ISME J. **7**:2248–2258.

Tveit, A., R. Schwacke, M. M. Svenning, and T. Urich. 2012. Organic carbon transformations in high-Arctic peat soils: Key functions and microorganisms. ISME J.

Tveit, A., R. Schwacke, M. M. Svenning, and T. Urich. 2013. Organic carbon transformations in high-Arctic peat soils: key functions and microorganisms. ISME J. **7**:299–311.

Urich, T., A. Lanzén, J. Qi, D. H. Huson, C. Schleper, and S. C. Schuster. 2008. Simultaneous assessment of soil microbial community structure and function through analysis of the meta-transcriptome. PLoS One **3**:e2527.

Urich, T., A. Lanzén, R. Stokke, R. B. Pedersen, C. Bayer, I. H. Thorseth, C. Schleper, I. H. Steen, and L. Øvreas. 2014. Microbial community structure and functioning in marine sediments associated with diffuse hydrothermal venting assessed by integrated meta-omics. Environ. Microbiol. **16**:2699-2710.

Urich, T., and C. Schleper. 2011. The "Double-RNA" Approach to Simultaneously Assess the Structure and Function of a Soil Microbial Community. Pages 587-596 Handbook of Molecular Microbial Ecology I. John Wiley & Sons, Inc.

Valster, R. M., B. A. Wullings, G. Bakker, H. Smidt, and D. van der Kooij. 2009. Free-living protozoa in two unchlorinated drinking water supplies, identified by phylogenic analysis of 18S rRNA gene sequences. Appl. Environ. Microbiol. **75**:4736-4746.

Valster, R. M., B. A. Wullings, R. van den Berg, and D. van der Kooij. 2011. Relationships between free-living protozoa, cultivable *Legionella* spp., and water quality characteristics in three drinking water supplies in the Caribbean. Appl. Environ. Microbiol. **77**:7321-7328.

Valster, R. M., B. A. Wullings, and D. van der Kooij. 2010. Detection of protozoan hosts for *Legionella pneumophila* in engineered water systems by using a biofilm batch test. Appl. Environ. Microbiol. **76**:7144-7153.

van Elsas, J. D., M. Chiurazzi, C. A. Mallon, D. Elhottová, V. Krištůfek, and J. F. Salles. 2012. Microbial diversity determines the invasion of soil by a bacterial pathogen. Proc. Natl. Acad. Sci. USA.

Van Wichelen, J., I. Van Gremberghe, P. Vanormelingen, A. E. Debeer, B. Leporcq, D. Menzel, G. A. Codd, J. P. Descy, and W. Vyverman. 2010. Strong effects of amoebae grazing on the biomass and genetic structure of a *Microcystis* bloom (Cyanobacteria). Environ. Microbiol. **12**:2797-2813.

Vannini, A., N. Bruni, A. Tomassini, S. Franceschini, and A. M. Vettraino. 2013. Pyrosequencing of environmental soil samples reveals biodiversity of the *Phytophthora* resident community in chestnut forests. FEMS Microbiol. Ecol. **85**:433–442.

Visvesvara, G. S., H. Moura, and F. L. Schuster. 2007. Pathogenic and opportunistic free living amoebae: *Acanthamoeba* spp., *Balamuthia mandrillaris*, *Naegleria fowleri*, and *Sappinia diploidea*. FEMS Immunol. Med. Mic. **50**:1-26.

Visvesvara, G. S., F. L. Schuster, and A. J. Martinez. 1993. *Balamuthia mandrillaris*, N. G., N. Sp., agent of amebic meningoencephalitis in humans and other animals. J. Eukaryot. Microbiol. **40**:504-514.

Vohník, M., Z. Burdíková, A. Vyhnal, and O. Koukol. 2011. Interactions between testate amoebae and saprotrophic microfungi in a Scots pine litter microcosm. Microb. Ecol. **61**:660-668.

Vos, M., A. B. Wolf, S. J. Jennings, and G. A. Kowalchuk. 2013. Micro-scale determinants of bacterial diversity in soil. FEMS Microbiol. Rev. **37**:936–954.

Wall, D. H., and J. Six. 2015. Give soils their due. Science **347**:695.

Walochnik, J., and J. Mulec. 2009. Free-living amoebae in carbonate precipitating microhabitats of karst caves and a new vahlkampfiid amoeba, *Allovahlkampfia spelaea* gen. nov., sp. nov. Acta Protozool. **48**:25-33.

Walochnik, J., A. Obwaller, and H. Aspöck. 2000. Correlations between morphological, molecular biological, and physiological characteristics in clinical and nonclinical isolates of *Acanthamoeba* spp. Appl. Environ. Microbiol. **66**:4408-4413.

Wanner, M. 1991. Zur Ökologie von Thekamöben (Protozoa: Rhizopoda) in süddeutschen Wäldern. Arch. Protistenkd. **140**:237-288.

Wanner, M., M. Elmer, M. Kazda, and W. E. R. Xylander. 2008. Community assembly of terrestrial testate amoebae: how is the very first beginning characterized? Microb. Ecol. **56**:43-54.

Weber, A. A. T., and J. Pawlowski. 2013. Can abundance of protists be inferred from sequence data: a case study of Foraminifera. PLoS One **8**:e56739.

Weekers, P. H. H., P. L. E. Bodelier, J. P. H. Wijen, and G. D. Vogels. 1993. Effects of grazing by the free-living soil amoebae *Acanthamoeba castellanii, Acanthamoeba polyphaga*, and *Hartmannella vermiformis* on various bacteria. Appl. Environ. Microbiol. **59**:2317-2319.

Weisse, T., N. Karstens, V. C. L. Meyer, L. Janke, S. Lettner, and K. Teichgräber. 2001. Niche separation in common prostome freshwater ciliates: the effect of food and temperature. Aquat. Microb. Ecol. **26**:167-179.

Weitere, M., T. Bergfeld, S. A. Rice, C. Matz, and S. Kjelleberg. 2005. Grazing resistance of *Pseudomonas aeruginosa* biofilms depends on type of protective mechanism, developmental stage and protozoan feeding mode. Environ. Microbiol. **7**:1593-1601.

Wikmark, O. G., C. Einvik, J. F. De Jonckheere, and S. D. Johansen. 2006. Short-term sequence evolution and vertical inheritance of the *Naegleria* twin-ribozyme group I intron. BMC Evol. Biol. **6**:39.

Wilkinson, D. M. 2008. Testate amoebae and nutrient cycling: peering into the black box of soil ecology. Trends Ecol. Evol. **23**:596-599.

Wilkinson, D. M., and E. A. D. Mitchell. 2010. Testate amoebae and nutrient cycling with particular reference to soils. Geomicrobiol. J. **27**:520-533.

Winding, A., R. Rønn, and N. B. Hendriksen. 1997. Bacteria and protozoa in soil microhabitats as affected by earthworms. Biol. Fertil. Soils **24**:133-140.

Wuyts, J., P. De Rijk, Y. Van de Peer, G. Pison, P. Rousseeuw, and R. De Wachter. 2000. Comparative analysis of more than 3000 sequences reveals the existence of two pseudoknots in area V4 of eukaryotic small subunit ribosomal RNA. Nucleic Acids Res. **28**:4698-4708.

Yamaoka, I., N. Kawamura, M. Mizuno, and Y. Nagatani. 1984. Scale formation in an amoeba, *Cochliopodium* sp. J. Eukaryot. Microbiol. **31**:267-272.

Yeates, G. W., and W. Foissner. 1995. Testate amoebae as predators of nematodes. Biol. Fertil. Soils **20**:1-7.

Zadrobilkova, E., G. Walker, and I. Cepicka. 2015. Morphological and molecular evidence support a close relationship between the free-living archamoebae *Mastigella* and *Pelomyxa*. Protist **166**:14-41.

Zwart, K. B., S. L. G. E. Burgers, J. Bloem, L. A. Bouwman, L. Brussaard, G. Lebbink, W. A. M. Didden, J. C. Y. Marinissen, M. J. Vreeken-Buijs, and P. C. de Ruiter. 1994. Population dynamics in the belowground food webs in two different agricultural systems. Agr. Ecosyst. Environ. **51**:187-198.

Acknowledgements

First and foremost I wish to thank my advisor Prof. **Michael Bonkowski**. Michael has given me tremendous support already during my studies and especially during my time as a PhD student. Especially by giving me the chance to attend innumerable workshops, conferences and working times abroad he broadened my mind- and eventually helped me to keep motivated throughout the entire time. I guess it is very rare that a PhD student has the chance to attend a kick-off meeting of a big EU project, where even post-docs are rare...

This leads me to my next big thanks- to the EU project **EcoFINDERS** that financed my work. Further, all the members of EcoFINDERS who were always supportive and open for interesting collaborations, resulting in a plethora of present and future joint works. For this thesis I specifically want to thank **Emily Tisserant**, **Marc Buée** and **Francis Martin** (INRA, Nancy, France) for the joint work on high-throughput sequencing approaches; **Junling Zhang** (China agricultural university, China) for providing soils from high altitudes in Tibet and the subsequent description of isolated heterolobosean amoebae; **Christian Mulder** (Rijksinstituut voor Volksgezondheid en Milieu, the Netherlands) for the great improvements of a work on nematophagous protists; Prof. **Bryan Griffiths** (Scotland's Rural College, Great Britain) for proofreading;

I am more than grateful to all the taxonomic experts outside the EcoFINDERS consortium without their help I would have been unable to become an expert in the field of Protistology, which enabled the fulfilment of the tasks performed within my PhD and EcoFINDERS; **Alexey Smirnov** and **Alexander Kudryavtsev** (St. Petersburg State University, Russia) for their patients in introducing me to amoebae especially when I worked with both in St. Petersburg- resulting in two chapters and most likely great future joint works; **David Bass** and **Cédric Berney** (Natural History Museum London, Great Britain) for joint taxonomic work on variosean amoebae, but also high-throughput work on cercozoan protists; **Johan de Jonckheere** (Université catholique de Louvain, Belgium) for teaching me to work with heterolobosean amoebae and the resulting publication; **Julia Walochnik** (Medical University Vienna, Austria) for the stimulating discussions during conferences and the resulting work on the diversity of *Acanthamoeba*.

I am indebted to **Tim Urich** (University of Vienna, Austria) for helping me getting into the field of high-throughput sequencing analyses, especially metatranscriptomics, which resulted in several joint projects; to **Cornelia Bondow** and **Jörg Römbke** (ECT Oekotoxikologie, Germany) for working on the joint project evaluating climate change on the soil protist community.

Thanks to my entire working group, especially **Maike Hünninghaus**, **Anna Maria Fiore-Donne** and **Robert Koller** for stimulating discussions and their contribution to some chapters of this thesis, our technicians **Anna Herzog**, **Conny Thielen** and **Irene Brockhaus**, as well as **Jamila Rosengarten** and **Kenneth Dumack** who did some of the experimental work as part of their bachelor theses.

I gratefully acknowledge Prof. **Hartmut Arndt** (University of Cologne, Germany) and **Francis Martin** for their great support as co-supervisors as they were always available for questions and feedback. Thanks a lot to Prof. **Frank Schäbitz** for chairing my defense.

Last but not least I want to specially thank **many unmentioned, but not less important people** involved in other side-projects and also those outside of science for "mental support" throughout the time of my PhD, often in form of highly welcome topic-changes; My **mum** by giving me the chance to take the "University-road" as the pre-requisite for my scientific career and especially my wife **Verena** who suffered a lot from regular work night- and weekend session.

Subpublications

Chapter 1: Geisen, S., J. Weinert, A. Kudryavtsev, A. Glotova, M. Bonkowski and A. V. Smirnov (2014). "Two new species of the genus *Stenamoeba* (Discosea, Longamoebia): cytoplasmic MTOC is present in one more amoebae lineage." Eur. J. Protistol. 50(2): 153-165.

Chapter 2: Geisen, S., A. Kudryavtsev, M. Bonkowski and A. Smirnov (2014). "Discrepancy between species borders at morphological and molecular levels in the genus *Cochliopodium* (Amoebozoa, Himatismenida), with the description of *Cochliopodium plurinucleolum* n. sp." Protist 165(3): 364-383.

Chapter 3: Berney, C., S. Geisen, J. Van Wichelen, F. Nitsche, P. Vanormelingen, M. Bonkowski and D. Bass (2015). "Expansion of the 'Reticulosphere': diversity of novel branching and network-forming amoebae helps to define Variosea (Amoebozoa)." Protist 166(2): 271-295.

Chapter 4: Geisen, S., M. Bonkowski, J. Zhang and J. F. De Jonckheere (2015). "Heterogeneity in the genus *Allovahlkampfia* and the description of the new genus *Parafumarolamoeba* (Vahlkampfiidae; Heterolobosea)." Eur. J. Protistol.

Chapter 5: Geisen, S., C. Bandow, J. Römbke and M. Bonkowski (2014). "Soil water availability strongly alters the community composition of soil protists." Pedobiologia 57(4-6): 205–213.

Chapter 6: Geisen, S., A. M. Fiore-Donno, J. Walochnik and M. Bonkowski (2014). "*Acanthamoeba* everywhere: high diversity of *Acanthamoeba* in soils." Parasitol. Res. 113(9): 3151-3158.

Chapter 8: Geisen, S., A. T. Tveit, I. M. Clark, A. Richter, M. M. Svenning, M. Bonkowski and T. Urich (2015). "Metatranscriptomic census of active protists in soils." ISME J.

Chapter 10: Geisen, S., J. Rosengarten, R. Koller, C. Mulder, T. Urich and M. Bonkowski (2015). "Pack hunting by a common soil amoeba on nematodes." Environ. Microbiol.